DATA REDUCTION AND ERROR ANALYSIS FOR THE PHYSICAL SCIENCES

Second Edition

Philip R. Bevington

Late Associate Professor of Physics
Case Western Reserve University

D. Keith Robinson

Professor of Physics
Case Western Reserve University

McGraw-Hill, Inc.

New York   St. Louis   San Francisco   Auckland   Bogotá   Caracas
Lisbon   London   Madrid   Mexico   Milan   Montreal   New Delhi
Paris   San Juan   Singapore   Sydney   Tokyo   Toronto

This book was set in Times Roman by Science Typographers, Inc.
The editors were Susan J. Tubb and John M. Morriss.
The cover was designed by Andrew Cantor.
Project supervision was done by Science Typographers, Inc.
R. R. Donnelley & Sons Company was printer and binder.

DATA REDUCTION AND ERROR ANALYSIS FOR THE PHYSICAL SCIENCES

3 4 5 6 7 8 9 0 DOC DOC 9 0 9 8 7 6 5 4 3

P/N 053187-0
PART OF
ISBN 0-07-911243-9

Library of Congress Cataloging-in-Publication Data

Bevington, Philip R.
  Data reduction and error analysis for the physical sciences
Philip R. Bevington, D. Keith Robinson—2nd ed.
    p.   cm.
  Includes index.
  ISBN 0-07-911243-9 (set only)
  1. Multivariate analysis.   2. Error analysis (Mathematics)
3. Least squares.   4. Data reduction.   I. Robinson, D. Keith
(date).   II. Title.
QA278.B63   1992
511'.43—dc20   91-27275

TRADEMARK ACKNOWLEDGMENTS
GRAF/DRIVE PLUS is a registered trademark of FLEMING SOFTWARE.
Turbo Pascal is a registered trademark of Borland International Inc.

ABOUT THE AUTHORS

The late Philip R. Bevington was a professor of physics at Case Western Reserve University. He graduated from Harvard University in 1954 and received his Ph.D. from Duke University in 1960. He taught at Duke University for five years and was an Assistant Professor at Stanford University from 1963 to 1968 before coming to Case Western Reserve University. He was involved in research in nuclear structure physics with Van de Graaff accelerators. While at Stanford he was active in computer applications for nuclear physics and was responsible for development of the SCANS system.

D. Keith Robinson is professor of physics at Case Western Reserve University in Cleveland, Ohio. He received his B.Sc. in physics from Dalhousie University in Canada in 1954 and his D.Phil. from Oxford University in 1960. He was a member of the staff at Brookhaven National Laboratory from 1960 until 1966 when he joined Case Western Reserve University. His research in experimental elementary particle physics has included studies of boson resonances, antiproton–proton interactions, and radiative production and decay of hyperons. He has been strongly involved in introducing computers into the undergraduate physics program at CWRU.

CONTENTS

Appendixes

PREFACE TO THE SECOND EDITION

The late Philip Bevington's little book on data and error analysis has served for more than 20 years as a text for students and a handbook for research scientists. This revision of Phil's book attempts to follow the format and style of the original while updating it to the 1990s. The intent is still to provide an introduction to the techniques of data analysis and error reduction in a book that can serve students as a handbook throughout their undergraduate and graduate careers.

In his 1969 preface, Phil referred to the "proliferation of computers and their use in research laboratories." Since that time, the developments in computing have been spectacular. Computers have become common in homes; colleges and universities have installed computer laboratories and facilities for students; and rapidly growing numbers of students in fields as diverse as English literature and the physical sciences have purchased their own computers. The very nature of mathematical computing has been changed by the introduction of new techniques, such as spreadsheets and elaborate mathematical packages, often with dazzling built-in graphics capabilities. However, a strong grounding in data and error analysis remains an essential component of science education, and conventional computer programming skills are still important in the physical sciences.

As in the original edition, the book begins with a discussion of measuring uncertainties and continues with probability distributions, error analysis, and maximum-likelihood estimates of means and uncertainties. A new topic, Monte Carlo simulations, has been added as Chapter 5 to provide students with an introduction to this powerful technique for planning experiments and evaluating the statistical significance of results.

In the later chapters, more advanced topics in data analysis are discussed. In Chapters 6 through 9, the least-squares method is applied to problems of increasing complexity, from straight line fits to x-y coordinate pairs through nonlinear fits that require iterative solutions. Special emphasis is placed on the very powerful Marquardt method for solving nonlinear fitting problems. Chapter 10 provides an introduction to the direct application of the maximum-likelihood technique to the analysis of data. In Chapter 11, the interpretation of χ^2 is discussed, as well as the use of correlation coefficients and the F test. A new section on confidence levels has been added.

Several of the appendixes of the original edition have been retained, and two new ones have been added. Appendix A is devoted to numerical methods including polynomial interpolation, numerical integration and differentiation, methods of finding the roots of nonlinear equations, and data smoothing. A new section on cubic splines and their application to curve plotting has been added. Appendix B repeats the original discussion of matrices and determinants. The original tables and graphs of Appendix C are reproduced as convenient references and illustrations of the properties of some well known distributions. The graphical presentation of data is discussed in Appendix D. Suggestions for preparing effective graphs are accompanied by routines for making simple histograms and plots. In Appendix E the computer programs that have served to illustrate the text are listed.

The problem sets at the end of each chapter have been expanded to include exercises that range in complexity from simple statistical calculations to minor projects such as least-squares fitting and Monte Carlo calculations. Answers to selected exercises are provided.

COMPUTER ROUTINES

The provision of simple computer routines for calculating functions and solving least-squares fitting problems was a most useful feature of the original book. These routines served both as illustrations of many of the topics discussed in the book, and as handy routines for incorporation into users' programs. This feature has been retained.

The computer routines in the new version are somewhat more complex and detailed than those in the original version, reflecting requests of readers for "drivers" for the fitting routines, as well as the stronger computing skills of today's students. A full complement of supporting routines is provided to facilitate the construction of complete programs for fitting curves to data and plotting the results. To compensate for the increased complexity, the sample programs are provided both as listings in Appendix E and in machine readable form on an IBM compatible diskette.

Because undergraduate science students are most likely to be familiar with structured languages such as Pascal or C, and may have one of those languages on their personal computers, the examples in this book are written in Turbo Pascal. The routines should be compatible with Versions 4 and later of Turbo Pascal. They should also be compatible with earlier versions and with other Pascal compilers, with the exception of the use of "units" as convenient collections of procedures and functions. Because REAL variables in Turbo Pascal have a precision of 11 or 12 significant digits, it was not necessary to use double or extended precision for any of the sample calculations. However, higher precision is strongly recommended for research calculations and for FORTRAN programs.

Serious scientific computing still requires proficiency in FORTRAN, a much maligned, but nevertheless invaluable computer language, so a supplement with programs in that language will be made available.

COMPUTER DISK

Routines on the accompanying diskette are organized in subdirectories corresponding to the chapters and appendixes in which they are first mentioned in the text. The disk includes an executable program, `Qdisplay` which can produce high resolution graphics plots on a dot-matrix or laser printer. Instructions for use are provided in the "READ.ME" file on the diskette.

ACKNOWLEDGMENTS

I am most indebted to the late Philip R. Bevington for his fine text, which has been the basis of this revision, and a strong guide in the choice of new material. I also wish to thank Joan Bevington Brown for her encouragement and for providing me with Phil's notes and correspondence. Careful early reviews of some of the revised sections of the book were very helpful to me as were conversations with colleagues. I am grateful to my chairman, William Gordon, for his support throughout this undertaking and to my undergraduate laboratory students for several of the examples and much inspiration.

I would also like to thank the following reviewers for their many helpful comments and suggestions: Ray F. Cowan, Massachusetts Institute of Technology; W. N. Hubin, Kent State University; James E. Lawler, University of Wisconsin; David H. Miller, Purdue University; Richard Prepost, University of Wisconsin; Robert L. Rasera, University of Maryland, Baltimore County.

Finally, I wish to thank my wife, Margi, for her remarkable patience and support.

D. Keith Robinson

PREFACE TO THE FIRST EDITION

The purpose of this text is to provide an introduction to the techniques of data reduction and error analysis commonly employed by individuals doing research in the physical sciences and to present them in sufficient detail and breadth to make them useful for students throughout their undergraduate and graduate studies. The presentation is developed from a practical point of view, including enough derivation to justify the results, but emphasizing the methods more than the theory.

The level of primary concern is that of junior and senior undergraduate laboratory where a thorough study of these techniques is most appropriate. The treatment is intended to be comprehensive enough to be suitable for use by graduate students in experimental research who would benefit from the generalized methods for linear and nonlinear least-squares fitting and from the summaries of definitions and techniques.

At the same time, the introduction to the material is made self-supporting in that no prior knowledge of the methods of statistical evaluation is assumed; the material of each section is developed from first principles. A discussion of differential calculus and manipulation of matrices and determinants is included in the appendixes to supplement their use in the text.

The emphasis, however, is toward the application of more general techniques than are usually presented in undergraduate laboratories. With the proliferation of computers and their use in research laboratories, it is important that sophisticated concepts of data reduction be introduced. Computer routines written explicitly for each section are used throughout to illustrate in a practical way both the concepts and the procedures discussed.

The first five chapters introduce the concepts of errors, uncertainties, probability distributions, and methods of optimizing the estimates of parameters characterizing observations of a single variable. Chapters 6 to 9 deal with the problem of fitting, analytically, complex functions to observations of more than one variable, including estimates of the resulting uncertainties and tests for optimizing the functional form of the fit. The last third of the text contains a description of techniques for searching for the best fit to data with arbitrary functions. Techniques for manipulating data or extracting information without fitting are also discussed.

COMPUTER PROGRAMS

The primary purpose of the computer routines is to clarify the presentation, but they are meant to be usable for calculations as well. They are written as subroutines and function subprograms to provide flexibility for the user in applying them to his own data. The format of the routines is similar to that of the IBM Scientific Subroutine package for the IBM 360 computer system, including considerable commentary to define the parameters used and to describe the flow of the routine.

The routines are written in Fortran IV, but they are compatible with Fortran II except for the use of double precision and missing suffixes in the names of library functions, which are noted in the program descriptions. The sizes of most of the arrays are to be specified by the user and they are dimensioned in the routines with a size of 1. For most versions of Fortran, the dimension size in a subprogram is a dummy argument for one-dimensional arrays and need not correspond to the actual size used. Two-dimensional arrays are always dimensioned explicitly; these arrays and those which are wholly contained within the routines are assigned dimensions such that the routines can handle up to 10 terms of a fitting function and up to 100 data points. These dimensions may be increased for larger input dimensions or decreased for more efficient use of memory. All input and output variables are specified as arguments of the calling statement.

These routines have been debugged in both Fortran IV and Fortran II versions on small and large computers. They are intended to be usable operating routines and are reasonably efficient. Their most important function, however, is to serve as a framework on which to build and modify routines to serve the specific needs of the user.

ACKNOWLEDGMENTS

I am deeply indebted to members of the Stanford University Nuclear Physics Group and especially to Drs. A. S. Anderson, T. R. Fisher, D. W. Heikkinen, R. E. Pixley, F. Riess, and G. D. Sprouse for many helpful comments, suggestions, and techniques. The use of the SCANS (Stanford Computers for the Analysis of Nuclear Structure) PDP-7 and PDP-9 computers for the development of computer routines and the assistance of Wylbur Wryght of the Stanford Computation Center are gratefully acknowledged. I have benefitted from comments by Prof. Jay Orear and especially from many invaluable suggestions and criticisms of Prof. Hugh D. Young.

I wish also to express my appreciation to Mrs. Pat Johnson for her care and diligence in typing and to Mrs. H. Mae Sprouse for her artistic work on the illustrations. Special appreciation is due to my wife, Joan, and our children, Ann and Mark, for enduring so patiently the division of my attention during the writing of this book.

Philip R. Bevington

DATA REDUCTION AND ERROR ANALYSIS
FOR THE PHYSICAL SCIENCES

CHAPTER
1

UNCERTAINTIES
IN MEASUREMENTS

1.1 MEASURING ERRORS

It is a well established fact of scientific investigation that the first time an experiment is performed the results often bear all too little resemblance to the "truth" being sought. As the experiment is repeated, with successive refinements of technique and method, the results gradually and asymptotically approach what we may accept with some confidence to be a reliable description of events. We may sometimes feel that nature is loath to give up her secrets without a considerable expenditure of effort on our part, and that first steps in experimentation are bound to fail. Whatever the reason, it is certainly true that for all physical experiments, errors and uncertainties exist that must be reduced by improved experimental techniques and repeated measurements, and that these errors must always be estimated to establish the validity of our results.

Error is defined by Webster as "the difference between an observed or calculated value and the true value." Usually, of course, we do not know the "true" value; otherwise there would be no reason for performing the experiment. We may know approximately what it should be, however, either from earlier experiments or from theoretical predictions. Such approximations can yield an indication as to whether our result is of the right order of magnitude, but we must always determine in a systematic way from the data and the experimental conditions themselves how much confidence we can have in our experimental results.

There is one class of errors that we can deal with immediately: errors that originate from mistakes or blunders in measurement or computation. Fortunately, these errors are usually apparent either as obviously incorrect data points or as results that are not reasonably close to expected values. They are classified as *illegitimate errors* and generally can be corrected by carefully repeating the operations. Rather, we are considering random fluctuations in our measurements, and systematic errors that limit the precision and accuracy of our results in more or less well defined ways.

Accuracy versus Precision

It is important to distinguish between the terms *accuracy* and *precision*. The *accuracy* of an experiment is a measure of how close the result of the experiment is to the true value. Therefore, it is a measure of the correctness of the result. The *precision* of an experiment is a measure of how well the result has been determined, without reference to its agreement with the true value. The precision is also a measure of the reproducibility of the result. The distinction between the accuracy and the precision of a set of measurements is illustrated in Figure 1.1. *Absolute precision* indicates the magnitude of the uncertainty in the result in the same units as the result. *Relative precision* indicates the uncertainty in terms of a fraction of the value of the result.

It is obvious that we must consider the accuracy and precision simultaneously for any experiment. It would be a waste of time and energy to determine a result with high precision if we knew that the result would be highly inaccurate. Conversely, a result cannot be considered to be extremely accurate if the precision is low. In general, when we quote the *uncertainty* in an experimental

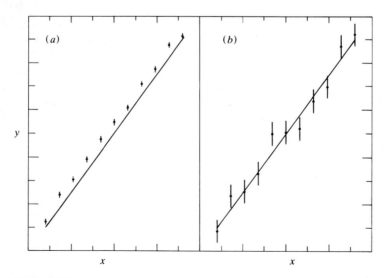

FIGURE 1.1
Illustration of the difference between precision and accuracy. (*a*) Precise but inaccurate data. (*b*) Accurate but imprecise data.

result, we are referring to the *precision* with which that result has been determined.

Systematic Errors

The *accuracy* of an experiment, as we have defined it, is generally dependent on how well we can control or compensate for *systematic errors*, errors that will make our results different from the "true" values with reproducible discrepancies. Errors of this type are not easy to detect and not easily studied by statistical analysis. They may result from faulty calibration of equipment or from bias on the part of the observer. They must be estimated from an analysis of the experimental conditions and techniques. A major part of the planning of an experiment should be devoted to understanding and reducing sources of systematic errors.

> **Example 1.1.** A student measures a table top with a steel meter stick and finds that the average of his measurements yields a result of 1.982 m for the length of the table. He subsequently learns that the meter stick was calibrated at 25°C and has an expansion coefficient of $0.0005°C^{-1}$. Because his measurements were made at a room temperature of 20°C, he multiplies his results by $1 + 0.0005 \times (20 - 25) = 0.9975$ so that his new determination of the length is 1.977 m.
>
> When the student repeats the experiment, he discovers that his technique for reading the meter stick was faulty in that he did not always read the divisions from directly above. By experimentation he determines that this consistently resulted in a reading that was 2 mm too short. The corrected result is 1.979 m.

In this experiment, the first result was given with a fairly high *precision*. The table top was found to be 1.982 m long, with a relative precision of about 1/2000, indicated by the fact that four significant figures were quoted. The corrections to this result were meant to improve the *accuracy* by compensating for known sources of deviation of the first result from the best estimate possible. These corrections did not improve the precision at all, but did in fact worsen it, because the corrections were themselves only estimates of the exact corrections.

Random Errors

The *precision* of an experiment is dependent on how well we can overcome *random errors*. These are the fluctuations in observations that yield results that differ from experiment to experiment and that require repeated experimentation to yield precise results. A given accuracy implies an equivalent precision and, therefore, also depends to some extent on random errors.

The problem of reducing random errors is essentially one of improving the experimental method and refining the techniques, as well as simply repeating the experiment. If the random errors result from instrumental uncertainties, they can be reduced by using more reliable and more precise measuring instruments. If the random errors result from statistical fluctuations associated with counting finite numbers of events, they may be reduced by counting more events. There are practical limits to these improvements. In the measurement of the length of the table of Example 1.1, the student might attempt to improve the precision of his measurements by using a magnifying glass to read the scale, or he might attempt to reduce statistical fluctuations in his measurements by repeating the measurement several times. In neither case would it be useful to reduce the random errors much below the systematic errors, such as those introduced by the calibration of the meter stick. The limits imposed by systematic errors are important considerations in planning and performing experiments.

Significant Figures and Roundoff

The precision of an experimental result is implied by the number of digits recorded in the result, although generally the uncertainty should be quoted specifically as well. The number of *significant figures* in a result is defined as follows:

1. The leftmost nonzero digit is the most significant digit.
2. If there is no decimal point, the rightmost nonzero digit is the least significant digit.
3. If there is a decimal point, the rightmost digit is the least significant digit, even if it is a 0.
4. All digits between the least and most significant digits are counted as significant digits.

For example, the following numbers each have four significant digits: 1234, 123,400, 123.4, 1001, 1000., 10.10, 0.0001010, 100.0. If there is no decimal point, there are ambiguities when the rightmost digit is 0. Thus, the number 1010 is considered to have only three significant digits even though the last digit might be physically significant. To avoid ambiguity, it is better to supply decimal points or to write such numbers in *scientific notation*, that is, as an argument in decimal notation multiplied by the appropriate power of 10. Thus, our example of 1010 would be written as 1010. or 1.010×10^3 if all four digits are significant.

   When quoting an experimental result, the number of significant figures should be approximately 1 more than that dictated by the experimental precision. The reason for including the extra digit is to avoid errors that might be caused by rounding errors in later calculations. If the result of the measurement of Example 1.1 is $L = 1.979$ m with an uncertainty of 0.012 m, this result could be quoted as $L = (1.979 \pm 0.012)$ m. However, if the first digit of the uncertainty is large, such as 0.082 m, then we should probably quote $L = (1.98 \pm 0.08)$ m. In other words, we let the uncertainty define the precision to which we quote our result.

When insignificant digits are dropped from a number, the last digit retained should be rounded off for the best accuracy. To round off a number to fewer significant digits than were specified originally, we truncate the number as desired and treat the excess digits as a decimal fraction. Then:

1. If the fraction is greater than $\frac{1}{2}$, increment the new least significant digit.
2. If the fraction is less than $\frac{1}{2}$, do not increment.
3. If the fraction equals $\frac{1}{2}$, increment the least significant digit only if it is odd.

The reason for rule 3 is that a fractional value of $\frac{1}{2}$ may result from a previous rounding up of a fraction that was slightly less than $\frac{1}{2}$ or a rounding down of a fraction that was slightly greater than $\frac{1}{2}$. For example, 1.249 and 1.251 both round to three significant figures as 1.25. If we were to round again to two significant figures, both would yield the same value, either 1.2 or 1.3, depending on our convention. Choosing to round up if the resulting last digit is odd and to round down if the resulting last digit is even, reduces systematic errors that would otherwise be introduced into the average of a group of such numbers. Note that with computer analysis of data, it is generally advisable to retain all available digits in intermediate calculations and round only the final results.

1.2 UNCERTAINTIES

Uncertainties in experimental results are of two main types: those that result from fluctuations in measurements and those associated with the theoretical description of our result. For example, if we measure the *length* of a table, we know that the result must be a number with units of length. Any uncertainties, aside from systematic errors, are associated with the fluctuations of our measurements from trial to trial. With an infinite number of measurements we might be able to estimate the length very precisely, but with a finite number of trials there will be a finite uncertainty. If, however, we were to describe the *shape* of an oval table, we would be faced with uncertainties both in the measurement of position of the edge of the table at various points and in the form of the equation to be used to describe the shape, whether it be circular, elliptical, or whatever. Thus, we shall be concerned in the following chapters with a comparison of the distribution of measured data points with the distribution predicted on the basis of a theoretical model. This comparison will help to indicate whether our method of extracting the results is valid or needs modification.

The term *error* signifies a deviation of the result from some "true" value. Usually we cannot know what the true value is, and we can consider only *estimates* of the errors inherent in the experiment. If we repeat an experiment, the results may well differ from those of the first attempt. We can express this difference as a *discrepancy* between the two results. Discrepancies arise because we can determine a result only with a given *uncertainty*.

A study of the distribution of the results of repeated measurements of the same quantity will lead to an understanding of the uncertainties in the measurements, and the uncertainties will serve as estimates of the errors. The quoted error is thus a measure of the spread of the distribution of repeated measurements. Because, in general, we shall not be able to quote the actual error of the results, we must develop a consistent method for determining and quoting the estimated error. We must also realize that the model from which we calculate theoretical parameters to describe the results of our experiment may not be the correct model to use. In the following chapters we shall discuss hypothetical parameters and probable distributions of errors pertaining to the "true" states of affairs, and we shall discuss methods of making experimental estimates of these parameters and the uncertainties associated with these determinations.

Minimizing Uncertainties and Best Results

Our preoccupation with error analysis is not confined just to the determination of the precision of our results. In general, we shall be interested in obtaining the maximum amount of useful information from the data on hand without being able either to repeat the experiment with better equipment or to reduce the statistical uncertainties by making more measurements. We shall be concerned, therefore, with the problem of extracting from the data the best estimates of theoretical parameters as well as of the random errors, and we shall want to understand the effect of these errors on our results, so that we can determine what confidence we can place in our final results. It is reasonable to expect that the most reliable results we can calculate from a given set of data will be those for which the estimated errors are the smallest. Thus, our development of techniques of error analysis will help to determine the optimum estimates of parameters to describe the data.

It must be noted, however, that our best efforts still will be only *estimates* of the quantities investigated.

1.3   PARENT AND SAMPLE DISTRIBUTIONS

If we make a measurement x_1 of a quantity x, we expect our observation to approximate the quantity, but we do not expect the experimental data point to be exactly equal to the quantity. If we make another measurement, we expect to observe a discrepancy between the two measurements because of random errors, and we do not expect either determination to be exactly correct, that is, equal to x. As we make more and more measurements, a pattern will emerge from the data. Some of the measurements will be too large, some will be too small. On the average, however, we expect them to be distributed around the correct value, assuming we can neglect or correct for systematic errors.

If we could make an infinite number of measurements, then we could describe exactly the distribution of the data points. This is not possible in practice, but we can hypothesize the existence of such a distribution that determines the probability of getting any particular observation in a single measurement. This distribution is called the *parent distribution*. Similarly, we can hypothesize that the measurements we have made are samples from the parent distribution and they form the *sample distribution*. In the limit of an infinite number of measurements, the sample distribution becomes the parent distribution.

> **Example 1.2.** A student makes 100 measurements of the length of a wooden block. His observations, corrected for systematic errors, range from about 18 to 22 cm, and many of the observations are identical. Figure 1.2 shows a *histogram* or frequency plot of a possible set of such measurements. The height of a data bar represents the number of measurements that fall between the two values indicated by the upper and lower limits of the bar on the abscissa of the plot. (See Appendix D.) If the distribution results from random errors in measurement, then it is very likely that it can be described in terms of the *Gaussian* or *normal error distribution*, the familiar bell-shaped curve of statistical analysis. (See Chapter 2.) A Gaussian curve based on these measurements is plotted as a solid line.

We could identify the solid curve, determined from the set of measurements displayed in the histogram, as the *sample distribution*. The measured data and the curve derived from them clearly do not agree exactly. The coarseness of

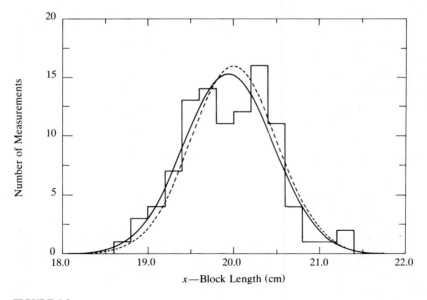

FIGURE 1.2
Histogram of measurements of the length of a block. The solid Gaussian curve was calculated from the mean ($\bar{x} = 19.9$ cm) and standard deviation ($s = 0.52$ cm) estimated from these measurements. The dashed curve represents the parent distribution with mean $\mu = 20.0$ cm and standard deviation $\sigma = 0.50$ cm.

the *experimental* histogram distinguishes it at once from the smooth *theoretical* Gaussian curve. We might imagine that, if the student were to make a great many measurements so that he could plot the histogram in finer and finer bins, that under ideal circumstances the histogram would eventually approach a smooth Gaussian curve. If we were to calculate the parameters from such a large sample, we could determine the *parent distribution*, represented by the dashed line in Figure 1.2.

The dashed curve in Figure 1.2 represents the probability of obtaining values of the variable x from a set of measurements. The area under the curve at the point x bounded by a range dx gives the number of events expected in that region from a 100-event sample. The area, divided by the total area of the plot, is the probability $P(x)\,dx$ that a randomly selected measurement will yield an observed value of x within the range $(x - dx/2) \le x < (x + dx/2)$. It is convenient to think in terms of a *probability density function*, which, for our sample parent population, is just the dashed curve of Figure 1.2 normalized to unit area.

Notation

A number of parameters of the parent distribution have been defined by convention. We shall use Greek letters to denote them, and Latin letters to denote experimental estimations of them.

In order to determine the parameters of the parent distribution, we shall assume that the results of experiments asymptotically approach the parent quantities as the number of measurements approaches infinity; that is, the parameters of the experimental distribution equal the parameters of the parent distribution *in the limit of an infinite number of measurements*. If we specify that there are N observations in a given experiment, then we can denote this by

$$(\text{parent parameter}) = \lim_{N \to \infty} (\text{experimental parameter})$$

If we make N measurements and label them x_1, x_2, x_3, and so forth, up to a final measurement x_N, then we can identify the sum of all these measurements as

$$\sum_{i=1}^{N} x_i \equiv x_1 + x_2 + x_3 + \cdots + x_N$$

where the left-hand side is interpreted as the sum of the observations x_i over the index i from $i = 1$ to $i = N$ inclusive. Because we shall be making frequent use of the sum over N measurements of various quantities, we shall simplify the notation by omitting the index whenever we are considering a sum where the index i runs from 1 to N;

$$\sum x_i \equiv \sum_{i=1}^{N} x_i$$

Mean, Median, and Mode

With the preceding definitions, the *mean* $\bar{x}$ of the experimental distribution is given as the sum of N determinations x_i of the quantity x divided by the number of determinations

$$\bar{x} \equiv \frac{1}{N} \sum x_i \tag{1.1}$$

and the mean μ of the parent population is defined as the limit

$$\mu \equiv \lim_{N \to \infty} \left(\frac{1}{N} \sum x_i \right) \tag{1.2}$$

The mean is therefore equivalent to the centroid or *average* value of the quantity x.

The *median* of the parent population $\mu_{1/2}$ is defined as that value for which, in the limit of an infinite number of determinations x_i, half the observations will be less than the median and half will be greater. In terms of the parent distribution, this means that the probability is 50% that any measurement x_i will be larger or smaller than the median

$$P(x_i < \mu_{1/2}) = P_i(x \ge \mu_{1/2}) = \tfrac{1}{2} \tag{1.3}$$

so that the median line cuts the area of the probability density distribution in half. Because of inconvenience in computation, the median is not often used as a statistical parameter.

The *mode*, or *most probable value* μ_{max}, of the parent population is that value for which the parent distribution has the greatest value. In any given experimental measurement, this value is the one that is most likely to be observed. In the limit of a large number of observations, this value will probably occur most often

$$P(\mu_{max}) \ge P(x \ne \mu_{max}) \tag{1.4}$$

The relationship of the mean, median, and most probable value to one another is illustrated in Figure 1.3. For a symmetrical distribution these parameters would all be equal by the symmetry of their definitions. For an asymmetric distribution such as that of Figure 1.3, the median generally falls between the most probable value and the mean. The most probable value corresponds to the peak of the distribution, and the areas on either side of the median are equal.

Deviations

The *deviation* d_i of any measurement x_i from the mean μ of the parent distribution is defined as the difference between x_i and μ:

$$d_i \equiv x_i - \mu \tag{1.5}$$

For computational purposes, deviations are generally defined with respect to

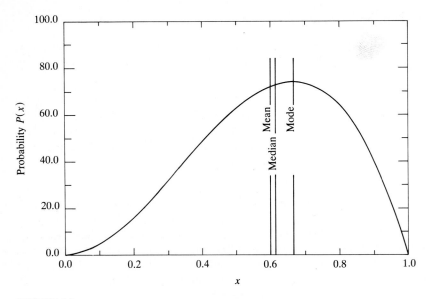

FIGURE 1.3
Asymmetric distribution illustrating the positions of the mean, median, and most probable value of the variable.

the mean, rather than the median or most probable value. If μ is the true value of the quantity, d_i is also the true error in x_i.

The average of the deviations $\bar{d}$ must vanish by virtue of the definition of the mean in Equation (1.2):

$$\lim_{N \to \infty} \bar{d} = \lim_{N \to \infty} \left[\frac{1}{N} \sum (x_i - \mu) \right] = \lim_{N \to \infty} \left(\frac{1}{N} \sum x_i \right) - \mu = 0 \qquad (1.6)$$

The *average deviation* α, therefore, is defined as the average of the absolute values of the deviations:

$$\alpha \equiv \lim_{N \to \infty} \left[\frac{1}{N} \sum |x_i - \mu| \right] \qquad (1.7)$$

The average deviation is a measure of the *dispersion* of the expected observations about the mean. The presence of the absolute value sign makes its use inconvenient for statistical analysis.

A parameter that is easier to use analytically and that can be justified fairly well on theoretical grounds to be a more appropriate measure of the dispersion of the observations is the *standard deviation* σ. The *variance* σ^2 is defined as the limit of the average of the squares of the deviations from the mean μ:

$$\sigma^2 \equiv \lim_{N \to \infty} \left[\frac{1}{N} \sum (x_i - \mu)^2 \right] = \lim_{N \to \infty} \left(\frac{1}{N} \sum x_i^2 \right) - \mu^2 \qquad (1.8)$$

and the standard deviation σ is the square root of the variance. Note that the second form of Equation (1.8) is often described as "the average of the squares minus the square of the average." The standard deviation is the root mean square of the deviations, and is associated with the *second moment* of the data about the mean. The corresponding expression for the standard deviation s of the sample population is given by

$$s^2 \equiv \frac{1}{N-1} \sum (x_i - \bar{x})^2 \qquad (1.9)$$

where the factor $N - 1$, rather than N, is required in the denominator to account for the fact that the parameter $\bar{x}$ has been determined from the data and not independently.

Significance

The mean μ and the standard deviation, as well as the median, the most probable value, and the average deviation, are all parameters that characterize the information we are seeking when we perform an experiment. Often we wish to describe our distribution in terms of just the mean and standard deviation. The mean may not be exactly equal to the datum in question if the parent distribution is not symmetrical about the mean, but it should have the same characteristics. If a more detailed description is desired, it may be useful to compute higher moments about the mean.

In general, the best we can say about the mean is that it is one of the parameters that specifies the probability distribution: It has the same units as the "true" value and, in accordance with convention, we shall consider it to be the best estimate of the "true" value under the prevailing experimental conditions.

The variance σ^2 and the standard deviation σ characterize the uncertainties associated with our experimental attempts to determine the "true" values. For a given number of observations, the uncertainty in determining the mean of the parent distribution is proportional to the standard deviation of that distribution. The standard deviation σ is, therefore, an appropriate measure of the uncertainty due to fluctuations in the observations in our attempt to determine the "true" value.

Although, in general, the distribution resulting from purely statistical errors can be described well by the two parameters, the mean and the standard deviation, we should be aware that, at distances of a few standard deviations from the mean of an experimental distribution, nonstatistical errors may dominate. In especially severe cases, it may be preferable to describe the spread of the distribution in terms of the average deviation, rather than the standard deviation, because the latter tends to deemphasize measurements that are far from the mean. There are also distributions for which the variance does not exist. The average deviation or some other quantity must be used as a parameter to indicate the spread of the distribution in such cases. In the following sections, however, we shall be concerned mainly with distributions that result from statistical errors and for which the variance exists.

1.4 MEAN AND STANDARD DEVIATION OF DISTRIBUTIONS

We can define the mean μ and the standard deviation σ in terms of the parent distribution $P(x)$ of the parent population. The probability function $P(x)$ is defined such that in the limit of a very large number of observations, the fraction dN of observations of the variable x that yield values between x and $x + dx$ is given by $dN = P(x)\,dx$.

The mean μ is the *expectation value* $\langle x \rangle$ of x, and the variance σ^2 is the expectation value $\langle (x - \mu)^2 \rangle$ of the square of the deviations of x from μ. The expectation value $\langle f(x) \rangle$ of any function of x is defined as the weighted average of $f(x)$, over all possible values of the variable x, with each value of $f(x)$ weighted by the probability density distribution $P(x)$.

Discrete Distributions

If the probability function is a discrete function $P(x)$ of the observed value x, we replace the sum over the individual observations Σx_i in Equation (1.2) by a sum over the value of the possible observations multiplied by the number of times these observations are expected to occur. If there are n possible different observable values of the quantity x, which we denote by x_j (where the index j runs from 1 to n with no two values of x_j equal), we should expect from a total of N observations to obtain each observable $NP(x_j)$ times. The mean can then be expressed as

$$\mu = \lim_{N \to \infty} \Sigma x_j = \lim_{N \to \infty} \frac{1}{N} \sum_{j=1}^{n} \left[x_j NP(x_j) \right]$$

$$= \lim_{N \to \infty} \sum_{j=1}^{n} \left[x_j P(x_j) \right] \tag{1.10}$$

Similarly, the variance σ in Equation (1.8) can be expressed in terms of the probability function $P(x)$:

$$\sigma^2 = \sum_{j=1}^{n} \left[(x_j - \mu)^2 P(x_j) \right] = \sum_{j=1}^{n} \left[x_j P(x_j) \right]^2 - \mu^2 \tag{1.11}$$

In general, the expectation value of any function $f(x)$ of x is given by

$$\langle f(x) \rangle = \sum_{j=1}^{n} \left[f(x_j) P(x_j) \right] \tag{1.12}$$

Continuous Distributions

If the probability function is a continuous smoothly varying function $P(x)$ of the observed value x, we replace the sum over the individual observations by an integral over all values of x multiplied by the probability $P(x)$. The mean μ becomes the first moment of the parent distribution

$$\mu = \int_{-\infty}^{\infty} x P(x) \, dx \tag{1.13}$$

and the variance σ^2 becomes the second central product moment

$$\sigma^2 = \int_{-\infty}^{\infty} (x - \mu)^2 P(x) \, dx = \int_{-\infty}^{\infty} x^2 P(x) \, dx - \mu^2 \tag{1.14}$$

The expectation value of any function of x is

$$\langle f(x) \rangle = \int_{-\infty}^{\infty} f(x) P(x) \, dx \tag{1.15}$$

What is the connection between the probability distribution of the parent population and an experimental sample we obtain? We have already seen that the uncertainties of the experimental conditions preclude a determination of the "true" values themselves. As a matter of fact, there are three levels of abstraction between the data and the information we seek:

1. From our experimental data points we can determine a sample frequency distribution that describes the way in which these particular data points are distributed over the range of possible data points. We use $\bar{x}$ to denote the mean of the data and $s^2(N - 1)/N$ to denote the sample variance. The shape and magnitude of the sample distribution vary from sample to sample.
2. From the parameters of the sample probability distribution we can estimate the parameters of the probability distribution of the parent population of possible observations. Our best estimate for the mean μ is the mean of the sample distribution $\bar{x}$, and the best estimate for the variance σ^2 is the compensated variance s^2. Even the shape of this parent distribution must be estimated or assumed.
3. From the estimated parameters of the parent distribution we estimate the results sought. In general, we shall assume that the estimated parameters of the parent distribution are equivalent to the "true" values, but the estimated parent distribution is a function of the experimental conditions as well as the "true" values, and these may not necessarily be separable.

Let us refer again to Figure 1.2, which shows a histogram of the measurements of the length of the block and two Gaussian curves. The calculation of the solid curve was based on the parameters $\bar{x} = 19.94$ cm and $s = 0.52$ cm determined experimentally from the data displayed in the histogram. The dashed curve was based on the parameters $\mu = 20.00$ cm and $\sigma = 0.50$ cm of the parent distribution, which could have been obtained by making a very large number of measurements.

The difference between the experimental mean $\bar{x}$ and the "true" mean μ is obvious on comparing the two curves. We should note, however, that even the definition of μ might be somewhat uncertain, because the edge of the block is not perfectly smooth and we should have to define exactly what we mean by the length. We are always restricted, in working with experimental data, to *estimating* the parameters of the parent population, and sometimes to defining what we mean by the *parent population*.

Nevertheless, by considering the data to be a sample from the parent population, we can estimate the shape and dispersion of the parent distribution to obtain useful information on the precision and reliability of our results. Thus, we find the sample mean $\bar{x}$ as an estimate of the mean μ in order to find the "true" value of length of the block, and we find the sample variance s as an estimate of the variance σ^2 in order to estimate the uncertainty in our value for μ.

SUMMARY

Systematic error: Reproducible inaccuracy introduced by faulty equipment, calibration, or technique.

Random error: Indefiniteness of result introduced by finite precision of measurement. Measure of fluctuation after repeated experimentation.

Uncertainty: Magnitude of error that is estimated to have been made in determination of results.

Accuracy: Measure of how close the result of an experiment comes to the "true" value.

Precision: Measure of how carefully the result is determined without reference to any "true" value.

Significant figures:

1. The leftmost nonzero digit is the most significant digit.
2. If there is no decimal point, the rightmost nonzero digit is the least significant digit.
3. If there is a decimal point, the rightmost digit is the least significant digit, even if it is zero.
4. All digits between the least and most significant digits are counted as significant digits.

Roundoff: Truncate the number to the specified number of significant digits and treat the excess digits as a decimal fraction.

1. If the fraction is greater than $\frac{1}{2}$, increment the least significant digit.

2. If the fraction is less than $\frac{1}{2}$, do not increment.

3. If the fraction equals $\frac{1}{2}$, increment the least significant digit only when it is odd.

Parent population: Hypothetical infinite set of data points of which the experimental data points are assumed to be a random sample.

Parent distribution: Probability distribution of the parent population from which the sample data are chosen.

Expectation value $\langle\ \rangle$: Weighted average of a function $f(x)$ over all values of x:

$$\langle f(x)\rangle = \lim_{N\to\infty}\left[\frac{1}{N}\Sigma f(x_i)\right] = \sum_{j=1}^{n}\left[f(x_j)P(x_j)\right] = \int_{-\infty}^{\infty}f(x)P(x)\,dx$$

Median $\mu_{1/2}$: $P(x_i < \mu_{1/2}) = P_i(x \geq \mu_{1/2}) = \frac{1}{2}$

Most probable value μ_{max}: $P(\mu_{max}) \geq P(x \neq \mu_{max})$

Mean: $\mu \equiv \langle x \rangle$

Average deviation: $\alpha \equiv \langle|x_i - \mu|\rangle$

Variance: $\sigma^2 \equiv \langle(x_i - \mu)^2\rangle = \langle x^2\rangle - \mu^2$

Standard deviation: $\sigma = \sqrt{\sigma^2}$

Sample mean: $\bar{x} = (1/N)\Sigma x_i$

Sample variance: $s^2 = 1/(N-1)\Sigma(x_i - \bar{x})^2$

EXERCISES

1.1. How many significant features are there in the following numbers?

(a) 976.45   (b) 84,000   (c) 0.0094   (d) 301.07
(e) 4.000   (f) 10   (g) 5280   (h) 400.
(i) 4.00×10^2 (j) 3.010×10^4

1.2. What is the most significant figure in Exercise 1.1? What is the least significant?

1.3. Round off each of the numbers in Exercise 1.1 to two significant digits.

1.4. Find the mean, median, and most probable value of x for the following data (from rolling dice).

i	x_i	i	x_i	i	x_i	i	x_i	i	x_i
1	3	6	8	11	12	16	6	21	5
2	7	7	9	12	8	17	7	22	10
3	3	8	7	13	6	18	8	23	8
4	7	9	5	14	6	19	9	24	8
5	12	10	7	15	7	20	8	25	8

1.5. Find the mean, median, and most probable grade from the following set of grades. Group them to find the most probable value.

i	x_i	i	x_i	i	x_i	i	x_i
1	73	11	73	21	69	31	56
2	91	12	46	22	70	32	94
3	72	13	64	23	82	33	51
4	81	14	61	24	90	34	79
5	82	15	50	25	63	35	63
6	46	16	89	26	70	36	87
7	89	17	91	27	94	37	54
8	75	18	82	28	44	38	100
9	62	19	71	29	100	39	72
10	58	20	76	30	88	40	81

1.6. Calculate the standard deviation of the data of Exercise 1.4.

1.7. Calculate the standard deviation of the data of Exercise 1.5.

1.8. Justify the second equality in Equations (1.8) and (1.14).

CHAPTER
2

PROBABILITY DISTRIBUTIONS

Of the many probability distributions that are involved in the analysis of experimental data, three play a fundamental role: the *binomial distribution*, the *Poisson distribution*, and the *Gaussian distribution*. Of these, the Gaussian or normal error distribution is undoubtedly the most important in statistical analysis of data. Practically, it is useful because it seems to describe the distribution of random observations for many experiments, as well as describing the distributions obtained when we try to estimate the parameters of most other probability distributions.

The Poisson distribution is generally appropriate for counting experiments where the data represent the number of items or events observed per unit interval. It is important in the study of random processes such as those associated with the radioactive decay of elementary particles or nuclear states, and is also applied to data that have been sorted into ranges to form a frequency table or a histogram.

The binomial distribution is generally applied to experiments in which the result is one of a small number of possible final states, such as the number of "heads" or "tails" in a series of coin tosses, or the number of particles scattered forward or backward relative to the direction of the incident particle in a particle physics experiment. Because both the Poisson and the Gaussian distributions can be considered as limiting cases of the binomial distribution, we shall devote some attention to the derivation of the binomial distribution from basic considerations.

2.1   BINOMIAL DISTRIBUTION

Suppose we toss a coin in the air and let it land. There is a 50% probability that it will land heads up and a 50% probability that it will land tails up. By this we mean that if we continue tossing a coin repeatedly, the fraction of times that it lands with heads up will asymptotically approach $\frac{1}{2}$, indicating that there was a probability of $\frac{1}{2}$ of doing so. For any given toss, the probability cannot determine whether or not it will land heads up; it can only describe how we should expect a large number of tosses to be divided into two possibilities.

Suppose we toss two coins at a time. There are now four different possible permutations of the way in which they can land: both heads up, both tails up, and two mixtures of heads and tails depending on which one is heads up. Because each of these permutations is equally probable, the probability for any choice of them is $\frac{1}{4}$ or 25%. To find the probability for obtaining a particular mixture of heads and tails, without differentiating between the two kinds of mixtures, we must add the probabilities corresponding to each possible kind. Thus, the total probability of finding either head up and the other tail up is $\frac{1}{2}$. Note that the sum of the probabilities for all possibilities $(\frac{1}{4} + \frac{1}{4} + \frac{1}{4} + \frac{1}{4})$ is always equal to 1 because *something* is bound to happen.

Let us extrapolate these ideas to the general case. Suppose we toss n coins into the air, where n is some integer. Alternatively, suppose that we toss one coin n times. What is the probability that exactly x of these coins will land heads up, without distinguishing which of the coins actually belongs to which group? We can consider the probability $P(x; n)$ to be a function of the number n of coins tossed and of the number x of coins that land heads up. For a given experiment in which n coins are tossed, this probability $P(x; n)$ will vary as a function of x. Of course, x must be an integer for any physical experiment, but we can consider the probability to be smoothly varying with x as a continuous variable for mathematical purposes.

Permutations and Combinations

If n coins are tossed, there are 2^n different possible ways in which they can land. This follows from the fact that the first coin has two possible orientations, for each of these the second coin also has two such orientations, for each of these the third coin also has two, and so on. Because each of these possibilities is equally probable, the probability for any one of these possibilities to occur at any toss of n coins is $1/2^n$.

How many of these possibilities will contribute to our observations of x coins with heads up? Imagine two boxes, one labelled "heads" and divided into x slots, and the other labelled "tails". We shall consider first the question of how many permutations of the coins result in the proper separation of x in one box and $n - x$ in the other; then we shall consider the question of how many combinations of these permutations should be considered to be different from each other.

In order to enumerate the number of *permutations* $Pm(n, x)$, let us pick up the coins one at a time from the collection of n coins and put x of them into the "heads" box. We have a choice of n coins for the first one we pick up. For our second selection we can choose from the remaining $n - 1$ coins. The range of choice is diminished until the last selection of the xth coin can be made from only $n - x + 1$ remaining coins. The total number of choices for coins to fill the x slots in the "heads" box is the product of the numbers of individual choices:

$$Pm(n, x) = n(n - 1)(n - 2) \cdots (n - x + 2)(n - x + 1) \qquad (2.1)$$

This expansion can be expressed more easily in terms of factorials

$$Pm(n, x) = \frac{n!}{(n - x)!} \qquad (2.2)$$

So far we have calculated the number of permutations $Pm(n, x)$ that will yield x coins in the "heads" box and $n - x$ coins in the "tails" box, with the provision that we have identified which coin was placed in the "heads" box first, which was placed in second, and so on. That is, we have *ordered* the x coins in the "heads" box. In our computation of 2^n different possible permutations of the n coins, we are only interested in which coins landed heads up or heads down, not which landed first. Therefore, we must consider contributions different only if there are different coins in the two boxes, not if the x coins within the "heads" box are permuted into different time orderings.

The number of different *combinations* $C(n, x)$ of the permutations in the preceding enumeration results from combining the $x!$ different ways in which x coins in the "heads" box can be permuted within the box. For every $x!$ permutations, there will be only one new combination. Thus, the number of different combinations $C(n, x)$ is the number of permutations $Pm(n, x)$ divided by the degeneracy factor $x!$ of the permutations:

$$C(n, x) = \frac{Pm(n, x)}{x!} = \frac{n!}{x!(n - x)!} = \binom{n}{x} \qquad (2.3)$$

This is the number of different possible combinations of n items taken x at a time, commonly referred to as $\binom{n}{x}$ or "n over x".

Probability

The probability $P(x; n)$ that we should observe x coins with heads up and $n - x$ with tails up is the product of the number of different combinations $C(n, x)$ that contribute to that set of observations multiplied by the probability for each of the combinations to occur, which we have found to be $(1/2)^n$.

Actually, we should separate the probability for each combination into two parts: one part is the probability $(\frac{1}{2})^x$ for x coins to be heads up; the other part is the probability $(\frac{1}{2})^{n-x}$ for the other $n - x$ coins to be tails up. The product of these two parts is the probability of the combination. In the general case (i.e., lopsided but identical coins), the probability p of success for each item (in this case landing heads up) is not equal to the probability $q = 1 - p$ for failure (landing tails up). The probability for each of the combinations of x coins heads up and $n - x$ coins tails up is $p^x q^{n-x}$.

With these definitions of p and q, the probability $P_B(x; n, p)$ for observing x of the n items to be in the state with probability p is given by the *binomial distribution*

$$P_B(x; n, p) = \binom{n}{x} p^x q^{n-x} = \frac{n!}{x!(n-x)!} p^x (1-p)^{n-x} \tag{2.4}$$

where $q = 1 - p$. The name for the binomial distribution comes from the fact that the coefficients $P_B(x; n, p)$ are closely related to the binomial theorem for the expansion of a power of a sum. According to the binomial theorem,

$$(p + q)^n = \sum_{x=0}^{n} \left[\binom{n}{x} p^x q^{n-x} \right] \tag{2.5}$$

The $(j + 1)$th term, corresponding to $x = j$, of this expansion, therefore, is equal to the probability $P_B(j; n, p)$. We can use this result to show that the binomial distribution coefficients $P_B(x; n, p)$ are normalized to a sum of 1. The right-hand side of Equation (2.5) is the sum of probabilities over all possible values of x from 0 to n and the left-hand side is just $1^n = 1$.

Mean and Standard Deviation

The mean of the binomial distribution is evaluated by combining the definition of μ in Equation (1.2) with the formula for the probability function of Equation (2.4):

$$\mu = \sum_{x=0}^{n} \left[x \frac{n!}{x!(n-x)!} p^x (1-p)^{n-x} \right] = np \tag{2.6}$$

We interpret this to mean that if we perform an experiment with n items and observe the number x of successes, after a large number of repeated experiments the average $\bar{x}$ of the number of successes will approach a mean value μ given by the probability for success of each item p times the number of items n. In the case of coin tossing where $p = \frac{1}{2}$, we should expect on the average to observe half the coins land heads up, which seems eminently reasonable.

The variance σ^2 of a binomial distribution is similarly evaluated by combining Equations (1.8) and (2.4):

$$\sigma^2 = \sum_{x=0}^{n} \left[(x - \mu)^2 \frac{n!}{x!(n-x)!} p^x (1-p)^{n-x} \right] = np(1-p) \tag{2.7}$$

The evaluation of these sums is left as an exercise. We are mainly interested in the results, which are remarkably simple.

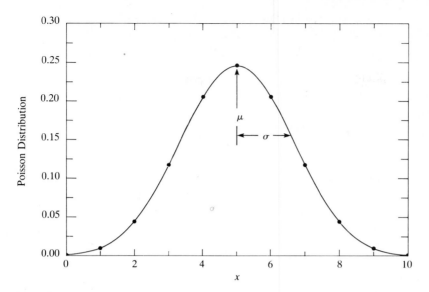

FIGURE 2.1
Binomial distribution for $\mu = 5.0$ and $p = \frac{1}{2}$ shown as a continuous curve although the function is only defined at the discrete points indicated by the round dots.

If the probability for a single success p is equal to the probability for failure $p = q = \frac{1}{2}$, then the distribution is symmetric about the mean μ, and the median $\mu_{1/2}$ and the most probable value are both equal to the mean. In this case, the variance σ_2 is equal to half the mean $\sigma^2 = \mu/2$. If p and q are not equal, the distribution is asymmetric with a smaller variance.

> **Example 2.1.** Suppose we toss 10 coins into the air a total of 100 times. With each coin toss we observe the number of coins that land heads up and denote that number by x_i, where i is the number of the toss; i ranges from 1 to 100 and x_i can be any integer from 0 to 10. The probability function governing the distribution of the observed values of x is given by the binomial distribution $P_B(x; n, p)$ with $n = 10$ and $p = \frac{1}{2}$. The parent distribution is not affected by the number N of repeated procedures in the experiment.

The parent distribution $P_B(x; 10, \frac{1}{2})$ is shown in Figure 2.1 as a smooth curve drawn through discrete points. The mean μ is given by Equation (2.6):

$$\mu = np = 10\left(\tfrac{1}{2}\right) = 5$$

the standard deviation σ is given by Equation (2.7):

$$\sigma = \sqrt{np(1-p)} = \sqrt{10\left(\tfrac{1}{2}\right)\left(\tfrac{1}{2}\right)} = \sqrt{2.5} \approx 1.58$$

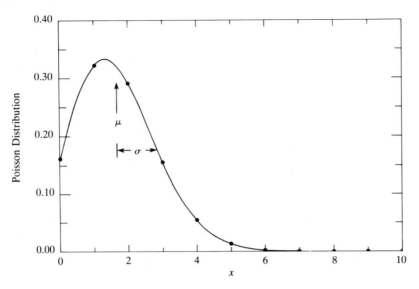

FIGURE 2.2
Binomial distribution for $\mu = 10/6$ and $p = 1/6$ shown as a continuous curve.

The curve is symmetric about its peak at the mean and the magnitudes of the points are such that the sum of the probabilities over all the points is equal to 1.

Example 2.2. Suppose we roll 10 dice. What is the probability that x of these dice will land with the 1 up? If we throw one die, the probability of its landing with 1 up is $p = \frac{1}{6}$. If we throw 10 dice, the probability for x of them landing with 1 up is given by the binomial distribution $P_B(x; n, p)$ with $n = 10$ and $p = \frac{1}{6}$:

$$P_B\left(x; 10, \frac{1}{6}\right) = \frac{10!}{x!(10-x)!}\left(\frac{1}{6}\right)^x\left(\frac{5}{6}\right)^{1-x}$$

This distribution is illustrated in Figure 2.2 as a smooth curve drawn through discrete points. The mean and standard deviation are

$$\mu = 10/6 \simeq 1.67$$

and
$$\sigma = \sqrt{10(1/6)(5/6)} \simeq 1.18$$

The distribution is not symmetric about the mean or about any other point. The most probable value is $x = 1$, but the peak of the smooth curve occurs for a slightly larger value of x.

Example 2.3. A particle physicist makes some preliminary measurements of the angular distribution of K mesons scattered from a liquid hydrogen target. She knows that there should be equal numbers of particles scattered forward and backward in the center-of-mass system of the particles. She measures 1000 interactions and finds that 472 scatter forward and 528 backward. What uncertainty should she quote in these numbers?

The uncertainty is given by the standard deviation from Equation (2.7),

$$\sigma = \sqrt{np(1-p)} = \sqrt{1000(\tfrac{1}{2})(\tfrac{1}{2})} = \sqrt{250} \simeq 15.8$$

Thus, she could quote

$$f_F = (472 \pm 15.8)/1000 = 0.472 \pm 0.15$$

for the fraction of particles scattered in the forward direction and

$$f_B = (528 \pm 15.8)/1000 = 0.528 \pm 0.15$$

for the fraction scattered backward.

Note that the uncertainties in the numbers scattering forward and backward must be the same because losses from one group must be made up in the other.

If the experimenter did not know the a priori probabilities of scattering forward and backward, she would have to estimate p and q from her measurements; that is,

$$p \simeq 472/1000 = 0.472$$

and
$$q \simeq 528/1000 = 0.528$$

She would then calculate

$$\sigma \simeq \sqrt{1000(0.472)(0.528)} = \sqrt{249.2} = 15.8$$

For probability p near 50%, the standard deviation is relatively insensitive to uncertainties in the experimental determination of p.

2.2   POISSON DISTRIBUTION

The Poisson distribution represents an approximation to the binomial distribution for the special case where the average number of successes is much smaller than the possible number; that is, when $\mu \ll n$ because $p \ll 1$. For such experiments the binomial distribution correctly describes the probability $P_B(x; n, p)$ of observing x events per time interval out of n possible events, each of which has a probability p of occurring, but the large number n of possible events makes exact evaluation from the binomial distribution impossible. Furthermore, in these experiments, neither the number n of possible events nor the probability p for each is usually known. What may be known instead is the average number of events μ expected in each time interval or its estimate $\bar{x}$. The Poisson distribution provides an analytical form appropriate to such investigations that describes the probability distribution in terms of just the variable x and the parameter μ.

Let us consider the binomial distribution in the limiting case of $p \ll 1$. We are interested in its behavior as n becomes infinitely large while the mean

$\mu = np$ remains constant. Equation (2.4) for the probability function of the binomial distribution may be written as

$$P_B(x; n, p) = \frac{1}{x!} \frac{n!}{(n-x)!} p^x (1-p)^{-x} (1-p)^n \tag{2.8}$$

If we expand the second term

$$\frac{n!}{(n-x)!} = n(n-1)(n-2) \cdots (n-x-2)(n-x-1) \tag{2.9}$$

we can consider it to be the product of x terms, each of which is very nearly equal to n because $x \ll n$ in the region of interest. This term asymptotically approaches n^x. The product of the second and third terms thus becomes $(np)^x = \mu^x$. The fourth term is approximately equal to $1 + px$, which tends to 1 as p tends to 0.

The last term can be rearranged by substituting μ/p for n to show that it asymptotically approaches $e^{-\mu}$:

$$\lim_{p \to 0} (1-p)^n = \lim_{p \to 0} \left[(1-p)^{1/p}\right]^{\mu} = \left(\frac{1}{e}\right)^{\mu} = e^{-\mu} \tag{2.10}$$

Combining these approximations, we find that the binomial distribution probability function $P_B(x; n, p)$ asymptotically approaches the *Poisson distribution* $P_p(x; \mu)$ as p approaches 0:

$$\lim_{p \to 0} P_B(x; n, p) = P_P(x; \mu) \equiv \frac{\mu^x}{x!} e^{-\mu} \tag{2.11}$$

Because this distribution is an approximation to the binomial distribution for $p \ll 1$, the distribution is asymmetric about its mean μ and will resemble that of Figure 2.2. Note that $P_B(x; \mu)$ does not become 0 for $x = 0$ and is not defined for negative values of x. This restriction is not troublesome for counting experiments because the number of counts per unit time interval can never be negative.

Deviation

The Poisson distribution can also be derived for the case where the number of events observed is small compared to the total possible number of events.[1] Assume that the average rate at which events of interest occur is constant over a given interval of time and that event occurrences are randomly distributed over that interval. Then, the probability dQ of observing no events in a time interval

---

[1] This derivation follows that of Orear, pages 21–22.

dt is given by

$$dQ(0;t,\tau) = -P(0;t,\tau)\frac{dt}{\tau} \tag{2.12}$$

where $P(x;t,\tau)$ is the probability of observing x events in the time interval dt, τ is a constant proportionality factor that is associated with the mean time between events, and the minus sign accounts for the fact that increasing the differential time interval dt decreases the probability proportionally. Integrating this equation yields the probability of observing no events within a time t to be

$$P(0;t,\tau) = P_0 e^{-t/\tau} \tag{2.13}$$

where P_0 is the constant of integration.

The probability $P(x;t,\tau)$ for observing x events in the time interval τ can be evaluated by integrating the differential probability

$$d^x Q(x;t,\tau) = \frac{e^{-t/\tau}}{x!} \prod_{i=1}^{x} \frac{dt_i}{\tau} \tag{2.14}$$

which is the product of the probabilities of observing each event in a different interval dt_i and the probability $e^{-t/\tau}$ of not observing any other events in the remaining time. The factor of $x!$ in the denominator compensates for the ordering implicit in the probabilities $dQ_i(1,t,\tau)$ as discussed in the preceding section on permutations and combinations.

Thus, the probability of observing x events in the time interval t is obtained by integration

$$P_P(x;\mu) = P(x;t,\tau) = \frac{e^{-t/\tau}}{x!}\left(\frac{t}{\tau}\right)^x \tag{2.15}$$

or

$$P_P(x;\mu) = \frac{\mu^x}{x!}e^{-\mu} \tag{2.16}$$

which is the expression for the Poisson distribution, where $\mu = t/\tau$ is the average number of events observed in the time interval t. Equation (2.16) represents a normalized probability function; that is, the sum of the function evaluated at each of the allowed values of the variable x is unity:

$$\sum_{x=0}^{\infty} P_P(x,\mu) = \sum_{x=0}^{\infty} \frac{\mu^x}{x!}e^{-\mu} = e^{-\mu} \sum_{x=0}^{\infty} \frac{\mu^x}{x!} = e^{-\mu}e^{\mu} = 1 \tag{2.17}$$

Mean and Standard Deviation

The Poisson distribution, like the binomial distribution, is a *discrete* distribution. That is, it is defined only at integral values of the variable x, although the parameter μ is a positive, real number. The mean of the Poisson distribution is actually the parameter μ that appears in the probability function $P_P(x;\mu)$ of

Equation (2.16). To verify this, we can evaluate the expectation value $\langle x \rangle$ of x:

$$\langle x \rangle = \sum_{x=0}^{\infty} \left(x \frac{\mu^x}{x!} e^{-\mu} \right) = \mu e^{-\mu} \sum_{x=1}^{\infty} \frac{\mu^{x-1}}{(x-1)!} = \mu e^{-\mu} \sum_{y=0}^{\infty} \frac{\mu^y}{y!} = \mu \quad (2.18)$$

To find the standard deviation σ, the expectation value of the square of the deviations can be evaluated:

$$\sigma^2 = \langle (x - \mu)^2 \rangle = \sum_{x=0}^{\infty} \left[(x - \mu)^2 \frac{\mu^x}{x!} e^{-\mu} \right] = \mu \quad (2.19)$$

Thus, the standard deviation σ is equal to the square root of the mean μ and the Poisson distribution has only a single parameter, μ.

Computation of the Poisson distribution by Equation (2.16) can be limited by the factorial function in the denominator. The problem can be avoided by using logarithms or by using the recursion relations

$$P(0; \mu) = e^{-\mu} \qquad P(x; \mu) = \frac{\mu}{x} P(x - 1; \mu) \quad (2.20)$$

This form has the disadvantage that, in order to calculate the function for particular values of x and μ, the function must be calculated at all lower values of x as well. However, if the function is to be summed from $x = 0$ to some upper limit to obtain the summed probability or to generate the distribution for a Monte Carlo calculation (Chapter 5), the function must be calculated at all lower values of x anyway.

> **Example 2.4.** Two students measure the background cosmic radiation in the physics laboratory as part of an experiment to determine the mean lifetime of two radioactive isotopes of silver. (See Example 8.1.) They record the number of counts in the detector in a series of 2-s intervals for 100 intervals and find that the mean number of counts is 1.69 counts per interval. From the mean they estimate the standard deviation to be $\sigma = \sqrt{1.69} = 1.30$, compared to $s = 1.29$ from a direct calculation with Equation (1.9).
>
> The students then repeat the exercise, this time recording the number of counts in 15-s intervals for 60 intervals and obtain a mean of 11.48 counts per 15-s interval and a standard deviation $\sigma = \sqrt{11.48} = 3.17$. The standard deviation calculated from the data with Equation (1.9) is $s = 3.39$.

Histograms of the two sets of data are shown in Figures 2.3 and 2.4. The calculated mean in each case was used as an estimate of the mean of the parent distribution, and a Poisson curve was calculated for each data set, based on the calculated means. These distributions are shown in Figures 2.3 and 2.4 as solid curves, although only the points at integral values of the abscissa are physically significant. The asymmetry of the distribution in Figure 2.3 is obvious, as is the fact that the mean μ does not coincide with the most probable value of x at the peak of the curve. The curve of Figure 2.4, on the other hand, is almost symmetric about its mean and the data are consistent with the curve. As

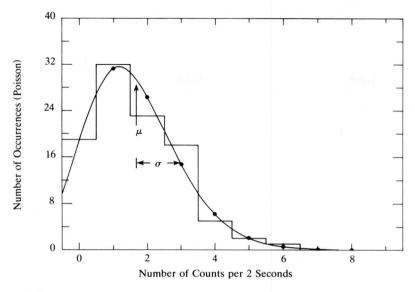

FIGURE 2.3
Histogram of counts in a cosmic ray detector. The Poisson distribution is an estimate of the parent distribution based on the measured mean $\bar{x} = 1.69$. It is shown as a continuous curve although the function is only defined at the discrete points indicated by the round dots. Only the circled calculation points are defined.

μ increases, the symmetry of the Poisson distribution increases and the distribution becomes indistinguishable from the Gaussian distribution.

Summed Probability

We may want to know the probability of obtaining a sample value of x between limits x_1 and x_2 from a Poisson distribution with mean μ. This probability is obtained by summing the values of the function calculated at the integral values of x between the two integral limits x_1 and x_2,

$$S_P(x_1, x_2; \mu) = \sum_{x_1}^{x_2} P_P(x; \mu) \tag{2.21}$$

where the limits of the sum are the integral parts of the numbers. More likely, we may want to find the probability of recording n or more events in a given interval when the mean number of events is μ. This is just the sum

$$S_P(n, \infty; \mu) = \sum_{x=n}^{\infty} P_P(x; \mu) = 1 - \sum_{x=0}^{n-1} P_P(x; \mu) = 1 - e^{-\mu} \sum_{x=0}^{n-1} \frac{\mu^x}{x!} \tag{2.22}$$

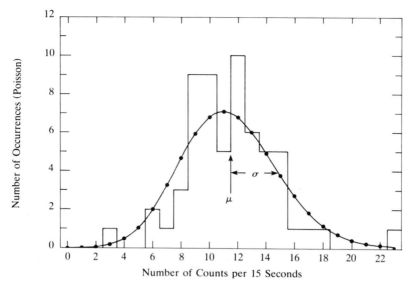

FIGURE 2.4

Histogram of counts in a cosmic ray detector. The Poisson distribution, shown as a continuous curve, is an estimate of the parent distribution based on the measured mean $\bar{x} = 11.48$. Only the circled calculated points are defined.

In Example 2.4, the mean number of counts recorded in a 15-s time interval was $\mu = 11.48$. In one of the intervals, 23 counts were recorded. From Equation (2.22), the probability of collecting 23 or more events in a single 15-s time interval is ~ 0.0018, and the probability of this occurring in any one of sixty 15-s time intervals is just the complement of the joint probability that 23 or more counts *not* be observed in any of the 60 time intervals, or $p \simeq 1 - (1 - 0.0018)^{60} \simeq 0.10$, or about 10%.

For large values of μ, the probability sum of Equation (2.22) may be approximated by an integral of the Gaussian function.

2.3 GAUSSIAN OR NORMAL ERROR DISTRIBUTION

The Gaussian distribution is an approximation to the binomial distribution for the special limiting case where the number of possible different observations n becomes infinitely large and the probability of success for each is finitely large so $np \gg 1$. It is also, as we observed, the limiting case for the Poisson distribution as μ becomes large.

There are several derivations of the Gaussian distribution from first principles, none of them as convincing as the fact that the distribution is reasonable, that it has a fairly simple analytic form, and that it is accepted by convention and experimentation to be the most likely distribution for most experiments. In addition, it has the satisfying characteristic that the most probable estimate of the mean μ from a random sample of observations x is the average of those observations $\bar{x}$.

Characteristics

The Gaussian probability function is defined as

$$P_G(x; \mu, \sigma) = \frac{1}{\sigma\sqrt{2\pi}} \exp\left[-\frac{1}{2}\left(\frac{x - \mu}{\sigma}\right)^2\right] \tag{2.23}$$

This is a continuous function describing the probability of obtaining the value x in a random observation from a parent distribution with parameters μ and σ, corresponding to the mean and standard deviation, respectively. Because the distribution is continuous, we must define an interval in which the value of the observation x will fall. The probability function is properly defined such that the probability $dQ_G(x; \mu, \sigma)$ that the value of a random observation will fall within an interval dx around x is given by

$$dQ_G(x; \mu, \sigma) = P_G(x; \mu, \sigma)\, dx \tag{2.24}$$

considering dx to be an infinitesimal differential, and the probability function is normalized, so that

$$\int_{x=-\infty}^{x=\infty} dQ_G(x; \mu, \sigma) \int_{x=-\infty}^{x=\infty} dQ_G = \int_{x=-\infty}^{x=\infty} P_G(x; \mu, \sigma)\, dx = 1 \tag{2.25}$$

The width of the curve is determined by the value of σ, such that for $x = \mu \pm \sigma$, the height of the curve is reduced to $e^{-1/2}$ of its value at the peak:

$$P_G(\mu \pm \sigma; \mu, \sigma) = e^{-1/2} P_G(\mu; \mu, \sigma) \tag{2.26}$$

The shape of the Gaussian distribution is shown in Figure 2.5. The curve displays the characteristic bell shape and symmetry about the mean μ.

We can characterize a distribution by its *full-width at half maximum* Γ, often referred to as the *half-width*, defined as the range of x between values at which the probability $P_G(x; \mu, \sigma)$ is half its maximum value:

$$P_G(\mu \pm 1/2\Gamma; \mu, \sigma) = 1/2 P_G(\mu; \mu, \sigma) \tag{2.27}$$

With this definition, we can determine from Equation (2.23) that

$$\Gamma = 2.354\sigma \tag{2.28}$$

As illustrated in Figure 2.5, tangents drawn along a portion of steepest descent of the curve intersect the curve at the $e^{-1/2}$ points $x = \mu \pm \sigma$ and intersect the x axis at the points $x = \mu \pm 2\sigma$.

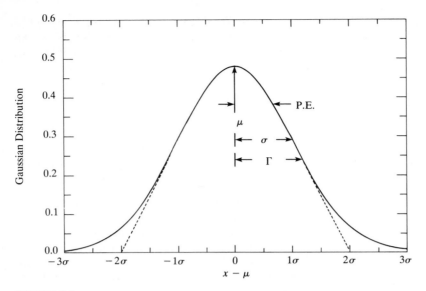

FIGURE 2.5
Gaussian probability distribution illustrating the relation of μ, σ, Γ, and P.E. to the curve. The curve has unit area.

Standard Gaussian Distribution

It is generally convenient to use a standard form of the Gaussian equation obtained by defining the dimensionless variable $z = (x - \mu)/\sigma$, because with this change of variable, we can write

$$P_G(z)\, dz = \frac{1}{\sqrt{2\pi}}\, \exp\left(-\frac{z}{2}\right)^2 dz \qquad (2.29)$$

Thus, from a single computer routine or a table of values of $P_G(z)$, we can find the Gaussian probability function $P_G(x; \mu, \sigma)$ for all values of the parameters μ and σ by changing the variable and scaling the function by $1/\sigma$ to preserve the normalization.

Mean and Standard Deviation

The parameters μ and σ in Equation (2.23) for the Gaussian probability density distribution correspond to the mean and standard deviation of the function. This equivalence can be verified by calculating μ and σ with Equations (1.13) and (1.14) as the expectation values for the Gaussian function of x and $(x - \mu)^2$, respectively.

For a finite data sample, which is expected to follow the Gaussian probability density distribution, the mean and standard deviation can be calculated directly from Equations (1.1) and (1.9). The resulting values of $\bar{x}$ and s

will be *estimates* of the mean μ and standard deviation σ. Values of $\bar{x}$ and s, obtained in this way from the original 100 measurements of the length of the block of Example 1.1, were used as estimates of μ and σ in Equation (2.23) to calculate the solid Gaussian curve in Figure 1.2. The curve was scaled to have the same area as the histogram. The curve does not represent the true parent distribution but rather a sample distribution based on our measurements.

Integral Probability

We are often interested in knowing the probability that a measurement will deviate from the mean by a specified amount Δx. The answer can be determined by evaluating numerically the integral

$$A_G(\Delta x; \mu, \sigma) = \frac{1}{\sigma\sqrt{2\pi}} \int_{\mu-\Delta x}^{\mu+\Delta x} \exp\left[-\frac{1}{2}\left(\frac{x-\mu}{\sigma}\right)^2\right] dx \qquad (2.30)$$

which gives the probability that any random value of x will deviate from the mean by *less than* $\pm\Delta x$. Because the probability function $P_G(x; \mu, \sigma)$ is normalized to unity, the probability that a measurement will deviate from the mean by *more* than Δx is just $1 - A_G(\Delta x; \mu, \sigma)$. Of particular interest are the probabilities associated with deviations of σ, 2σ, and so forth, from the mean, corresponding to 1, 2, and so on standard deviations. We may also be interested in the probable error (P.E.), defined to be the absolute value of the deviation $|x - \mu|$ such that the probability for the deviation of any random observation $|x_i - \mu|$ is less than $\frac{1}{2}$. That is, half the observations of an experiment would be expected to fall within the boundaries denoted by $\mu \pm$ P.E.

If we use the standard form of the Gaussian distribution of Equation (2.29), we can calculate the integrated probability $A_G(z)$ in terms of the dimensionless variable $z = (x - \mu)/\sigma$,

$$A_G(z) = \frac{1}{\sqrt{2\pi}} \int_{-\Delta z}^{\Delta z} e^{-z^2/2} dz \qquad (2.31)$$

where $\Delta z = \Delta x/\sigma$ measures the deviation from the mean in units of the standard deviation σ.

The integral of Equation (2.31) cannot be evaluated analytically, so in order to obtain the probability $A_G(\Delta x, \mu, \sigma)$ it is necessary either to expand the Gaussian function in a Taylor's series and integrate the series term-by-term, or to integrate numerically. With modern computers, numerical integration is fast and accurate, and reliable results can be obtained from a simple quadratic integration (Appendix A).

Tables and Graphs

The Gaussian probability function $P_G(z)$ and the integral probability $A_G(z)$ are tabulated and plotted in Appendices C.1 and C.2, respectively. From the integral probability table of Appendix C.2, we note that the probabilities are about 68% and 95% that a given measurement will fall within 1 and 2 standard deviations of the mean, respectively. Similarly, by considering the 50% probability limit we can see that the probable error is given by P.E. $= 0.6745\sigma$.

Comparison of Gaussian and Poisson Distributions

A comparison of the Poisson and Gaussian curves reveals the nature of the Poisson distribution. It is the appropriate distribution for describing experiments in which the possible values of the data are strictly bounded on one side but not on the other. The Poisson curve of Figure 2.3 exhibits the typical Poisson shape. The Poisson curve of Figure 2.4 differs little from the corresponding Gaussian curve of Figure 2.5, indicating that for large values of the mean μ, the Gaussian distribution becomes an acceptable description of the Poisson distribution. Because, in general, the Gaussian distribution is more convenient to calculate than the Poisson distribution, it is often the preferred choice. However, one should remember that the Poisson distribution is only defined at 0 and positive integral values of the variable x, whereas the Gaussian function is defined at all values of x.

2.4   LORENTZIAN DISTRIBUTION

There are many other distributions that appear in scientific research. Some are phenomenological distributions, created to parameterize certain data distributions. Others are well grounded in theory. One such distribution in the latter category is the Lorentzian distribution, similar but unrelated to the binomial distribution. The Lorentzian distribution is an appropriate distribution for describing data corresponding to resonant behavior, such as the variation with energy of the cross section of a nuclear or particle reaction or absorption of radiation in the Mössbauer effect.

The *Lorentzian distribution* function $P_L(x; \mu, \Gamma)$, also called the *Cauchy distribution*, is defined as

$$P_L(x; \mu, \Gamma) = \frac{1}{\pi} \frac{\Gamma/2}{(x - \mu)^2 + (\Gamma/2)^2} \tag{2.32}$$

This distribution is symmetric about its mean μ with a width characterized by its half-width Γ. The most striking difference between it and the Gaussian distribution is that it does not diminish to 0 as rapidly; the behavior for large deviations is proportional to the inverse square of the deviation, rather than exponentially related to the square of the deviation.

As with the Gaussian distribution, the Lorentzian distribution function is a continuous function, and the probability of observing a value x must be related to the interval within which the observation may fall. The probability

$dQ_L(x; \mu, \Gamma)$ for an observation to fall within an infinitesimal differential interval dx around x is given by the product of the probability function $P_L(x; \mu, \Gamma)$ and the size of the interval dx;

$$dQ_L(x; \mu, \Gamma) = P_L(x; \mu, \Gamma) \, dx \tag{2.33}$$

The normalization of the probability density function $P_L(x; \mu, \Gamma)$ is such that the integral of the probability over all possible values of x is unity:

$$\int_{-\infty}^{\infty} P_L(x; \mu, \Gamma) \, dx = \frac{1}{\pi} \int_{-\infty}^{\infty} \frac{1}{1 + z^2} \, dz = 1 \tag{2.34}$$

where $z = (x - \mu)/(\Gamma/2)$.

Mean and Half-Width

The mean μ of the Lorentzian distribution is given as one of the parameters in Equation (2.32). It is obvious from the symmetry of the distribution that μ must be equal to the mean as well as to the median and to the most probable value.

The standard deviation is not defined for the Lorentzian distribution as a consequence of its slowly decreasing behavior for large deviations. If we attempt to evaluate the expectation value for the square of the deviations

$$\sigma^2 = \left\langle (x - \mu)^2 \right\rangle = \frac{1}{\pi} \frac{\Gamma^2}{4} \int_{-\infty}^{\infty} \frac{z^2}{1 + z^2} \, dz \tag{2.35}$$

we find that the integral is unbounded: the integral does not converge for large deviations. Although it it possible to calculate a *sample standard deviation* by evaluating the average value of the square of the deviations from the sample mean, this calculation has no meaning and will not converge to a fixed value as the number of samples increases.

The width of the Lorentzian distribution is instead characterized by the *full-width at half maximum* Γ, generally called the *half-width*. This parameter is defined such that when the deviation from the mean is equal to $\frac{1}{2}$, the half-width, $x - \mu = \pm\Gamma/2$, the probability function $P_L(x; \mu, \Gamma)$ is half its value at the maximum. Thus, the half-width Γ is the full width of the curve measured at the level of half the maximum probability. We can verify that this identification of Γ with the full-width at half maximum is correct by substituting $x = \mu \pm \Gamma/2$ into Equation (2.32).

The Lorentzian and Gaussian distributions are shown for comparison in Figure 2.6, for $\mu = 10$ and $\Gamma = 2.354$ (corresponding to $\sigma = 1$ for the Gaussian function). Both distributions are normalized according to their definitions in Equations (2.23) and (2.32). For both curves, the value of the maximum probability is inversely proportional to the half-width Γ. This results in a peak value of $2/\pi\Gamma \simeq 0.270$ for the Lorentzian distribution and a peak value of $1/\sqrt{2\pi} \simeq 0.399$ for the Gaussian distribution.

Except for the normalization, the Lorentzian distribution is equivalent to the dispersion relation that is used, for example, in describing the cross section

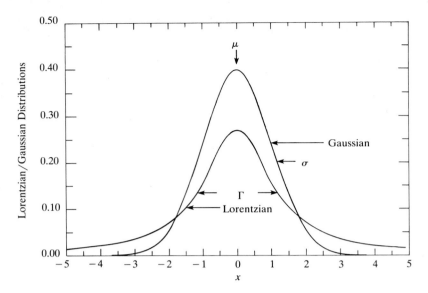

FIGURE 2.6
Comparison of normalized Lorentzian and Gaussian distributions, with $\Gamma = 2.354\sigma$.

of a nuclear reaction for a Breit-Wigner resonance:

$$\sigma = \pi\lambda^2 \frac{\Gamma_1\Gamma_2}{(E - E_0)^2 + (\Gamma/2)^2} \qquad (2.36)$$

SUMMARY

Binomial distribution: Describes the probability of observing x successes out of n tries when the probability for success in each try is p:

$$p_B(x; n, p) = \binom{n}{x}p^x q^{n-x} = \frac{n!}{x!(n-x)!}p^x(1-p)^{n-x}$$

$$\mu = np \qquad \sigma^2 = np(1-p)$$

Poisson distribution: Limiting case of the binomial distribution for large n and constant μ; appropriate for describing small samples from large populations.

$$P_P(x; \mu) = \frac{\mu^x}{x!}e^{-\mu}, \qquad \sigma^2 = \mu$$

Gaussian distribution: Limiting case of the binomial distribution for large n and finite p; appropriate for smooth symmetric distributions.

$$P_G(x;\mu,\sigma) = \frac{1}{\sigma\sqrt{2\pi}} \exp\left[-\frac{1}{2}\left(\frac{x-\mu}{\sigma}\right)^2\right]$$

Half-width $\Gamma = 2.354\sigma$; probable error P.E. $= 0.6745\sigma$.
Standard form:

$$P_G(z)\,dz = \frac{1}{\sqrt{2\pi}} \exp\left(-\frac{z}{2}\right)^2 dz$$

Lorentzian distribution: Distribution relation

$$P_L(x;\mu,\Gamma) = \frac{1}{\pi}\,\frac{\Gamma/2}{(x-\mu)^2 + (\Gamma/2)^2}$$

EXERCISES

2.1. Evaluate the following:

$$(a)\ \binom{6}{3} \qquad (b)\ \binom{4}{2} \qquad (c)\ \binom{10}{3} \qquad (d)\ \binom{52}{4}$$

2.2. Evaluate the binomial distribution $P_B(x;n,p)$ for $n=6$, $p=\frac{1}{2}$, and $x=0$ to 6. Sketch the distribution and identify the mean and standard deviation. Repeat for $p=\frac{1}{6}$.

2.3. The probability distribution of the sum of the points showing on a pair of dice is given by

$$P(x) = \frac{x-1}{36} \qquad 2 \le x \le 7$$

$$= \frac{13-x}{36} \qquad 7 \le x \le 12$$

Find the mean, median, and standard deviation of the distribution.

2.4. Show that the sum in Equation (2.6) reduces to $\mu = np$. *Hint:* Define $y = x-1$ and $m = n-1$ and use the fact that

$$\sum_{y=0}^{m}\left[\frac{m!}{y!(m-y)!}p^y(1-p)^{m-y}\right] = \sum_{y=0}^{m} P_B(y;m,p) = 1$$

2.5. On a certain kind of slot machine there are 10 different symbols that can appear in each of three windows. The machine pays off different amounts when either one, two, or three lemons appear. What should be the payoff ratio for each of the three possibilities if the machine is honest and there is no cut for the house?

2.6. Show that the sum in Equation (2.7) reduces to $\sigma^2 = np(1-p)$. *Hint:* Define $y = x-1$ and $m = n-1$ and use the results of Exercise 2.4.

2.7. At rush hour on a typical day, 25.0% of the cars approaching a fork in the street turn left and 75.0% turn right. On a particular day, 283 cars turned left and 752 turned right. Find the predicted uncertainty in these numbers and the probability that these measurements were not made on a "typical day"; that is, find the probability of obtaining a result that is as far or farther from the mean than the result measured on the particular day.

2.8. In a certain physics course, 7.3% of the students failed and 92.7% passed, averaged over many semesters.
   (a) What is the expected number of failures in a particular class of 32 students, drawn from the same population?
   (b) What is the probability that five or more students will fail?

2.9. Evaluate and plot the two Poisson distributions of Example 2.4. Plot on each graph the corresponding Gaussian distribution with the same mean and standard deviation.

2.10. Verify that, for the Poisson distribution, if μ is an integer, the probability for the $x = \mu$ is equal to probability for $x = \mu - 1$, $P_P(\mu, \mu) = P_P(\mu - 1; \mu)$.

2.11. Show that the sum in Equation (2.19) reduces to $\sigma^2 = \mu$. *Hint:* Use Equation (1.8) to simplify the expression. Define $y = x - 1$ and show that the sum reduces to $\mu\langle y + 1\rangle = \mu^2$.

2.12. Members of a large collaboration that operates a giant proton-decay detector in a salt mine near Cleveland, Ohio, detected a burst of 8 neutrinos in their apparatus coincident with the optical observation of the explosion of the Supernova 1987A.
   (a) If the average number of neutrinos detected in the apparatus is 2 per day, what is the probability of detecting a fluctuation of 8 or more in one day?
   (b) In fact, the 8 neutrinos were all detected within a 10-min period. What is the probability of detecting a fluctuation of 8 or more neutrinos in a 10-min period if the average rate is 2 per 24 hours?

2.13. In a scattering experiment to measure the polarization of an elementary particle, a total of $N = 1000$ particles were scattered from a target. Of these, 670 were observed to be scattered to the right and 330 to the left. Assume that there is no uncertainty in $N = N_R + N_L$.
   (a) Based on the experimental estimate of the probability, what is the uncertainty in N_R? In N_L?
   (b) The asymmetry parameter is defined as $A = (N_R - N_L)/(N_R + N_L)$. Calculate the experimental asymmetry and its uncertainty.
   (c) Assume that the asymmetry has been predicted to be $A = 0.400$ and recalculate the uncertainties in (a) and (b) using the predicted probability.

2.14. A problem arises when recording data with electronic counters in that the system may saturate when rates are very high, leading to a "dead time". For example, after a particle has passed through a detector, the equipment will be "dead" while the detector recovers and the electronics stores away the results. If a second particle passes through the detector in this time period, it will not be counted.
   (a) Assume that a counter has a dead time of 200 ns (200×10^{-9} s) and is exposed to a beam of 1×10^6 particles per second so that the mean number of particles hitting the counter in the 200-ns time slot is $\mu = 0.2$. From the Poisson probability for this process, find the efficiency of the counter; that is, the ratio of the average number of particles counted to the average number that pass through the counter in the 200-ns time period.
   (b) Repeat the calculation for beam rates of 2, 4, 6, 8, and 10×10^6 particles per second, and plot a graph of counter efficiency as a function of beam rate.

2.15. Show by numerical calculation that, for the Gaussian probability distribution, the full-width at half maximum Γ is related to the standard deviation by $\Gamma = 2.354\sigma$ [Equation (2.28)].

2.16. The probability that an electron is at a distance r from the center of the nucleus of a hydrogen atom is given by

$$P(r) = Cr^2e^{-r/R}$$

Find the mean radius $\bar{r}$ and the standard deviation. Find the value of the constant C.

2.17. Show that a tangent to the Gaussian function is steepest at $x = \mu \pm \sigma$, and therefore intersects the curve at the $e^{-1/2}$ points. Show also that these tangents intersect the y axis at $x = \mu \pm 2\sigma$.

CHAPTER
3

ERROR ANALYSIS

In Chapter 1 we discussed methods for extracting from a set of data points estimates of the mean and standard deviation that describe, respectively, the desired result and the uncertainties in the results. In this chapter we shall further consider how to estimate uncertainties in our measurements, the sources of the uncertainties, and how to combine uncertainties in separate measurements to find the error in a result calculated from those measurements.

3.1 INSTRUMENTAL AND STATISTICAL UNCERTAINTIES

Instrumental Uncertainties

If the quantity x has been measured with a physical instrument, the uncertainty in the measurement generally comes from fluctuations in readings of the instrumental scale, either because the settings are not exactly reproducible due to imperfections in the equipment, or because of human imprecision in observing settings, or a combination of both. Such uncertainties are called *instrumental* because they arise from a lack of perfect precision in the measuring instruments (including the observer). We can include in this category experiments that deal with measurements of such characteristics as length, mass, voltage, current, and so forth. These uncertainties are often independent of the actual value of the quantity being measured.

Instrumental uncertainties are generally determined by examining the instruments and considering the measuring procedure to estimate the reliability of the measurements. In general, one should attempt to make readings to a fraction of the smallest scale division on the instrument. For example, with a good mercury thermometer, it is often easy to estimate the level of the mercury to a least count of one-half of the smallest scale division and possibly even to one-fifth of a division. The measurement is generally quoted to plus or minus one-half of the least count, and this number represents an estimate of the standard deviation of a single measurement. Recalling that, for a Gaussian distribution, there is a 68% probability that a random measurement will lie within 1 standard deviation of the mean, we observe that our object in estimating errors is not to place outer limits on the range of the measurement, which is impossible, but to set a particular *confidence level* that a repeated measurement of the quantity will fall this close to the mean or closer. Often we choose the standard deviation, the 68% confidence level, but other levels are used as well. We shall discuss the concept of confidence levels in Chapter 11.

If it is possible to make repeated measurements, then an estimate of the standard deviation can be calculated from the spread of these measurements as discussed in Chapter 1. The resulting estimate of the standard deviation corresponds to the expected uncertainty in a single measurement. In principle, this *internal* method of determining the uncertainty should agree with that obtained by the *external* method of considering the equipment and the experiment itself, and in fact, any significant discrepancy between the two suggests a problem, such as a misunderstanding of some aspect of the experimental procedure. However, when reasonable agreement is achieved, then the standard deviation calculated internally from the data generally provides the better estimate of the uncertainties.

Statistical Uncertainties

If the measured quantity x represents the number of counts in a detector per unit time interval for a random process, then the uncertainties are called *statistical* because they arise, not from a lack of precision in the measuring instruments, but from overall statistical fluctuations in the collections of finite numbers of counts over finite intervals of time. For statistical fluctuations, we can estimate analytically the standard deviation for each observation, without having to determine it experimentally. If we were to make the same measurement repeatedly, we should find that the observed values were distributed about their mean in a Poisson distribution (as discussed in Section 2.2) instead of a Gaussian distribution. We can justify the use of this distribution intuitively by considering that we should expect a distribution that is related to the binomial distribution, but that is consistent with our boundary conditions that we can collect any positive number of counts, but no fewer than zero counts, in any time interval.

One immediate advantage of the Poisson distribution is that the standard deviation is automatically determined:

$$\sigma = \sqrt{\mu} \tag{3.1}$$

The relative uncertainty, the ratio of the standard deviation to the average rate,

$\sigma/\mu = 1/\sqrt{\mu}$, decreases as the number of counts received per interval increases. Thus our relative uncertainties are smaller when our counting rates are higher.

The value for μ to be used in Equation (3.1) for determining the standard deviation σ is, of course, the value of the mean counting rate from the parent population, of which each measurement x is only an approximate sample. In the limit of an infinite number of determinations, the average of all the measurements would very closely approximate the parent value, but often we cannot make more than one measurement of each value of x, much less an infinite number. Thus, we are forced to use $\sqrt{x}$ as an estimate of the standard deviation of a single measurement.

Example 3.1. Consider an experiment in which we count gamma rays emitted by a strong radioactive source. We cannot determine the counting rate instantaneously because no counts will be detected in an infinitesimal time interval. But we can determine the number of counts x detected over a time interval Δt, and this should be representative of the average counting rate over that interval. Assume that we have recorded 5212 counts in a 1-s time interval. The distribution of counts is random in time and follows the Poisson probability function, so our estimate of the standard deviation of the distribution is $\sigma = \sqrt{5212}$. Thus, we should record our result for the number of counts x in the time interval Δt as 5212 ± 72 and the relative error is

$$\frac{\sigma_x}{x} = \frac{\sqrt{x}}{x} = \frac{1}{\sqrt{x}} \simeq \frac{1}{72} = 0.021 = 1.4\%$$

There may also be instrumental uncertainties contributing to the overall uncertainties. For example, we can determine the time intervals with only finite precision. However, we may have some control over these uncertainties and can often organize our experiment so that the statistical errors are dominant. Suppose that the major instrumental error in our example is the uncertainty $\sigma_t = 0.01$ s in the time interval $\Delta t = 1.00$ s. The relative uncertainty in the time interval is thus

$$\frac{\sigma_t}{\Delta t} = \frac{0.01}{1.00} = 0.001 = 1.0\%$$

This relative instrumental error in the time interval will produce a 1.0% relative error in the number of counts x. Because the instrumental uncertainty is comparable to the statistical uncertainty, it might be wise to attempt a more precise measurement of the interval or to increase its length. If we increase the counting time interval from 1 s to 4 s, the number of counts x will increase by about a factor of 4 and the relative statistical error will therefore decrease by a factor of 2 to about 0.7%, whereas the instrumental uncertainty will decrease by a factor of 4 to 0.25%, as long as the instrumental uncertainty σ_t remains constant at 0.01 s.

3.2 PROPAGATION OF ERRORS

We often want to determine a dependent variable x that is a function of one or more different measured variables. We must know how to propagate or carry over the uncertainties in the measured variables to determine the uncertainty in the dependent variable.

> **Example 3.2.** Suppose we wish to find the volume V of a box of length L, width W, and height H. We can measure each of the three dimensions to be L_0, width W_0, and height H_0 and combine these measurements to yield a value for the volume:
>
> $$V_0 = L_0 W_0 H_0 \qquad (3.2)$$
>
> How do the uncertainties in the estimates L_0, W_0, and H_0, affect the resulting uncertainties in the final result V_0?

If we knew the actual errors, $\Delta L = L - L_0$ and so forth, in each dimension, we could obtain an estimate of the error in the final result V_0 by expanding V about the point (L_0, W_0, H_0) in a Taylor series. The first term in the Taylor expansion gives

$$V \simeq V_0 + \Delta L \left(\frac{\partial V}{\partial L} \right)_{W_0 H_0} + \Delta W \left(\frac{\partial V}{\partial W} \right)_{L_0 H_0} + \Delta H \left(\frac{\partial V}{\partial H} \right)_{L_0 W_0} \qquad (3.3)$$

from which we can find $\Delta V = V - V_0$. The terms in parentheses are the partial derivatives of V, with respect to each of the dimensions, L, W, and H, evaluated at the point L_0, W_0, H_0. They are the proportionality constants between changes in V and infinitesimally small changes in the corresponding dimensions. The partial derivative of V with respect to L, for example, is evaluated with the other variables W and H held fixed at the values W_0 and H_0 as indicated by the subscript. This approximation neglects higher-order terms in the Taylor expansion, which is equivalent to neglecting the fact that the partial derivatives are not constant over the ranges of L, W, and H given by their errors. If the errors are large, we must include in this definition at least second partial derivatives ($\partial^2 V / \partial L^2$, etc.) and partial cross derivatives ($\partial^2 V / \partial L \, \partial W$, etc.), but we shall omit these from the discussion that follows.

For our example of $V = LWH$, Equation (3.3) gives

$$\Delta V \simeq W_0 H_0 \, \Delta L + L_0 H_0 \, \Delta W + L_0 W_0 \, \Delta H \qquad (3.4)$$

which we could evaluate if we knew the uncertainties ΔL, ΔW, and ΔH.

Uncertainties

In general, however, we do not know the actual errors in the determination of the dependent variables (or if we do, we should make the necessary corrections). Instead, we may be able to estimate the error in each measured quantity, or to estimate some characteristic, such as the standard deviation σ, of the probability distribution of the measured qualities. How can we combine the standard deviation of the individual measurements to estimate the uncertainty in the result?

Suppose we want to determine a quantity x that is a function of at least two measured variables, u and v. We shall determine the characteristics of x from those of u and v and from the fundamental dependence

$$x = f(u, v, \ldots) \tag{3.5}$$

Although it may not always be exact, we shall assume that the most probable value for x is given by

$$\bar{x} = f(\bar{u}, \bar{v}, \ldots) \tag{3.6}$$

The uncertainty in the resulting value for x can be found by considering the spread of the values of x resulting from combining the individual measurements $u_i, v_i, \ldots$ into individual results x_i:

$$x_i = f(u_i, v_i, \ldots) \tag{3.7}$$

In the limit of an infinite number of measurements, the mean of the distribution will coincide with the average $\bar{x}$ given in Equation (3.6) and we can use the definition of Equation (1.8) to find the variance σ_x^2 (which is the square of the standard deviation σ_x):

$$\sigma_x^2 = \lim_{N \to \infty} \left[\frac{1}{N} \sum (x_i - \bar{x}_i)^2 \right] \tag{3.8}$$

Just as we expressed the deviation of V in Equation (3.4) as a function of the deviations in the dimensions L, W, and H, so we can express the deviations $x_i - \bar{x}$ in terms of the deviations $u_i - \bar{u}, v_i - \bar{v}, \ldots$ of the observed parameters

$$x_i - \bar{x} \simeq (u_i - \bar{u})\left(\frac{\partial x}{\partial u}\right) + (v - \bar{v})\left(\frac{\partial x}{\partial v}\right) + \cdots \tag{3.9}$$

where we have omitted specific notation of the fact that each of the partial derivatives is evaluated with all the other variables fixed at their mean values.

Variance and Covariance

Combining Equations (3.8) and (3.9) we can express the variance σ_x^2 for x in terms of the variances $\sigma_u^2, \sigma_v^2, \ldots$ for the variables $u, v, \ldots$, which were actually measured:

$$\sigma_x^2 \simeq \lim_{N \to \infty} \frac{1}{N} \sum \left[(u_i - \bar{u})\left(\frac{\partial x}{\partial u}\right) + (v_i - \bar{v})\left(\frac{\partial x}{\partial v}\right) + \cdots \right]^2$$

$$\simeq \lim_{N \to \infty} \frac{1}{N} \sum \left[(u_i - \bar{u})^2\left(\frac{\partial x}{\partial u}\right)^2 + (v_i - \bar{v})^2\left(\frac{\partial x}{\partial v}\right)^2 \right.$$

$$\left. + 2(u_i - \bar{u})(v_i - \bar{v})\left(\frac{\partial x}{\partial u}\right)\left(\frac{\partial x}{\partial v}\right) + \cdots \right] \tag{3.10}$$

The first two terms of Equation (3.10) can be expressed in terms of the variances σ_u^2 and σ_v^2 given by Equation (1.8):

$$\sigma_u^2 = \lim_{N \to \infty} \left[\frac{1}{N} \sum (u_i - \bar{u}_i)^2 \right] \qquad \sigma_v^2 = \lim_{N \to \infty} \left[\frac{1}{N} \sum (v_i - \bar{v}_i)^2 \right] \quad (3.11)$$

In order to express the third term of Equation (3.10) in a similar form, we introduce the *covariances* σ_{uv}^2 between the variables u and v defined analogous to the variances of Equation (3.11):

$$\sigma_{uv}^2 \equiv \lim_{N \to \infty} \left[\frac{1}{N} \sum \left[(u_i - \bar{u})(v_i - \bar{v}) \right] \right] \quad (3.12)$$

With these definitions, the approximation for the standard deviation σ_x for x given in Equation (3.10) becomes

$$\sigma_x^2 \simeq \sigma_u^2 \left(\frac{\partial x}{\partial u} \right)^2 + \sigma_v^2 \left(\frac{\partial x}{\partial v} \right)^2 + \cdots + 2\sigma_{uv}^2 \left(\frac{\partial x}{\partial u} \right) \left(\frac{\partial x}{\partial v} \right) + \cdots \quad (3.13)$$

This is the *error propagation equation*.

The first two terms in the equation are averages of squares of deviations weighted by the squares of the partial derivatives, and may be considered to be the averages of the squares of the deviations in x produced by the uncertainties in u and in v, respectively. In general, these terms dominate the uncertainties. If there are additional variables besides u and v in the determination of x, their contributions to the variance of x will have similar terms.

The third term is the average of the cross terms involving products of deviations in u and v weighted by the product of the partial derivatives. If the fluctuations in the measured quantities u and $v, \ldots$ are uncorrelated, then, on the average, we should expect to find equal distributions of positive and negative values for this term, and we should expect the term to vanish in the limit of a large random selection of observations. This is often a reasonable approximation and Equation (3.13) then reduces to

$$\sigma_x^2 \simeq \sigma_u^2 \left(\frac{\partial x}{\partial u} \right)^2 + \sigma_v^2 \left(\frac{\partial x}{\partial v} \right)^2 + \cdots \quad (3.14)$$

with similar terms for additional variables. In general, we shall use Equation (3.14) for determining the effects of measuring uncertainties on the final result and shall neglect the covariant terms. However, as we shall see in Chapter 7, the covariant terms often make important contributions to the uncertainties in parameters determined by fitting curves to data by the least-squares method.

3.3 SPECIFIC ERROR FORMULAS

The expressions of Equations (3.13) and (3.14) were derived for the general relationship of Equation (3.5) giving x as an arbitrary function of u and $v, \ldots$. In the following specific cases of functions $f(u, v, \ldots)$, the parameters a and b are defined as positive constants and u and v are variables.

Simple Sums and Differences

If the dependent variable x is related to a measured quantity u by the relation

$$x = u \pm a \qquad (3.15)$$

then the partial derivative $\partial x/\partial u = 1$ and the uncertainty in x is just

$$\sigma_x = \sigma_u \qquad (3.16)$$

and the relative uncertainty is given by

$$\frac{\sigma_x}{x} = \frac{\sigma_u}{x} = \frac{\sigma_u}{u \pm a} \qquad (3.17)$$

Note that if we are dealing with a difference $(x = u - a)$, the uncertainty in x can be greater than x.

Example 3.3. In an experiment to count particles emitted by a decaying radioactive source, we measure $N_1 = 723$ counts in a 15-s time interval at the beginning of the experiment and $N_2 = 19$ counts in a 15-s time interval later in the experiment. The events are random and obey Poisson statistics so that we know that the uncertainties in N_1 and N_2 are just their square roots. Assume that we have made a very careful measurement of the background radiation in the absence of the radioactive source and obtained a value $B = 14.2$ counts with negligible error for the same time interval Δt. Because we have averaged over a long time period, the mean number of background counts in the 15-s interval is not an integral number.

For the first time interval, the corrected number of counts is

$$x_1 = N_1 - B = 723 - 14.2 = 708.8 \text{ counts}$$

The uncertainty in x_1 is given by

$$\sigma_{x_1} = \sigma_{N_1} = \sqrt{723} \simeq 26.9 \text{ counts}$$

and the relative uncertainty is

$$\frac{\sigma_x}{x} = \frac{26.9}{708} = 0.038 \simeq 3.8\%$$

For the second time interval, the corrected number of events is

$$x_2 = N_2 - B = 19 - 14.2 \simeq 4.8 \text{ counts}$$

The uncertainty in x is given by

$$\sigma_{x_2} = \sigma_{N_2} = \sqrt{19} \simeq 4.4 \text{ counts}$$

and the relative uncertainty in x is

$$\frac{\sigma_x}{x} \simeq \frac{4.4}{4.8} = 0.91$$

Weighted Sums and Differences.

If x is the weighted sum of u and v,

$$x^2 = au \pm bv \qquad (3.18)$$

the partial derivatives are simply the constants

$$\left(\frac{\partial x}{\partial u}\right) = a \qquad \left(\frac{\partial x}{\partial v}\right) = \pm b \qquad (3.19)$$

and we obtain

$$\sigma_x^2 = a^2\sigma_u^2 + b^2\sigma_v^2 \pm ab\sigma_{uv}^2 \qquad (3.20)$$

Note the possibility that the variance σ_x^2 might vanish if the covariance σ_{uv}^2 has the proper magnitude and sign. This could happen in the unlikely event that the fluctuations were completely correlated so that each erroneous observation of u was exactly compensated for by a corresponding erroneous observation of v.

Example 3.4. Suppose that, in the previous example, the background radiation B had not been averaged over a long time period but was simply measured for 15 s to give $B = 14$ with standard deviation $\sigma_B = \sqrt{14} \simeq 3.7$ counts. Then the uncertainty in x would be given by

$$\sigma_x^2 = \sigma_N^2 + (-\sigma_B)^2 = N + B$$

because the uncertainties in N and B are equal to their square roots.

For the first time interval, we would calculate

$$x_1 = (723 - 14) \pm \sqrt{723 + 14} = 709 \pm 27.1 \text{ counts}$$

and the relative uncertainty is

$$\frac{\sigma_x}{x} = \frac{27.1}{709} \simeq 0.038$$

For the second time interval, we would calculate

$$x_2 = (19 - 14) \pm \sqrt{19 + 14} = 5 \pm 5.7 \text{ counts}$$

and the relative uncertainty is

$$\frac{\sigma_x}{x} = \frac{5.7}{5} \simeq 1.1$$

Multiplication and Division.

If x is the weighted product of u and v,

$$x = \pm auv \qquad (3.21)$$

the partial derivatives of each variable are functions of the other variable,

$$\left(\frac{\partial x}{\partial u}\right) = \pm av \qquad \left(\frac{\partial x}{\partial u}\right) = \pm bu \tag{3.22}$$

and the standard deviation of x becomes

$$\sigma_x = (av\sigma_u) + (au\sigma_v) + 2auv\sigma_{uv} \tag{3.23}$$

which can be expressed more symmetrically as

$$\frac{\sigma_x^2}{x^2} = \frac{\sigma_u^2}{u^2} + \frac{\sigma_v^2}{v^2} + 2\frac{\sigma_{uv}^2}{uv} \tag{3.24}$$

Similarly, if x is obtained through division,

$$x = \pm \frac{au}{v} \tag{3.25}$$

the variance for x is given by

$$\frac{\sigma_x^2}{x^2} = \frac{\sigma_u^2}{u^2} + \frac{\sigma_v^2}{v^2} - 2\frac{\sigma_{uv}^2}{uv} \tag{3.26}$$

Example 3.5. The area of a triangle is equal to half the product of the base times the height $A = bh/2$. If the base and height have values $b = 5.0 \pm 0.1$ cm and $h = 10.0 \pm 0.3$ cm, the area is $A = 25.0$ cm^2 and the uncertainty in the area is given by

$$\frac{\sigma_A^2}{A^2} = \frac{\sigma_b^2}{b^2} + \frac{\sigma_h^2}{h^2} \tag{3.27}$$

or

$$\sigma_A^2 = A^2\left(\frac{\sigma_b^2}{b^2} + \frac{\sigma_h^2}{h^2}\right)$$

$$= 25^2(\text{cm}^4)\left(\frac{0.1^2}{5^2} + \frac{0.3^2}{10^2}\right)(\text{cm}^2/\text{cm}^2)$$

$$\simeq 0.81 \text{ cm}^2$$

Although the absolute uncertainty in the height is 3 times the absolute uncertainty in the base, the relative uncertainty σ_h is only $1\frac{1}{2}$ times as large and its contribution to the variance of the area is only $(1\frac{1}{2})^2$ as large.

Powers

If x is obtained by raising the variable u to a power

$$x = au^{\pm b} \tag{3.28}$$

the derivative of x with respect to u is

$$\left(\frac{\partial x}{\partial u}\right) = \pm abu^{\pm b-1} = \pm \frac{bx}{u} \tag{3.29}$$

and relative error in x becomes

$$\frac{\sigma_x}{x} = \pm b \frac{\sigma_u}{u} \tag{3.30}$$

For the special cases of $b = +1$, we have

$$x = au \qquad \sigma_x = a\sigma_u$$

so

$$\frac{\sigma_x}{x} = \frac{\sigma_u}{u} \tag{3.31}$$

For $b = -1$, we have

$$x = \frac{a}{u} \qquad \sigma_x = -\frac{a\sigma_u}{u^2}$$

so

$$\frac{\sigma_x}{x} = -\frac{\sigma_u}{u} \tag{3.32}$$

The negative sign indicates that, in division, a positive error in u will produce a corresponding negative error in x.

> **Example 3.6.** The area of a circle is proportional to the square of the radius $A = \pi r^2$. If the radius is determined to be $r = 10.0 \pm 0.3$ cm, the area is $A = 100\pi$ cm^2 with an uncertainty given by
>
> $$\frac{\sigma_A}{A} = 2\frac{\sigma_r}{r}$$
>
> or $\quad \sigma_A = 2A\frac{\sigma_r}{r} = 2\pi(10.0\text{ cm})^2(0.3\text{ cm})/(10.0\text{ cm}) = 6\text{ cm}^2$

Exponentials

If x is obtained by raising the natural base to a power proportional to u,

$$x = ae^{\pm bu} \tag{3.33}$$

the derivative of x with respect to u is

$$\frac{\partial x}{\partial u} = \pm abe^{\pm bu} = \pm bx \tag{3.34}$$

and the relative uncertainty becomes

$$\frac{\sigma_x}{x} = \pm b\sigma_u \tag{3.35}$$

If the constant that is raised to the power is not equal to e, the expression can be rewritten as

$$x = a^{\pm bu} \tag{3.36}$$

$$= (e^{\ln a})^{\pm bu} = e^{\pm(b\ln a)u}$$

$$= e^{\pm cu} \quad \text{with } c = b\ln a$$

where ln indicates the natural logarithm. Solving in the same manner as before we obtain

$$\frac{\sigma_x}{x} = \pm c\sigma_u = \pm(b \ln a)\sigma_u \tag{3.37}$$

Logarithms

If x is obtained by taking the logarithm of u,

$$x = a \ln(\pm bu) \tag{3.38}$$

the derivative with respect to u is

$$\frac{\partial x}{\partial u} = \frac{a}{u} \tag{3.39}$$

so

$$\sigma_x = a\frac{\sigma_u}{u} \tag{3.40}$$

These relations can be useful for making quick estimates of the uncertainty in a calculated quantity caused by the uncertainty in a measured variable. For a simple product or quotient of the measured variable u with a constant, a 1% error in u causes a 1% error in x. If u is raised to a power b, the resulting error in x becomes b% for a 1% uncertainty in u. Even if the complete expression for x involves other measured variables, $x = f(u, v, \ldots)$ and is considerably more complicated than these simple examples, it is often possible to use these relations to make approximate estimates of uncertainties.

3.4 APPLICATION OF ERROR EQUATIONS

Even for relatively simple calculations, such as those encountered in undergraduate laboratory experiments, blind application of the general error propagation expression [Equation (3.14)] can lead to very lengthy and discouraging equations, especially if the final results depend on several different measured quantities. Often the error equations can be simplified by neglecting terms that make negligible contributions to the final uncertainty, but this requires a certain amount of practice.

Approximations

The student should attempt to make quick, approximate estimates of the various contributions to the uncertainty in the final result by considering separately the terms in Equation (3.14). A convenient rule of thumb is to neglect terms that make final contributions that are less than 10% of the largest contribution. (Like all rules of this sort, one should be wary of special cases. Several smaller contributions to the final uncertainty can sum to be as important as one larger uncertainty.)

Example 3.7. Suppose that the area of a rectangle $A = LW$ is to be determined from the following measurements of the lengths of two sides:

$$L = 22.1 \pm 0.1 \text{ cm} \qquad W = 7.3 \pm 0.1 \text{ cm}$$

The relative contribution of σ_L to the error in L will be

$$\frac{\sigma_{A_L}}{A} = \frac{\sigma_L}{L} = \frac{0.1}{22.1} \simeq 0.005$$

and the corresponding contribution of σ_W will be

$$\frac{\sigma_{A_W}}{A} = \frac{\sigma_W}{W} = \frac{0.1}{7.3} \simeq 0.014$$

The contribution from σ_L is thus about one-third of that from σ_W. However, when the contributions are combined, we obtain

$$\sigma_A = A\sqrt{0.014^2 + 0.005^2}$$

which can be expanded to give

$$\sigma_A \simeq 0.014A\left(1 + \frac{1}{2}\left(\frac{0.005}{0.014}\right)^2\right) \simeq 0.014A(1 + 0.06) = 0.015A$$

Thus, the effective contribution from σ_L is only about 6% of the effective contribution from σ_W and could safely be neglected in this calculation.

Computer Calculation of Uncertainties

Finding analytic forms for the partial derivatives is sometimes quite difficult. One should always break Equation (3.14) into separate components and not attempt to find one complete equation that incorporates all error terms. In fact, if the analysis is being done by computer, it may not even be necessary to find the derivatives explicitly. The computer can find numerically the variations in the dependent variable caused by variations in each independent, or measured, variable.

Suppose that we have a particularly complicated equation, or set of equations, relating our final result x to the individually measured variables u, v, and so forth. Let us assume that the actual equations are programmed as a computer function `Calculate`, which returns the single variable x when called with arguments corresponding to the measured parameters

```
x = Calculate(u,v,w...)
```

We shall further assume that correlations are small so that the covariances may be ignored. Then, to find the variations of x with the measured quantities u, v, and so forth, we can make successive calls to the function of the form

```
dXu = Calculate(u + du,v,w, ··· ) - x,
dXv = Calculate(u,v + dv,w, ··· ) - x,
dXw = Calculate(u,v,w + dw, ··· ) - x,
```
    etc.

where d u, d v, d w, and so forth are the standard deviations σ_u, σ_v, σ_w, and so on. The resulting contributions to the uncertainty in x are combined in quadrature as

$$dX = sqrt(\ sqr(dXu) + sqr(dXv) + sqr(dXw) + \cdots)$$

Note that it would not be correct to incorporate all the variations into one equation such as

$$dX = Calculate(u + du, v + dv, w + dw, \cdots) - x$$

because this would imply that the errors d u, d v, and so on were actually known quantities, rather than independent, estimated variations of the measured quantities, corresponding to estimates of the widths of the distributions of the measured variables.

SUMMARY

Covariance: $\sigma_{uv}^2 = \langle (u - \bar{u})(v - \bar{v}) \rangle$.
Propagation of errors: Assume $x = f(u, v)$:

$$\sigma_x^2 = \sigma_u^2 \left(\frac{\partial x}{\partial u} \right)^2 + \sigma_v^2 \left(\frac{\partial x}{\partial v} \right)^2 + 2\sigma_{uv}^2 \left(\frac{\partial x}{\partial u} \right) \left(\frac{\partial x}{\partial v} \right)$$

For u and v uncorrelated, $\sigma_{uv}^2 = 0$.
Specific formulas:

$$x^2 = au \pm bv \qquad \sigma_x^2 = a^2\sigma_u^2 + b^2\sigma_v^2 \pm ab\sigma_{uv}^2$$

$$x = \pm auv \qquad \frac{\sigma_x^2}{x^2} = \frac{\sigma_u^2}{u^2} + \frac{\sigma_v^2}{v^2} + 2\frac{\sigma_{uv}^2}{uv}$$

$$x = \pm \frac{au}{v} \qquad \frac{\sigma_x^2}{x^2} = \frac{\sigma_u^2}{u^2} + \frac{\sigma_v^2}{v^2} - 2\frac{\sigma_{uv}^2}{uv}$$

$$x = au^{\pm b} \qquad \frac{\sigma_x}{x} = \pm b\frac{\sigma_u}{u}$$

$$x = ae^{\pm bu} \qquad \frac{\sigma_x}{x} = \pm b\sigma_u$$

$$x = a^{\pm bu} \qquad \frac{\sigma_x}{x} = \pm (b \ln a)\sigma_u$$

$$x = a \ln(\pm bu) \qquad \sigma_x = a\frac{\sigma_u}{u}$$

EXERCISES

3.1. Find the uncertainty σ_x in x as a function of the uncertainties σ_u and σ_v in u and v for the following functions:
(a) $x = 1/2(u + v)$ (b) $x = 1/2(u - v)$ (c) $x = 1/u^2$
(d) $x = uv^2$ (e) $x = u^2 + v^2$

3.2. If the diameter of a round table is determined to within 1%, how well is its area known? Would it be better to determine its radius to within 1%?

3.3. The resistance R of a cylindrical conductor is proportional to its length L and inversely proportional to its cross-sectional area $A = \pi r^2$. Which should be determined with higher precision, r or L, to optimize the determination of R? How much higher?

3.4. The initial activity N_0 and the lifetime τ of a radioactive source are known with uncertainties of 1% each. For estimation of the activity $N = N_0 e^{-t/\tau}$, the error in the initial activity N_0 dominates for small t and vice versa for large t (compared with τ). For what value of t/τ do the errors in N_0 and r contribute equally to the uncertainty in N?

3.5. Snell's law relates the angle of refraction θ_2 of a light ray travelling in a medium of index of refraction n_2 to the angle of incidence θ_1 of a ray travelling in a medium of index n_1 through the equation $n_2 \sin \theta_2 = n_1 \sin \theta_1$. Find n_2 and its uncertainty from the following measurements:

$$\theta_1 = (22.03 \pm 0.2)° \qquad \theta_2 = (14.45 \pm 0.2)° \qquad n_1 = 1.0000$$

Assume that there is no uncertainty in n_1.

3.6. The change in frequency produced by the Doppler shift when a sound source of frequency ν is moving with velocity v toward a fixed observer is given by $\Delta\nu = u\nu/(u - v)$, where u is the velocity of sound. Find the uncertainty in $\Delta\nu$ resulting from the measuring uncertainties in u, ν, and v. Assume the values $u = (332 \pm 5)$ m/s, $\nu = (1000 \pm 1)$ Hz, and $v = (0.123 \pm 0.003)$ m/s.

3.7. The radius of a circle can be calculated from measurements of the length L of a chord and the distance h from the chord to the circumference of the circle from the equation $R = L^2/2h + h/2$. Calculate the radius and its uncertainty from the following values of L and h.
(*a*) $L = (125.0 \pm 5.0)$ cm, $h = (0.51 \pm 0.22)$ cm.
(*b*) $L = (125.0 \pm 5.0)$ cm, $h = (57.4 \pm 1.2)$ cm.
Was it necessary to use the second term to calculate R in both (*a*) and (*b*)? Explain.

3.8. After measuring the speed of sound u several times, a student concludes that the standard deviation of his measurements is $\sigma = 12$ m/s. Assume that the uncertainties are random and that the experiment is not limited by systematic effects and determine how many measurements would be required to give a final uncertainty in the mean of ± 2.0 m/s.

3.9. Students in the undergraduate laboratory recorded the following counts in 1-min intervals from a radioactive source. The nominal mean decay rate from the source is 3.7 decays per minute.

Decays per minute	0	1	2	3	4	5	6	7	8	9	10
Frequency of occurrence	1	9	20	24	19	11	11	0	3	1	1

(*a*) Find the mean decay rate and its standard deviation. Compare the standard deviation to the value expected from the Poisson distribution for the mean value that you obtained.
(*b*) Plot a histogram of the data and show Poisson curves of both the parent and observed distributions.

3.10. Find by numerical integration the probability of observing a value from the Gaussian distribution that is:

    (*a*) More than 1 standard deviation (σ) from the mean.

    (*b*) More than 2 standard deviations from the mean.

    (*c*) More than 3 standard deviations from the mean.

3.11. Find by numerical integration the probability of observing a value from the Lorentzian distribution that is:

    (*a*) More than 1 half-width ($\Gamma/2$) from the mean.

    (*b*) More than 2 half-widths from the mean.

    (*c*) More than 3 half-widths from the mean.

CHAPTER 4

ESTIMATES
OF MEAN
AND ERRORS

4.1 METHOD OF LEAST SQUARES

In Chapter 2 we defined the mean μ of the parent distribution and noted that the most probable estimate of the mean μ of a random set of observations is the average $\bar{x}$ of the observations. The justification for that statement is based on the assumption that the measurements are distributed according to the Gaussian distribution. In general, we expect the distribution of measurements to be either Gaussian or Poisson, but because these distributions are indistinguishable for most physical situations we can assume the Gaussian distribution is obeyed.

Method of Maximum Likelihood

Assume that, in an experiment, we have observed a set on N data points that are randomly selected from the infinite set of the parent population, distributed according to the parent distribution. If the parent distribution is Gaussian with mean μ and standard deviation σ, the probability dQ_i for making any single observation x_i within an interval dx is given by

$$dQ_i = P_i \, dx \tag{4.1}$$

with probability function, $P_i \equiv P_G(x_i; \mu, \sigma)$,

$$P_i = \frac{1}{\sigma\sqrt{2\pi}} \exp\left[-\frac{1}{2}\left(\frac{x_i - \mu}{\sigma}\right)^2\right] \tag{4.2}$$

Because, in general, we do not know the mean μ of the distribution for a physical experiment, we must estimate it from some experimentally derived parameter. Let us call the estimate μ'. What formula for deriving μ' from the data will yield the maximum likelihood that the parent distribution had a mean equal to μ?

If we hypothesize a trial distribution with a mean μ' and standard deviation $\sigma' = \sigma$, the probability of observing the value x_i is given by the probability function

$$P_i(\mu') = \frac{1}{\sigma\sqrt{2\pi}} \exp\left[-\frac{1}{2}\left(\frac{x_i - \mu'}{\sigma}\right)^2\right] \tag{4.3}$$

Considering the entire set of N observations, the probability for observing that particular set is given by the product of the individual probability functions, $P_i(\mu')$,

$$P(\mu') = \prod_{i=1}^{N} P_i(\mu') \tag{4.4}$$

where the symbol $\prod$ denotes the product of the N probabilities $P_i(\mu')$.

The product of the constants multiplying the exponential in Equation (4.3) is the same as the product to the Nth power, and the product of the exponentials is the same as the exponential of the sum of the arguments. Therefore, Equation (4.4) reduces to

$$P(\mu') = \left(\frac{1}{\sigma\sqrt{2\pi}}\right)^N \exp\left[-\frac{1}{2}\sum\left(\frac{x_i - \mu'}{\sigma}\right)^2\right] \tag{4.5}$$

According to the *method of maximum likelihood*, if we compare the probabilities $P(\mu')$ of obtaining our set of observations from various parent populations with different means μ' but with the same standard deviation $\sigma' = \sigma$, the probability is greatest that the data were derived from a population with $\mu' = \mu$; that is, the most likely population from which such a set of data might have come is assumed to be the correct one.

Calculation of the Mean

The method of maximum likelihood states that the most probable value for μ' is the one that gives the maximum value for the probability $P(\mu')$ of Equation (4.5). Because this probability is the product of a constant times an exponential to a negative argument, maximizing the probability $P(\mu')$ is equivalent to minimizing the argument X of the exponential,

$$X = -\frac{1}{2}\sum\left(\frac{x_i - \mu'}{\sigma}\right)^2 \tag{4.6}$$

To find the minimum value of a function X we set the derivative of the function to 0,

$$\frac{dX}{d\mu'} = -\frac{d}{d\mu'} \frac{1}{2} \sum \left(\frac{x_i - \mu'}{\sigma} \right)^2 = 0 \tag{4.7}$$

and obtain

$$\frac{dX}{d\mu'} = -\frac{1}{2} \sum \frac{d}{d\mu'} \left(\frac{x_i - \mu'}{\sigma} \right)^2 = \sum \left(\frac{x_i - \mu'}{\sigma^2} \right) = 0 \tag{4.8}$$

which, because σ is a constant, gives

$$\mu' = \bar{x} \equiv \frac{1}{N} \sum x_i \tag{4.9}$$

Thus, the maximum likelihood method for estimating the mean by maximizing the probability $P(\mu')$ of Equation (4.5) shows that the most probable value of the mean is just the average $\bar{x}$ as defined in Equation (1.1).

Estimated Error in the Mean

What uncertainty σ is associated with our determination of the mean μ' in Equation (4.9)? We have assumed that all data points x_i were drawn from the same parent distribution and were thus obtained with an uncertainty characterized by the same standard deviation σ. Each of these data points contributes to the determination of the mean μ' and therefore each data point contributes some uncertainty to the determination of the final results. A histogram of our data points would follow the Gaussian shape, peaking at the value μ' and exhibiting a width corresponding to the standard deviation σ. Clearly we are able to determine the mean to much better than $\pm\sigma$, and our determination will improve as we increase the number of measured points N and are thus able to improve the agreement between our experimental histogram and the smooth Gaussian curve.

In Chapter 3 we developed the error propagation equation [see Equation (3.13)] for finding the contribution of the uncertainties in several terms contributing to a single result. Applying this relation to Equation (4.9) to find the variance σ_μ^2 of the mean μ', we obtain

$$\sigma_\mu^2 = \sum \left[\sigma_i^2 \left(\frac{\partial \mu'}{\partial x_i} \right)^2 \right] \tag{4.10}$$

where the variance σ_i^2 in each measured data point x_i is weighted by the square of the effect $\partial \mu'/\partial x_i$ that that data point has on the result. This approximation neglects correlations between the measurements x_i as well as second- and higher-order terms in the expansion of the variance σ_μ^2, but it should be a reasonable approximation as long as none of the data points contributes a major portion of the final result.

If the uncertainties of the data points are all equal $\sigma_i = \sigma$, the partial derivatives in Equation (4.10) are simply

$$\frac{\partial \mu'}{\partial x_i} = \frac{\partial}{\partial x_i} \left(\frac{1}{N} \sum x_i \right) = \frac{1}{N} \tag{4.11}$$

and combining Equations (4.10) and (4.11), we obtain

$$\sigma_\mu^2 = \sum \left[\sigma_i^2 \left(\frac{1}{N} \right)^2 \right] = \frac{\sigma^2}{N} \tag{4.12}$$

for the estimated error in the mean σ_μ. Thus, the standard deviation of our determination of the mean μ' and, therefore, the precision of our estimate of the quantity μ, improves as the square root of the number of measurements.

The standard deviation σ of the parent population can be estimated from a consideration of the measuring equipment and conditions, or internally from the data, according to Equation (1.8):

$$\sigma \simeq s = \sqrt{\frac{1}{N-1} \sum (x_i - \bar{x})^2} \tag{4.13}$$

which gives for the uncertainty σ_μ in the determination of the mean

$$\sigma_\mu = \frac{\sigma}{\sqrt{N}} \simeq \frac{s}{\sqrt{N}} \tag{4.14}$$

In principle, the value of σ obtained from Equation (4.13) should be consistent with the estimate made from the experimental equipment.

> **Example 4.1.** We return to the student's measurement of the length of a block (Example 1.2). Let us assume that the length had been established previously by other, careful measurements to be 20.000 cm. The student measures the block 100 times and concludes, from a consideration of his ruler, the geometry of the block, and his own ability, that the uncertainty in each of his individual measurements is approximately ± 0.5 cm. All measurements are made with the same experimental apparatus, and the finite precision of this apparatus, characterized by σ, results in a spread of observations grouped around the known length $x = 20.000$ cm. A histogram of these data is displayed in Figure 1.2.

Because the uncertainties in all the data points are equal, $\sigma_i = \sigma$, the student estimates the length of the block according to Equation (4.9) to be $\mu \simeq x \equiv 19.942$ cm with a standard deviation from Equation (4.13) of 0.522. From Equation (4.14), he estimates the uncertainty in his determination of the mean to be $\sigma_\mu = \sigma / \sqrt{N} = 0.522 / \sqrt{100} = 0.0522$. He quotes his result as $L = 19.942 \pm 0.052$ cm.

We note that the measured mean length is about 1.1 standard deviations away from the known length of the block. Checking the integral of the Gaussian probability equation in Table C.2, we observe that we should expect a measurement to be within 1.1σ about 73% of the time, or, on the other hand, we should expect to exceed 1.1σ about 27% of the time.

A Warning About Statistics

Equation (4.12) might suggest that the error in the mean of a set of measurements x_i can be reduced indefinitely by repeated measurements of x_i. We should be aware of the limitations of this equation before assuming that an experimental result can be improved to any desired degree of accuracy if we are willing to do enough work. There are three main limitations to consider, those of available time and resources, those imposed by systematic errors, and those imposed by nonstatistical fluctuations.

The first of these limitations is a very practical one. It may not be possible to take enough repeated measurements to make a significant improvement in the standard deviation of the result. The student of Example 1.2 may be able to make 100 measurements of the length of a block, but it is unlikely that he would have the patience to make 4 times as many measurements to cut the uncertainty by a factor of 2. Similarly, an experiment at a particle accelerator may be assigned 1000 hours of beam time. It is highly improbable that 16,000 hours would be made available to improve the precision of the result by a factor of 4.

Every experiment is subject to systematic errors at some level. Even after every possible effort has been made to understand the experimental equipment, and correct for all known defects and errors of calibration, there comes a point at which further knowledge is unobtainable. Some of the difficulties in obtaining an accurate measurements of the length of a block are obvious. The calibration of our meter stick may not be known to be a very high precision, and any irregularities of the block itself may bias the measurements. At the particle accelerator, the physicist may find limits imposed by uncertainty about the dead times of counters and the angular acceptance of his apparatus.

The phrase "nonstatistical fluctuations" can hide a multitude of sins, or at least problems, in our experiments. It is a rare experiment that follows the Gaussian distribution beyond 3 or 4 standard deviations. More likely, some unexplained data points will appear far from the mean. Such points may imply the existence of other contaminating points within the central probability region, masked by the large body of good points. A thorough study of background effects and sources of possible contaminating is obviously required, but at some level, these effects are bound to limit the accuracy of the experiment.

What are we to make of those unexpected points that appear in our data plots well beyond their level of probability? Some may arise from a chance careless measurement. Did our attention wander at the instant when we should have recorded the data point? Did we accidentally interchange two digits in writing down our meter stick measurement? Perhaps we can understand and make corrections for some of these effects. Other anomalies in the data may be caused by equipment malfunction. Did our electronic detector respond to a particularly striking clash of metal from the local all-powerful rock radio station? Did our trusty computer decide to service a student's video game rather than respond to an urgent data interrupt? And was the distribution that we chose to represent our data the correct one for this experiment?

We may be able to make corrections for these problems, once we are aware of their existence, but there are always others. At some level, things will happen that we can not understand, and for which we can not make corrections, and these "things" will cause data to appear where statistically no data should exist and data points to vanish that should have been there. The moral is, be aware and do not trust statistics in the tails of the distributions.

Elimination of Data Points

There will be occasions when we feel justified in eliminating or correcting outlying data points. For example, suppose that, among the measurements of the block length in Example 1.2, one had been recorded as 91.2 cm. The student would likely conclude that he had meant to write 19.2 cm and either ignore or correct the point.

What if one measurement had been recorded as 22.2 cm? Should any action be taken? The point is about 4 standard deviations away from the mean of all the data points, and referring to Table C.2 we see that there is about a 0.06% probability of obtaining in a single measurement a value that is that far from the mean. Thus, in a hundred measurements we should expect to collect about $100 \times 0.0006 = 0.06$ such events.

The established condition for discarding data in such circumstances is known as Chauvenet's criterion, which states that we should discard a data point if we expect less than half an event to be further from the mean than the suspect point. If our sample point exceeds this requirement and, as long as we are convinced that our data do indeed follow the Gaussian distribution, we may discard the point with reasonable confidence and recalculate the mean and standard deviation.

Removing an outlying point has a greater effect on the standard deviation than on the mean of a data sample, because the standard deviation depends on the squares of the deviations. Deleting one such point will lead to a smaller standard deviation and perhaps another point or two will now become candidates for rejection. We should be very cautious about changing data unless we are confident that we understand the source of the problem we are seeking to correct, and repeated point deletion is generally not recommended. The importance of keeping good records of any changes to the data sample should also be emphasized.

Weighting the Data—Nonuniform Uncertainties

In developing the probability $P(\mu')$ of Equation (4.5) from the individual probabilities $P_i(\mu')$ of Equation (4.3), we assumed that the data points were all extracted from the same parent population. In some circumstances, however, some data points might be measured with better or worse precision than others. We can express this quantitatively by assuming parent distributions with the same mean μ but with different standard deviations σ_i.

If we assign to each data point x_i, its own standard deviation σ_i, representing the precision with which that particular data point was measured, Equation (4.5) for the probability $P(\mu')$ that the observed set of N data points come from parent distributions with means $\mu_i = \mu'$ and standard deviations σ_i becomes

$$P(\mu') = \prod_{i=1}^{n} \left(\frac{1}{\sigma_i \sqrt{2\pi}} \right) \exp\left[-\frac{1}{2} \sum \left(\frac{x_i - \mu'}{\sigma_i} \right)^2 \right] \qquad (4.15)$$

Using the method of maximum likelihood, we must maximize this probability, which is equivalent to minimizing the argument in the exponential

$$-\frac{1}{2} \frac{d}{d\mu'} \sum \left(\frac{x_i - \mu'}{\sigma_i} \right)^2 = \sum \left(\frac{x_i - \mu'}{\sigma_i^2} \right) = 0 \qquad (4.16)$$

The most probable value is therefore the weighted average of the data points

$$\mu' = \frac{\sum(x_i/\sigma_i^2)}{\sum(1/\sigma_i^2)} \qquad (4.17)$$

where each data point x_i in the sum is weighted inversely by its own variance σ_i^2.

Error in the Weighted Mean

If the uncertainties of the data points are not equal, we evaluate $\partial\mu'/\partial x_i$ from the expression of Equation (4.17) for the mean μ':

$$\frac{\partial\mu'}{\partial x_i} = \frac{\partial}{\partial x_i} \frac{\sum(x_i/\sigma_i^2)}{\sum(1/\sigma_i^2)} = \frac{1/\sigma_i^2}{\sum(1/\sigma_i^2)} \qquad (4.18)$$

Substituting this result into Equation (4.10) yields a general formula for the uncertainty of the mean σ:

$$\sigma_\mu^2 = \sum \frac{1/\sigma_i^2}{[\sum(1/\sigma_i^2)]^2} = \frac{1}{\sum(1/\sigma_i^2)} \qquad (4.19)$$

Relative Uncertainties

It may be that the relative values of σ_i are known, but the absolute magnitudes are not. For example, if one set of data is acquired with one scale range and another set with a different scale range, the σ_i may be equal within each set but differ by a known factor between the two sets, as would be the case if σ_i were proportional to the scale range. In such a case, the *relative* values of the σ_i should be included as weighting factors in the determination of the mean μ and its uncertainty, and the *absolute* magnitudes of the σ_i can be estimated from the dispersion of the data points around the mean.

Let us define weighting factors w_i such that

$$kw_i = 1/\sigma_i^2, \tag{4.20}$$

where k is an unknown scaling constant and the σ_i are the standard deviations associated with each measurement. We assume that the weights w_i are known but that the absolute values of the standard deviations σ_i are not. Then, Equation (4.17) can be written

$$\mu' = \frac{\Sigma\left(x_i/\sigma_i^2\right)}{\Sigma\left(1/\sigma_i^2\right)} = \frac{\Sigma kw_i x_i}{\Sigma kw_i} = \frac{\Sigma w_i x_i}{\Sigma w_i} \tag{4.21}$$

and the result depends only on the relative weights and not on the absolute magnitudes of the σ_i.

To find the error in the estimate μ' of the mean we must calculate a weighted *average variance* of the data:

$$\sigma^2 = \frac{\Sigma w_i(x_i - \mu')^2}{\Sigma w_i} \times \frac{N}{(N-1)} = \left(\frac{\Sigma w_i x_i^2}{\Sigma w_i} - \mu'^2\right) \times \frac{N}{(N-1)} \tag{4.22}$$

where the last factor corrects for the fact that the mean μ' was itself determined from the data. We may recognize the expression in brackets as the difference between the weighted average of the squares of our measurements x_i and the square of the weighted average. The variance of the mean can then be determined by substituting the expression for σ^2 from Equation (4.22) into Equation (4.14):

$$\sigma_\mu^2 = \frac{\sigma^2}{N} \tag{4.23}$$

If they are required, the value of the scaling constant k and of the values of the separate variances σ_i can be estimated by equating the two expressions of σ_μ of Equations (4.14) and (4.19) and replacing $1/\sigma_i$ by kw_i to give

$$\frac{\sigma^2}{N} = \frac{1}{\Sigma\left(1/\sigma_i^2\right)} = \frac{1}{k\Sigma w_i} \tag{4.24}$$

so

$$k = \frac{N}{\sigma^2} \frac{1}{\Sigma w_i} \tag{4.25}$$

and therefore

$$\sigma_i^2 = \frac{1}{kw_i} = \frac{\sigma^2 \Sigma w_i}{Nw_i} \tag{4.26}$$

Example 4.2. A student performs an experiment to determine the voltage of a standard cell. The student makes 40 measurements with the apparatus and finds a result $\bar{x}_1 = 1.022$ V with a spread $s_1 = 0.01$ V in the observations. After looking over her data she realizes that she could improve the equipment to decrease the uncertainty by a factor of 2.5 ($s_2 = 0.004$ V) so she makes 10 more measurements that yield a result $\bar{x}_2 = 1.018$ V.

The mean of all these observations is given by Equation (4.17):

$$\mu \simeq \frac{\dfrac{40(1.022)}{0.01^2} + \dfrac{10(1.018)}{0.004^2}}{\dfrac{40}{0.01^2} + \dfrac{10}{0.004^2}} \; V$$

$$= \frac{4.00(1.022) + 6.25(1.018)}{4.00 + 6.25} \; V$$

$$= 1.0196 \; V$$

The uncertainty in the mean σ is given by Equation (4.19):

$$\sigma_\mu \simeq \left(\frac{40}{0.01^2} + \frac{10}{0.004^2} \right)^{-1/2} = 0.00099 \; V$$

The result should be quoted as $\mu = 1.0196 \pm 0.0010$ V although $\mu = 1.020 \pm 0.001$ V would also be acceptable. Carrying the fourth place (which is completely undefined) after the decimal point just eliminates any possible rounding errors if these data should later be merged with data from other experiments.

The precision of the final result is better than that for either part of the experiment. The uncertainties in the estimates of the means μ_1 and μ_2 determined from the two sets of data independently are given by Equation (4.14):

$$\sigma_1 = \frac{0.01}{\sqrt{40}} \; V = 0.0016 \; V \qquad \sigma_2 = \frac{0.004}{\sqrt{10}} \; V = 0.0013 \; V$$

A comparison of these values illustrates the fact that taking more measurements decreases the resulting uncertainty only as the square root of the number of observations, which for this case is not so important as decreasing σ_i.

What if the student did not know the absolute uncertainties in her measurements, but only that the uncertainties had been improved by a factor of 2.5 by the professor's suggestion? She could obtain the estimate of the mean directly from Equation (4.21) by replacing $1/\sigma_1^2$ by the weight $w_i = 1$, and $1/\sigma_2^2$ by the weight $W_i = 2.5^2$, to give

$$\mu \simeq \frac{40(1)(1.022) \; V + 10(2.5^2)(1.018) \; V}{40(1) + 10(2.5)^2} = 1.0196 \; V$$

To find the error in the mean the student could calculate σ from her data by Equation (4.22) and use Equation (4.23) to determine σ_μ.

Discarding Data

Even though the student made four times as many observations at the lower precision (higher uncertainty), the high precision contribution is over 1.5 times as effective in determining the mean as is the high uncertainty contribution. The student should probably consider ignoring the low precision data entirely and using only the high precision data. Why should we ever throw away data that are not known to be bad? Additionally, because in this case the earlier data are weighted so as to be rather unimportant to the result, what is the point in neglecting them and thereby wasting all the effort that went into collecting those first 40 data points?

These are questions that arise again and again in experimental science as one works to find the elusive parameters of the parent distribution. The answer lies in the fact that experiments tend to be improved over time and often the earliest data taking period is best considered a training period for the experimenters and a "shakedown" period for the equipment. Why risk contaminating the sample with data of uncertain results when they contribute so little to the final result? The relative standard deviations of the two data sets can serve as a guide. If the spread of the later distribution shows marked improvement over that of the earlier data, then we should seriously consider throwing away the earlier data unless we are certain of their reliability. There is no hard and fast rule that defines when a group of data should be ignored—common sense must be applied. However, we should make an effort to overcome the natural bias toward using all data simply to recover our investment of time and effort. A greater profit may be gained by using the cleaner sample alone.

4.2 STATISTICAL FLUCTUATIONS

For many experiments the standard deviations σ_i can be determined more accurately from a knowledge of the estimated parent distribution than from the data or from other experiments. If the observations are known to follow the Gaussian distribution, the standard deviation σ is a free parameter and must be determined experimentally. If, however, the observations are known to be distributed according to the Poisson distribution, the standard deviation is equal to the square root of the mean.

As discussed in Chapter 2, the Poisson distribution is appropriate for describing the distribution of the data points in counting experiments where the observations are the numbers of events detected per unit time interval. In such experiments, there are fluctuations in the counting rate from observation to observation that result solely from the intrinsically random nature of the process and are independent of any imprecision in measuring the time interval or of any inexactness in counting the number of events occurring in the interval. Because the fluctuations in the observations result from the statistical nature of the process, they are classified as *statistical fluctuations*, and the resulting errors in the final determinations are classified as *statistical errors*.

In any given time interval there is a finite chance of observing *any* positive (or zero) integral number of events. The probability for observing any specific number of counts is given by the Poisson probability function, with mean μ_t, where the subscript t indicates that these are average values for the time interval of length Δt. Thus, if we make N measurements of the number of counts in time intervals of fixed length Δt, we expect that a histogram of the number of counts x_i recorded in each time interval would follow the Poisson distribution for mean μ_t.

Mean and Standard Deviation

For values of the mean μ_t greater than about 10, the Gaussian distribution closely approximates the shape of the Poisson distribution. Therefore, we can use the formula of Equation (4.9) for estimating the mean with the assumption that all data points were extracted for the same parent population and thus have the same uncertainties:

$$\mu_t \simeq \bar{x}_t = \frac{1}{N} \sum x_i \qquad (4.27)$$

Here the x_i are the numbers of events detected in the N time intervals Δt, and the assumption that the data were all drawn from the same parent population is equivalent to assuming that the lengths of the time intervals were the same for all measurements.

According to Equation (2.19), the variance σ^2 for a Poisson distribution is equal to the mean μ:

$$\sigma_t^2 = \mu_t \simeq \bar{x}_t \qquad (4.28)$$

The uncertainty in the mean σ_{t_μ} is obtained by combining Equations (4.12) and (4.28):

$$\sigma_{t_\mu} = \frac{\sigma_t}{\sqrt{N}} = \sqrt{\frac{\mu_t}{N}} \simeq \sqrt{\frac{\bar{x}_t}{N}} \qquad (4.29)$$

We usually wish to find the mean number of counts per unit time, which is just

$$\mu = \frac{\mu_t}{\Delta t} \quad \text{with} \quad \sigma_\mu = \frac{\sigma_{t_\mu}}{\Delta t} = \sqrt{\frac{\mu}{N \Delta t}} \qquad (4.30)$$

As we might expect, the uncertainty in the mean number of counts per unit time σ_μ is inversely proportional to the square roots of both the time interval Δt and the number of measurements N.

In some experiments, as in Example 4.2, data may be obtained with varying uncertainties. For purely statistical fluctuations, this implies that counts were recorded for different lengths of the time intervals Δt_i. If we wish to find the mean number of counts μ per unit time from data points that were obtained in different time periods, and therefore with different uncertainties, there are two possible ways to proceed. If we have the raw data counts (the x_i's) and we know they are all independent, then we can simply add all the x_i's and divide the sum by the sum of the time intervals:

$$\mu = \frac{\Sigma x_i}{\Sigma \Delta t_i} \quad \text{and} \quad \sigma^2 = \mu$$

The more likely situation is that we know only the means μ_j and corresponding standard deviations σ_j of the means, obtained from the experiments. For example, when dealing with published experimental data, we should assume that the errors incorporate instrumental as well as statistical uncertainties. With such data, the safest procedure is to apply Equations (4.17) and (4.19) to evaluate the weighted mean μ of the individual means μ_i and the standard deviation σ of the mean:

$$\mu \simeq \frac{\Sigma\left(\mu_j/\sigma_j^2\right)}{\Sigma\left(1/\sigma_j^2\right)} \quad \text{and} \quad \sigma_\mu = \frac{1}{\Sigma\left(1/\sigma_j^2\right)} \tag{4.31}$$

Example 4.3. The activity of a radioactive source is measured $N = 10$ times with a time interval $\Delta t = 1$ min. The data are given in Table 4.1. The average of these data points is $\bar{x} = 15.1$ counts per minute. The spread of the data points is characterized by $\sigma = 3.9$ counts per minute calculated from the mean according to Equation (4.27). The uncertainty in the mean is calculated according to Equation (4.29) to be $\sigma_{\bar{x}} \simeq 1.2$ counts per minute.

If we were to combine the data into one observation $x' = \Sigma x_i$ from one 10-min interval, we would obtain the same result. The activity is $x' = 151$ counts per 10 minutes $= 15.1$ counts per minute as before. The uncertainty in the result is given by the standard deviation of the single data point $\sigma_{x'} = \sqrt{151} \simeq 12.3$ counts per 10 minutes $= 1.2$ counts per minute.

Suppose that we made an additional measurement for a 10-min period and obtained $x'' \cong 147$ counts. We could combine x' and x'' exactly as before to obtain a total

$$\bar{x}_T = x' + x'' = (151 + 147)/(10 + 10) = 14.9 \text{ counts per minute}$$

with an uncertainty

$$\sigma_{\bar{x}_T} = \sqrt{298}/20 = 0.87 \text{ counts per minute}$$

which is smaller than $\sigma_{\bar{x}}$ by a factor of $\sqrt{2}$. Alternatively, we could combine the original data points according to Equation (4.17) and calculate the uncertainty in the final result σ_T from combining the uncertainties of the individual data points according to Equation (4.19).

Note that, although we could have simplified matters by recording all the data as one experimental point, $x = 298$ counts per 20 minutes, by so doing, we

TABLE 4.1
Experimental data for the activity of a radioactive source from the experiment of Example 4.3†

Interval Δt_i (min)	Counts x_i	
1	19	
1	11	
1	24	$\bar{x} = \dfrac{1}{N}\Sigma x_i = 151$ counts per 10 minutes
1	16	$= 15.1$ counts per minute
1	11	
1	15	
1	22	$\sigma \simeq \sqrt{\bar{x}} = 3.9$ counts per minute
1	9	
1	9	$\sigma_{\bar{x}} \simeq \dfrac{\sigma}{\sqrt{N}} = 1.2$ counts per minute
1	$\underline{15}$	
	Sum = 151	
$\underline{10}$	$\underline{147}$	$\sigma_{10} = \sqrt{147}$ counts per 10 minutes
		$= 1.2$ counts per minute
Total 20	298	$\bar{x}_{20} = (151 + 147)/(10 + 10)$
		$= 298/20 = 14.9$ counts per minute
		$\sigma_{20} = \sqrt{298}$ counts per 20 minutes
		$= 0.9$ counts per minute

†The data tabulated are the number of counts x_i detected in each time interval Δt_i.

would lose all independent information about the shape of the distribution that could be used as a partial check on the validity of the experiment.

4.3 χ^2 TEST OF A DISTRIBUTION

Once we have calculated the mean and standard deviation from our data, we may be in a position to say even more about the parent population. If we can be fairly confident of the type of parent distribution that describes the spread of the data points (e.g., Gaussian or Poisson distribution), then we can describe the parent distribution in detail and predict the outcome of future experiments from a statistical point of view.

Because we are concerned with the behavior of the probability function $P(x_i)$ as a function of the observed values of x_i, a complete discussion will be postponed until Chapter 11 following the development of procedures for comparing data with complex functions. Let us for now use the results of Chapter 11 without derivation. The test that we shall describe here is the χ^2 test for goodness of fit.

Probability Distribution

If measurements x_i are made of the quantity x, we can truncate the data to a common least count and group the observations into frequencies of identical observations to make a histogram. Let us assume that j runs from 1 to n so there are n possible different values of x_j, and let us call the frequency of observations, or number of counts in each histogram bin, $h(x_j)$ for each different measured value of x_j. If the probability for observing the value x_j in any random measurement is denoted by $P(x_j)$, then the expected number of such observations is $y(x_j) = NP(x_j)$ where N is the total number of measurements. Figures 4.1 and 4.2 show the same six-bin histogram, drawn from a Gaussian parent distribution with mean $\mu = 5.0$ and standard deviation $\sigma = 1$, corresponding to 100 total measurements. The parent distribution $y(x_j) = NP(x_j)$ is illustrated by the large Gaussian curve.

For each measured value x_j, there is a standard deviation $\sigma_j(h)$ associated with the uncertainty in the observed frequency $h(x_j)$. This is not the same as the uncertainty σ_i associated with the difference $x_i - \mu$ between the value of an individual measurement x_i and the mean μ of the data comprising the histogram. If we were to set out to determine the distribution function $P(x_j)$ very precisely by taking a number of sets of data ($k = 1 \cdots n_k$), then we would make several measurements $h_k(x_j)$ for each frequency $y(x_j)$. For each value of

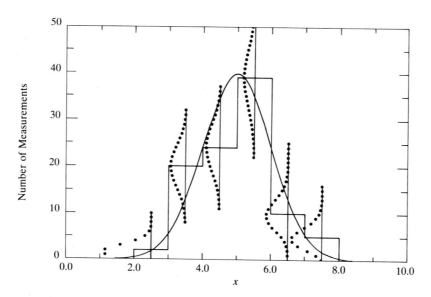

FIGURE 4.1

Histogram, drawn from a Gaussian parent distribution with mean $\mu = 5.0$ and standard deviation $\sigma = 1$, corresponding to 100 total measurements. The parent distribution $y(x_j) = NP(x_j)$ is illustrated by the large Gaussian curve. The smaller dotted curves represent the Poisson distribution of events in each bin, based on the sample data.

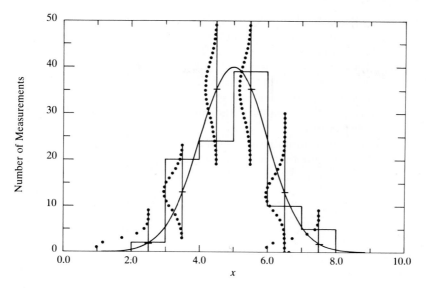

FIGURE 4.2
The same histogram as shown in Figure 4.1 with dotted curves representing the Poisson distribution of events in each bin, based on the parent distribution.

x_j, there is a distribution curve $p_j(y_k)$ that describes the probability of obtaining the value of the frequency $h_k(x_j)$ in the kth trial experiment when the expected value is $y(x_j)$. The spread of these measurements for each value of j is characterized by $\sigma_j(h)$. This spread is illustrated in Figure 4.2 by dotted Poisson curves representing $p_j(y_k)$, centered in each bin at the frequency $\mu_j = y(x_j)$ predicted by the parent population with standard deviation $\sigma_j(h) = \sqrt{\mu_j}$. In an actual experiment, we generally would not know the parameters, or perhaps even the form, of the parent distribution. Thus we should estimate the distributions of events in each bin from the data, as illustrated in Figure 4.1.

Definition of χ^2

With the preceding definitions for n, N, x_j, $h(x_j)$, $P(x_j)$, and $\sigma_j(h)$, the definition of χ^2 from Chapter 11 is

$$\chi^2 \equiv \sum_{j=1}^{n} \frac{\left[h(x_j) - NP(x_j)\right]^2}{\sigma_j(h)^2} \tag{4.32}$$

In most experiments, however, we do not know the values of $\sigma_j(h)$ because we make only one set of measurements $f(x_j)$. Fortunately, these uncertainties can be estimated from the data directly without measuring them explicitly.

If we consider the data of Figure 4.2, we observe that for each value of x_j, we have extracted a proportionate random sample of the parent population for that value. The fluctuations in the observed frequencies $h(x_j)$ come from the statistical probabilities of making random selections of finite numbers of items and are distributed according to the Poisson distribution with $y(x_j)$ as mean. Although the distribution of frequencies $y(x_j)$ in Figure 4.2 is Gaussian, the probability functions for the spreads of the measurements of each frequency $p_j(h_k)$ are Poisson distributions.

For the Poisson distribution, the variance $\sigma_j(h)^2$ is equal to the mean $y(x_j)$ of the distribution, and thus we can estimate $\sigma_j(h)$ from the data to be $\sigma_j(h) = \sqrt{NP(x_j)} \simeq \sqrt{h(x_j)}$. Equation (4.32) simplifies to

$$x^2 = \sum_{j=1}^{n} \frac{[h(x_j) - NP(x_j)]^2}{NP(x_j)} \simeq \sum_{j=1}^{n} \frac{[h(x_j) - NP(x_j)]^2}{h(x_j)} \qquad (4.33)$$

Test of χ^2

As defined in Equations (4.32) and (4.33), χ^2 is a statistic that characterizes the dispersion of the observed frequencies from the expected frequencies. The numerator of Equation (4.32) is a measure of the spread of the observations. The denominator is a good measure of the expected spread. We might imagine that for good agreement, the average spread of the data would correspond to the expected spread, and thus we should get a contribution of about 1 from each frequency, or $\chi^2 \simeq n$ from the entire distribution. As we shall see, this is almost correct.

However, we do not always get ideal agreement between the data and our expectations, and in order to test the goodness of fit of the observed frequencies to the assumed probability distribution, we must know how χ^2 is distributed; that is, we need to know the probability of observing our calculated values of χ^2 from a random sample of data. If our value of χ^2 is reasonably probable, then we can have confidence in our assumed distribution and vice versa.

If the observed frequencies were to agree exactly with the predicted frequencies $h(x_j) = NP(x_j)$, then we would find $\chi^2 = 0$. From our understanding of probability, we should realize that this is not a very likely outcome of an experiment. For any physical experiment where predicted and observed frequencies will not be equal, we should expect a value $\chi^2 \simeq n$. In fact, the true expectation value for χ^2 is

$$\langle \chi^2 \rangle = \nu = n - n_c \qquad (4.34)$$

where ν is the number of degrees of freedom and is equal to the number n of sample frequencies y minus the number n_c of constraints or parameters that have been calculated from the data to describe the probability function, $NP(x_j)$. For our example, even if $P(x_j)$ is chosen completely independently of the measured distribution $h(x_j)$, there is still the normalizing factor N corresponding to the total number of events in the distribution, so that the expectation value of χ^2 must at best be $\langle \chi^2 \rangle = n - 1$.

It is convenient to define the *reduced chi-square* as $\chi_\nu^2 \equiv \chi^2/\nu$, with expectation value $\langle \chi_\nu^2 \rangle = 1$. Values of χ_ν^2 much larger than 1 result from large deviations from the assumed distribution and may indicate poor measurements, incorrect assignment of uncertainties, or an incorrect choice of probability function. Very small values of χ_ν^2 are equally unacceptable and may imply some misunderstanding of the experiment. Rather than consider the probability of obtaining any particular value of χ^2 or χ_ν^2 (which is infinitesimally small), we shall use an integral test to determine the probability of observing a value of χ_ν^2 equal to or greater than the one we calculated. This is similar to our consideration of the probability that a measurement of a variable deviate by more than a certain amount from the mean.

Table C.4 gives the probability that a random sample of data points drawn from the assumed probability distribution would yield a value of χ^2 as large as or larger than the observed value in a given experiment with ν degrees of freedom.

If the probability is reasonably close to 1, then the assumed distribution describes the spread of the data points well. If the probability is small, either

TABLE 4.2
χ^2 analysis of the data of Example 4.1†

Length	Observed frequency h_j	From parent distribution			From sample distribution		
		y_j	σ_j	$\dfrac{y_j - h_j}{\sigma_j}$	y_j	σ_j	$\dfrac{y_j - h_j}{\sigma_j}$
18.7	1	0.54	0.74	-0.62	0.90	0.95	-0.10
18.9	3	1.42	1.19	-1.33	2.08	1.44	-0.63
19.1	4	3.16	1.78	-0.47	4.16	2.04	0.08
19.3	7	5.99	2.45	-0.41	7.17	2.68	0.06
19.5	13	9.68	3.11	-1.07	10.68	3.27	-0.71
19.7	14	13.33	3.65	-0.18	13.72	3.70	-0.07
19.9	11	15.64	3.95	1.17	15.23	3.90	1.08
20.1	12	15.64	3.95	0.92	14.60	3.82	0.68
20.3	16	13.33	3.65	-0.73	12.08	3.48	-1.13
20.5	11	9.68	3.11	-0.42	8.64	2.94	-0.80
20.7	4	5.99	2.45	0.81	5.33	2.31	0.58
20.9	1	3.16	1.78	1.21	2.84	1.69	1.09
21.1	1	1.42	1.19	0.35	1.31	1.14	0.27
21.3	2	0.54	0.74	-1.98	0.52	0.72	-2.06
		$\chi_\nu^2 = 12.82/13 = 0.99$			$\chi_\nu^2 = 10.31/13 = 0.79$		

†Parameters of the parent Gaussian distribution are $\mu = 20.00$ cm and $\sigma = 0.50$ cm; parameters for the sample distribution are $\mu = 19.94$ and $\sigma = 0.52$.

the assumed distribution is not a good estimate of the parent distribution or the data sample is not representative of the parent distribution. There is no yes-or-no answer to the test; in fact, we should expect to find a probability of about 0.5 with $\chi_\nu^2 \simeq 1$, because statistically the observed values of χ^2 should exceed the norm half the time. But in most cases, the probability is either reasonably large or unreasonably small, and the test is fairly conclusive. A further discussion of the statistical significance of the χ^2 probability function will be given in Chapter 11.

Let us consider again the data of Example 1.2, which are listed in Table 4.2. To test the agreement between the data and the predicted distribution, we have calculated the function $y(x_j) = NP(x_j)$ at each value of x_j from the mean and standard deviation of the parent distribution, and from the observed mean and standard deviation of the data. The results are listed in Table 4.2 with the uncertainties $\sigma_j = \sqrt{y(x_j)}$, and the individual contributions (before squaring) to χ^2, $[h(x_j) - NP(x_j)]/\sigma(h)_j$. The calculated values of χ^2 from the comparison between the data and each distributions are the sums of the squares of these last quantities.

Because we have 14 data points and one parameter, the normalization constant N determined from the data, the expectation value of χ^2 is $\nu = 14 - 1 = 13$. The values of χ^2 we obtained, 12.82 and 10.31, are reasonably close to the expected value. Referring to Table C.4, we observe that, for 13 degrees of freedom, the probability of obtaining in repeated experiments a value of $\chi^2/\nu \geq 0.99$ is $\sim 50\%$, and the probability of obtaining a value of $\chi^2/\nu \geq 0.79$ is $\sim 67\%$.

Generalizations of the χ^2 Test

In the preceding example we knew the parent distributions and were therefore able to determine the uncertainties $\sigma_j(h)$ from the predicted probability. In most cases, where the actual parameters of the probability function are being determined in the calculation, we must use an estimate of the parent population based on these parameters and must estimate the uncertainties in the $y(x_j)$'s from the data themselves. To do this we must replace the uncertainties in columns four and seven of Table 4.2 with the square roots of the observed frequencies in column two.

Furthermore, although our example was clearly based on a simple probability function, the χ^2 test is often generalized to compare data obtained in any type of experiment to the prediction of a model. The uncertainties in the measurements may be instrumental or statistical or a combination of both, and the uncertainty $\sigma_j(h)^2$ in the denominator of Equation (4.32) may represent a Gaussian error distribution rather than the Poisson distribution. In fact, several of the histogram bins in our example contained small numbers of counts, and thus, the statistical application of the test was not strictly correct, because we assume Gaussian statistics in the χ^2 calculation. However, the test still provides us with a reproducible method of evaluating the quality of our data, and if we are concerned with statistical accuracy, we can merge the low-count bins to satisfy the Gaussian statistics requirement.

Another application of the chi-squared test is in comparing two sets of data to attempt to decide whether or not they were drawn from the same parent population. Suppose that we have measured two distributions, $g(x_j)$ and $h(x_j)$, and wish to determine the probability that the two sets were not drawn from the same parent probability distribution $P(x_j)$. Clearly, we could apply the χ^2 test separately to the two sets of data and determine separately χ^2 probabilities that each set was not associated with the supposed parent population $P(x_j)$. However, we can also make a direct test, independent of the parent population, by writing

$$\chi^2 = \sum_{j=1}^{n} \frac{\left[g(x_j) - h(x_j)\right]^2}{\sigma^2(g) + \sigma^2(h)} \tag{4.35}$$

The denominator $\sigma^2(g) + \sigma^2(h)$ is just the variance of the difference $g(x_j) - h(x_j)$. As in the previous examples, the expectation value of χ^2 depends on the relation between the two parts of the numerator, $g(x_j)$ and $h(x_j)$. If the two parts, corresponding to the distributions of the two data sets, were obtained completely independently of one another, then the number of degrees of freedom equals n and $\langle\chi^2\rangle = n$. If one of the distributions $g(x_j)$ or $h(x_j)$ has been normalized to the other, then the number of degrees of freedom is reduced by 1 and $\langle\chi^2\rangle = n - 1$. Again, we interpret the χ^2 probability in a negative sense. If the value of χ^2/ν is large, and therefore the probability given in Table C.4 is low, we may conclude that the two sets of data were drawn from different distributions. However, for a low value of χ^2 and therefore high probability, we cannot draw the opposite conclusion that the two data sets $g(x_j)$ and $h(x_j)$ were drawn from the same distribution. There is always the possibility that there are indeed two different but closely similar distributions and that our data are not sufficiently sensitive to detect the difference between the two.

SUMMARY

Weighted mean:

$$\bar{x} = \frac{\Sigma(x_i/\sigma_i)}{\Sigma(1/\sigma_i^2)} \xrightarrow[\sigma_i = \sigma]{} \frac{1}{N}\Sigma x_i$$

Variance of mean:

$$\sigma_\mu = \frac{1}{\Sigma(1/\sigma_j^2)} \xrightarrow[\sigma_i = \sigma]{} \frac{\sigma^2}{N}$$

Instrumental uncertainties: Fluctuations in measurements due to finite precision of measuring instruments:

$$\sigma^2 \simeq s^2 = \frac{1}{N-1} \sum (x_i - \bar{x})^2$$

Statistical fluctuations: Fluctuations in observations resulting from statistical probability of taking random samples of finite numbers of items:

$$\sigma^2 = \mu \simeq \bar{x}$$

χ^2 *test:* Comparison of observed frequency distribution $h(x_j)$ of possible observations x_j versus predicted distribution $NP(x_j)$, where N is the number of data points and $P(x_j)$ is the theoretical probability distribution:

$$\chi^2 \equiv \sum_{j=1}^{n} \frac{\left[h_k(x_j) - NP(x_j) \right]^2}{\sigma_j^2}$$

Degrees of freedom ν: Number of data points minus the number of parameters to be determined from the data points.
Reduced χ^2: $\chi_\nu^2 = \chi^2/\nu$. For χ^2 tests, χ_ν^2 should be approximately equal to 1.
Graphs and tables of χ^2: Table C.4 gives the probability that a random sample of data when compared to its *parent distribution* would yield values of χ_ν^2 as large as or larger than the observed value.

EXERCISES

4.1. From the data of Exercise 1.4, calculate the uncertainty in the mean σ_μ. Is the actual error reasonable?

4.2. Repeat Exercise 4.1 for the data of Exercise 1.5.

4.3. Read the data of Example 2.4 from Figures 2.3 and 2.4. Recalculate the curves and calculate χ^2 and χ_ν^2 for the agreement between the curves and the histograms. Use only bins with 5 or more counts.

4.4. Work out the intermediate steps in Equation (4.19).

4.5. A student measures the period of a pendulum and obtains the following values.

Trial	1	2	3	4	5	6	7	8
Value	1.35	1.34	1.32	1.36	1.33	1.34	1.37	1.35

(*a*) Find the mean and standard deviation of the measurements and the standard deviation of the mean.
(*b*) Estimate the probability that another measurement will fall within 0.02 s of the mean.

4.6. (*a*) Find the mean and the standard deviation of the mean of the following numbers under the assumption that they were all drawn from the same parent population.

(b) In fact, data points 1 through 20 were measured with uniform uncertainty σ, whereas data points 21 through 30 were measured more carefully so that the uniform uncertainty was only $\sigma/2$. Find the mean and standard deviation of the mean under these conditions.

Trial	$x(\sigma)$	Trial	$x(\sigma)$	Trial	$x(\sigma/2)$
1	2.40	11	1.94	21	2.59
2	2.45	12	1.55	22	2.65
3	2.47	13	2.12	23	2.55
4	3.13	14	2.17	24	2.07
5	2.92	15	3.06	25	2.61
6	2.85	16	1.97	26	2.61
7	2.05	17	2.23	27	2.54
8	2.52	18	3.20	28	2.76
9	2.94	19	2.24	29	2.37
10	1.89	20	2.60	30	2.57

4.7. A counter is set to count gamma rays from a radioactive source. The total number of counts, including background, recorded in each 1-min interval is listed in the accompanying table. An independent measurement of the background in a 5-min interval gave 58 counts. From these data find:
(a) The mean background in a 1-min interval and its uncertainty.
(b) The corrected counting rate from the source alone and its uncertainty.

Trial	1	2	3	4	5	6	7	8	9	10
Total counts	125	130	105	126	128	119	137	131	115	116

4.8. The 1990 edition of the *Particle Data Tables* lists the following experimental measurements of the mean lifetime of the K_s^0 meson. Find the weighted mean of the data and the uncertainty in the mean.

i	1	2	3	4	5
τ_i	0.8920 ± 0.0044	0.881 ± 0.009	0.8913 ± 0.0032	0.9837 ± 0.0048	0.8958 ± 0.0045

4.9. Eleven students in an undergraduate laboratory combined their measurements of the mean lifetime of an excited state. Their individual measurements are tabulated.

Student	1	2	3	4	5	6	7	8	9	10	11
τ (s)	34.3	32.2	35.4	33.5	34.7	33.5	27.9	32.0	32.4	31.0	19.8
σ_τ	1.6	1.2	1.5	1.4	1.6	1.5	1.9	1.2	1.4	1.8	1.3

Find the maximum likelihood estimate of the mean and its uncertainty.

4.10. Assume that you have a box of resistors that have a Gaussian distribution of resistances with mean value $\mu = 100\ \Omega$ and standard deviation $\sigma = 20\ \Omega$ (i.e., 20% resistors). Suppose that you wish to form a subgroup of resistors with $\mu = 100\ \Omega$ and standard deviation of $5\ \Omega$ (i.e., 5% resistors) by selecting all resistors with resistance between the two limits $r_1 = \mu - a$ and $r_2 = \mu + a$.
(a) Find the value of a.
(b) What fraction of the resistors would satisfy the condition?

4.11. A student plotted a 10-bin histogram of 200 measured values of the period of a simple pendulum. From the mean and standard deviation of his measurements he calculated a Gaussian curve that he scaled so that the area of the curve was the same as that of the histogram. He then applied the χ^2 test to compare his data to the curve.

(*a*) What could he hope to learn from such a test?

(*b*) What value of χ^2 should he expect to obtain for such a fit?

(*c*) Refer to Table C.4 and find the values of χ^2 corresponding to the 10% and 90% probabilities for this fit.

4.12. Plot a histogram of the course grades listed in Exercise 1.5 in 10-point bins. Plot a Gaussian curve based on the mean and standard deviation of the data, normalized to the area of the histogram. Apply the χ^2 test and check the associated probability from Table C.4.

CHAPTER
5

MONTE CARLO
TECHNIQUES

5.1 INTRODUCTION

We have seen in the preceding chapter the importance of probability distributions in the analysis of data samples, and have observed that we are usually interested in the integrals or sums of such distributions over specified ranges. Although we have considered only experiments that are described by a single distribution, most experiments involve a combination of many different probability distributions. Consider, for example, a simple scattering experiment to measure the angular distribution of particles scattered from protons in a fixed target. The magnitude and direction of the momentum vector of the incident particles, the probability that a particle will collide with a proton in the target, and the resulting momentum vectors of the scattered particles can all be described in terms of probability distributions. The final experimental result can be treated in terms of a multiple integration over all these distributions.

Analytical evaluation of such an integral is rarely possible, so numerical methods must be used. However, even the simplest first-order numerical integration can become very tedious for a multidimensional integral. A one-dimensional integral of a function can be determined efficiently by evaluating the function N times on a regular grid, where the number of samples N depends on the structure of the function and the required accuracy. (See Appendix A.) A two-dimensional integral requires sampling in two dimensions and, for accuracy comparable to that of the corresponding one-dimensional problem, requires something like N^2 samples. A three-dimensional integral requires something like N^3 samples. For integrals with many dimensions, the number of grid points at which the function must be calculated becomes excessively large.

Before we continue with methods of extracting parameters from data, let us look at the Monte Carlo method, a way of evaluating these multiple integrals that depends on random sampling from probability density distributions, rather than regular grid-based sampling techniques. The Monte Carlo method provides the experimental scientist with one of the most powerful tools available for planning experiments and analyzing data. Basically, Monte Carlo is a method of calculating multiple integrals by random sampling. Practically, it provides a method of simulating experiments and creating models of experimental data. With a Monte Carlo calculation, we can test the statistical significance of data with relatively simple calculations, which require neither a deep theoretical understanding of statistical analysis nor sophisticated programming techniques.

The name *Monte Carlo* comes from the city on the Mediterranean with its famous casino, and a Monte Carlo calculation implies a statistical method of studying problems based on the use of random numbers, similar to those generated in the casino games of chance. One might reasonably ask whether the study of science can be aided by such associations, but in fact, with Monte Carlo techniques, very complicated scientific and mathematical problems can be solved with considerable ease and precision.

Example 5.1. Suppose that we wish to find the area of a circle of radius r_c but have forgotten the equation. We might inscribe the circle within a square of known area A_s and cover the surface of the square uniformly with small markers, say grains of rice. We find the ratio of the number of grains that lie within the circle to those that cover the square, and determine the area of the circle A_c from the relation

$$A_c = A_s N_c / N_s,$$

(5.1)

where N_c and N_s are the numbers of grains of rice within the boundaries of the circle and of the square, respectively.

What would be the accuracy of this determination; that is, how close should we expect our answer to agree with the true value for the area of a circle? Clearly it would depend on the number and size of the rice grains relative to the size of the square, and on the uniformity of both the grains and their distribution over the square. What if we decided that instead of attempting to cover the square uniformly, we would be content with a random sampling obtained by tossing the rice grains from a distance so that they landed randomly on the square, with every location equally probable? Then we would obtain an interesting result: Our problem would reduce to a simple binomial calculation as long as we did not overpopulate the square but kept the density of rice grains low so that position of any grain on the square was not influenced by the presence of other grains. We should find that, for a fixed number of grains N_s thrown onto the square, the uncertainty σ in the measurement of the circular area would be given by the standard deviation for the binomial distribution with probability $p = A_c / A_s$,

$$\sigma = \sqrt{N_s p(1 - p)} = \sqrt{N_c(1 - p)}$$

(5.2)

Thus, if we were to increase the number of rice grains N_c by a factor of 4, the relative error in our determination of the area of the circle would decrease by a factor of 2.

Replacing the tossed rice grains by a set of computer generated random numbers is an obvious improvement. Let us inscribe our circle of unit radius in a square of side length 2, and generate $N = 100$ pairs of random numbers between -1 and $+1$ to determine the area. Then the probability of a "hit" is just the ratio of the area of the circle to the area of a square, or $p = \pi/4$, so in 100 tries, the mean number of hits will be $\mu = 100p = 78.5$, and the standard deviation, from Equation (5.2), will be $\sigma = \sqrt{Np(1-p)} = \sqrt{100(\pi/4)(1-\pi/4)} = 4.1$. For our measurements of the area of the circle with 100 tries we should expect to obtain from Equation (5.1) $A_c = A_s \times N_c/N_s = (78.5 \pm 4.1) \times 2^2/100 = 3.14 \pm 0.16$.

Figure 5.1 shows a typical distribution of hits from one "toss" of 100 pairs of random numbers. In this example there were 73 hits, so we should estimate the area and its uncertainty from Equations (5.1) and (5.2) to be $A = 2.92 \pm 0.18$. To determine the uncertainty, we assumed that we did not know the a priori probability $p = \pi/4$ and, therefore, we used our experimental estimate $p \simeq 73/100$.

Figure 5.2 shows a histogram of the circle area estimates obtained in 100 independent Monte Carlo runs, each with 100 pairs of random numbers (or a total of 10,000 "tosses"). The Gaussian curve was calculated from the mean, $A = 3.127$, and standard deviation, $\sigma = 0.156$, of the 100 estimated areas.

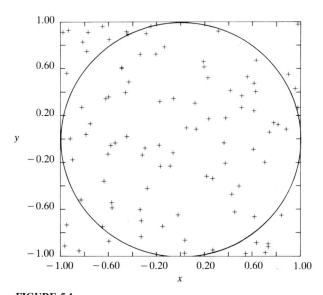

FIGURE 5.1
Estimation of the area of a circle by the Monte Carlo method. The plot illustrates a t distribution of hits from one "toss" of 100 pairs of random numbers uniformly distributed be -1.00 and $+1.00$.

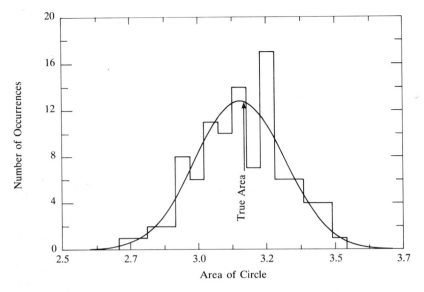

FIGURE 5.2
Histogram of the circle area estimates obtained in 100 independent Monte Carlo runs, each with 100 pairs of random numbers. The Gaussian curve was calculated from the mean $A = 3.127$ and standard deviation $\sigma = 0.156$ of the 100 estimated areas.

Obviously, the area determination problem of Example 5.1 is much too simple to require a Monte Carlo calculation. However, for problems involving integrations of many variables and for those with complicated integration limits, the Monte Carlo technique is invaluable, with its straightforward sampling and its relatively simple determination of the uncertainties.

5.2 RANDOM NUMBERS

A successful Monte Carlo calculation requires a reliable set of *random numbers*, but truly random numbers for use in calculations are hard to obtain. One might think of a scheme based upon measuring the times between cosmic ray hits in a detector, or on some physical process such as the generation of noise in an electronic circuit. Such numbers would be random in the sense that it would be impossible to predict the value of the next number from previous numbers but they are hardly convenient to use in extended calculations, and some might not have the necessary uniformity required for a Monte Carlo calculation.

In fact, it is generally preferable to use *pseudorandom numbers*, numbers generated by a computer algorithm designed to produce a sequence of apparently uncorrelated numbers that are uniformly distributed over a predefined range. In addition to the convenience of being able to generate these numbers within the Monte Carlo program itself, pseudorandom numbers have another important advantage over truly random numbers for Monte Carlo calculations. A Monte Carlo program may use a great many random numbers, and the path of the calculation through the program will depend on the numbers chosen in each run. With truly random numbers, every run of a Monte Carlo calculation would follow a different path and produce different results. Such a program would be very difficult to debug. With pseudorandom numbers, we can repeat a calculation with the same sequence of numbers, and search for any particular problems that may be hidden in the code.

There are other advantages too. If we are studying the sensitivity of a calculation to variations in a selected parameter, we can reduce the variance of the *difference* between results calculated with two trial values of the parameter by using the same random number sequence for those parts of the calculation which are independent of the parameter in question. Finally, a pseudorandom number generator can be written to be *portable*; that is, the sequence of numbers produced by the algorithm is independent of computer hardware and language, so that a given program will produce the same results when run on different computers. In view of these advantages and the fact that we rarely, if ever, encounter situations where truly random numbers are required, we shall henceforth use the term *random numbers* to denote *pseudorandom numbers*.

In general, our random number generator must satisfy the following basic criteria:

1. The distribution of the numbers should be uniform within a specified range and should satisfy statistical tests for randomness, such as lack of predictability and of correlations among neighboring numbers.
2. The calculation should produce a large number of unique numbers before repeating the cycle.
3. The calculation should be very fast.

A simple multiplication method is often used to generate random numbers, or *uniform deviates*, as they are often called. An integer starting value or *seed* r_0 and two integer constants are chosen. Successive random numbers are derived from the recursion relation

$$r_{i+1} = (a \times r_i) \bmod m \tag{5.3}$$

where the mod operation corresponds to dividing the product in parentheses by the integer m to obtain the remainder. With appropriate choices of constants a and m, we can obtain a finite sequence of numbers that appear to be randomly selected between 1 and $m - 1$. The length of the sequence is determined by the choice of constants and is limited by the computer word size. For example, if we choose $m = 37$ and $a = 5$, Equation (5.3) gives us the cycle of 36 nicely mixed up numbers, listed in Table 5.1. Random number generators included with computer languages are often based on some variation of this multiplication technique. Careful and thorough statistical studies must be made to be sure that

TABLE 5.1
Pseudorandom numbers†

i	r_i	i	r_i	i	r_i	i	r_i
1	1	10	6	19	36	28	31
2	5	11	30	20	32	29	7
3	25	12	2	21	12	30	35
4	14	13	10	22	23	31	27
5	33	14	13	23	4	32	24
6	17	15	28	24	20	33	9
7	11	16	29	25	26	34	8
8	18	17	34	26	19	35	3
9	16	18	22	27	21	36	15

†The generating equation is $r_{i+1} = (a \times r_i) \bmod m$, with $a = 5$ and $m = 37$.
The cycle repeats $a_{37} = a_1$, $a_{38} = a_2$, and so forth.

an untested random number generator produces an acceptable sequence of numbers.

Because the numbers generated by Equation (5.3) are not truly random, we might worry that our calculations are affected by hidden correlations in successively generated numbers. We can improve the randomness of our sample by *shuffling* the numbers. We generate two sequences of numbers with different generators a and m; one sequence is stored in an array and a number from the second sequence is used as an index to select numbers from the first sequence. For large programs that employ many random numbers, this method is limited by storage space, although *local shuffling* within a block of random numbers can be used.

Even a modest Monte Carlo program can require many random numbers, and to assure the statistical significance of results, we must be certain that the calculation does not use more than the maximum number generated by the algorithm before the sequence repeats. The sample generator of Equation (5.3) cannot produce more than $m - 1$ different values of r_i. The actual cycle length may be less than this range, depending on the choice of constants. The cycle length can be increased by employing two or more independent sequences such that the resulting cycle length is proportional to the product of the lengths of the component cycles.

A generator developed by Wichmann and Hill,[1] based on a simple linear combination of numbers from three independent sequences, is said to have a very long cycle ($\sim 7 \times 10^{12}$) and appears to be well tested. Because the algorithm uses three seeds, it is a little longer and slower than one or two seed algorithms, but its long repeat cycle, portability, and lack of correlations seem to make it a convenient, worry-free generator for most purposes. The algorithm is listed in Appendix E.

---

[1]The authors include a thorough and very useful discussion of the tests applied to a random number sequence, and of the development and testing of the published algorithm.

Although the fact that pseudorandom number generators always produce the same sequences of numbers from the same seeds is an advantage in program debugging, it may be a disadvantage in production running. For example, a simulation program developed for use as a science museum display could be very uninteresting if it repeated the same sequence of events every time it was run. If unpredictable seeds are required, they can easily be derived from the least counts of the computer clock. Commercial routines often include such a method of randomizing the starting seeds. On the other hand, if we wish to run a simulation program several times and to combine the results of the several different runs, the safest method to assure the statistical independence of the separate runs is to record the last values of the seeds at the end of each run and use these as starting seeds for the next run.

A thorough discussion of random number generation and of the Monte Carlo technique is given in Knuth.

5.3   RANDOM NUMBERS FROM PROBABILITY DISTRIBUTIONS

Transformation Method

Most number generators scale their output to provide real numbers uniformly distributed between 0 and 1. In general, however, we require numbers drawn from specific probability distributions. Let us define uniform deviates $p(r)$ drawn from a standard probability density distribution that is uniform between $r = 0$ and $r = 1$:

$$p(r) = \begin{cases} 1 & \text{for } 0 \leq r < 1 \\ 0 & \text{otherwise} \end{cases} \tag{5.4}$$

The distribution is *normalized* so that

$$\int_{-\infty}^{\infty} p(r)\, dr = \int_0^1 1\, dr = 1 \tag{5.5}$$

We shall refer to $p(r)$ as the *uniform distribution*.

Suppose that we require random deviates from a different normalized probability density distribution $P(r)$, which is defined to be uniform between $x = -1$ and 1; that is, the distribution

$$P(x) = \begin{cases} \frac{1}{2} & \text{for } -1 \leq x < 1 \\ 0 & \text{otherwise} \end{cases} \tag{5.6}$$

If we choose a random deviate r between 0 and 1 from the uniform distribution of Equation (5.4), it is obvious that we can calculate another random deviate x as a function of r:

$$x = f(r) = 2r - 1 \tag{5.7}$$

which will be uniformly distributed between -1 and $+1$. This is an example of a simple linear transformation.

To pick a random sample x from the distribution Equation (5.6), we started with a random deviate r drawn from the uniform distribution of Equation (5.4) and found a function $f(r)$ that gave the required relation between x and r. Let us find a general relation for obtaining a random deviate x from any probability density distribution $P(x)$, in terms of the random deviate r drawn from the uniform probability distribution $p(r)$.

Conservation of probability requires that

$$|p(r)\,dr| = |P(x)\,dx| \tag{5.8}$$

and, therefore, we can write

$$\int_{r=-\infty}^{r} p(r)\,dr = \int_{x=-\infty}^{x} P(x)\,dx \quad \text{or} \quad \int_{r=0}^{r} 1\,dr = \int_{x=-\infty}^{x} P(x)\,dx \tag{5.9}$$

which gives the general result

$$r = \int_{x=-\infty}^{x} P(x)\,dx \tag{5.10}$$

Thus, to find x, selected randomly from the probability distribution $P(x)$, we generate a random number r from the uniform distribution and find the value of the limit x that satisfies the integral equation (5.10).

Example 5.2. Consider the distribution described by the equation

$$P(x) = \begin{cases} A(1 + ax^2) & \text{for } -1 \le x < 1 \\ 0 & \text{otherwise} \end{cases} \tag{5.11}$$

where $P(x)$ is positive or zero everywhere within the specified range, and the normalizing constant A is chosen so that

$$\int_{-1}^{1} P(x)\,dx = 1 \tag{5.12}$$

We have

$$r = \int_{-\infty}^{x} P(x)\,dx = \int_{-1}^{x} A(1 + ax^2)\,dx \tag{5.13}$$

which gives

$$r = A(x + ax^3/3 + 1 + a/3) \tag{5.14}$$

and therefore, to find x we must solve the third-degree equation (5.14).

The procedure we have described is referred to as the *transformation method* of generating random deviates from probability distributions. In general, neither the integral equation (5.13) nor the solution of the resulting equation (5.14) can be obtained analytically, so numerical calculations are necessary.

The following steps are required to generate random deviates from a specific probability distribution by the transformation method with a numerical integration:

1. Decide on the range of x. Some probability density functions are defined in a finite range, as in Equation (5.6); others, such as the Gaussian function, extend to infinity. For numerical calculations, reasonable finite limits must be set on the range of the variable.
2. Normalize the probability function. If it is necessary to impose limits on the range of the variable x, then the function must be renormalized to assure that the integral is unity over the newly defined range. *The normalization integral should be calculated numerically by the same routine that is used to find y.*
3. Generate a random variable r drawn from the uniform distribution $p(r)$.
4. Integrate the normalized probability function $P(x)$ from negative infinity (or its defined lower limit) to the value $x = x$, where x satisfies Equation (5.10).

Because the Monte Carlo method usually requires the generation of large numbers of individual events, it is essential to have available fast numerical interpolation and integration routines. To reduce computing time, it is often efficient to set up tables of repeatedly used solutions or integrals within the initializing section of a Monte Carlo program. For example, to pick a random deviate x from the distribution of Equation (5.11), we could do the integral of Equation (5.13) numerically at the beginning of our program, and set up a table of values of r versus x. Then, when we require a random number from the distribution, we generate a random number r and search the table for the corresponding value of x. In general, the search should be followed by an interpolation within the table (Appendix A) to avoid introducing excessive graininess into the resulting distribution. It would be even more convenient, but a little trickier, to produce a table of x versus r, so that the required value of x could be obtained from an index derived from r. In all cases of precalculated tables, it is important to consider the resolution required in the generated variable, because this will determine the intervals at which data must be stored, and therefore the size of the table, and the time required for a search.

Rejection Method

Although the *transformation method* is probably the most useful method for obtaining random deviates drawn from particular distributions, the *rejection method* is often the easiest to use. This was the method that we used in Example 5.1 to find the area of a circle, by generating random numbers uniformly over the surface of the circle and rejecting all except those that fell within the circumference.

Example 5.3. Suppose we wish to obtain random deviates between $x = -1$ and $x = +1$, drawn from the distribution function

$$P(x) = 1 + ax^2 \tag{5.15}$$

which is just the unnormalized distribution of Equation (5.11). To use the rejection method, we begin by generating a random deviate x' uniformly distributed between -1 and $+1$, corresponding to the allowed range of x, and a second random deviate y' uniformly distributed between 0 and $(1 + a)$, corresponding to the allowed range of $P(x)$. We can see that x' and y' must be given by

$$x' = -1 + 2r_i \quad \text{and} \quad y' = (1 + a)r_{i+1} \tag{5.16}$$

where r_i and r_{i+1} are successively generated random values of r drawn from the uniform distribution.

We count an event as a "hit" if the point (x', y') falls between the curve defined by $P(x)$ and the x axis, that is, if $y' < P(x')$, and a "miss" if it falls above the curve. In the limit of a large number of trials, the entire plot, including the area between the curve and the x axis, will be uniformly populated by this operation and our selected samples will be the x coordinates of the "hits," or the values of x', drawn randomly from the distribution $P(x)$. Note that with this method it is not necessary to normalize the distribution to form a true probability function. It is sufficient that the distribution be positive and well behaved within its allowed range.

The advantage of the rejection method over the transformation method is its simplicity. An integration is not required—only the probability function itself must be calculated. A disadvantage of the method is often its low efficiency. In a complex Monte Carlo program only a small fraction of the events may survive the complete calculation to become successful "hits" and the generation and subsequent rejection of so many random numbers may be very time consuming. To reduce this problem, it is advisable to place the strictest possible limits on the random coordinates used to map out the distribution function when using the rejection method.

5.4   SPECIFIC DISTRIBUTIONS

Gaussian Distribution

Almost any Monte Carlo calculation that simulates experimental measurements will require the generation of deviates drawn from a Gaussian distribution, or *Gaussian deviates*. A common application is simulation of measuring uncertainties by *smearing* variables. Fortunately, because of the convenient scaling properties of the Gaussian function, it is only necessary to generate Gaussian deviates from the standard distribution

$$P_G(z)\, dz = \frac{1}{\sqrt{2\pi}} \exp\left[-\frac{z^2}{2} \right] dz, \tag{5.17}$$

with mean 0 and standard deviation 1, and to scale to different means μ and standard deviations σ by calculating

$$x = \sigma z + \mu \tag{5.18}$$

There are several different ways of obtaining random samples of the variable z from the distribution $P_G(z)$ of Equation (5.17). The two most obvious are the rejection and transformation methods discussed previously. Because the Gaussian function is defined between $-\infty$ and $+\infty$, these methods require that limits be placed on the range of z. For most calculations in which the Gaussian function is being used to simulate smearing of data caused by measuring errors, a range of $\pm 3\sigma$ should be satisfactory because all but $\sim 0.3\%$ of normally distributed events lie within this range.

Because the Gaussian function cannot be integrated analytically, numerical integrations are required for the transformation method. Decisions must be made on the order of integration and the step size as well as on the limits. A first- or second-order numerical integration (Appendix A) is generally satisfactory, with a linear interpolation to find an approximation to the value of x in Equation (5.10) at the required value of the integral.

An interesting method for generating Gaussian deviates is based on the fact that if we repeatedly calculate the means of groups of numbers drawn randomly from any distribution, the distribution of those means tends to a Gaussian as the number of means increases. Thus, if we calculate many times the sums of N uniform deviates, drawn from the uniform distribution, we should expect the sums to fall into a truncated Gaussian distribution, bounded by 0 and N, with mean value $N/2$. If we generate N values of r from the distribution of Equation (5.4) and calculate

$$r_G = \sum_{i=1}^{N} r_i - N/2 \tag{5.19}$$

the variable r_G will be drawn from an approximately Gaussian distribution with mean $\mu = 0$ and standard deviation $\sigma = \sqrt{N/12}$. We should note that the maximum range of r_G will be limited to $\mu \pm N/2$ or $\mu \pm \sigma\sqrt{3N}$. For $N = 2$, the sum is a triangle function and as N increases, the distribution quickly takes on a Gaussian-like shape. Values of N as small as $N = 4$ are suitable for low statistics calculations. With $N = 4$, we have $\sigma = \sqrt{1/3} \simeq 0.058$ and the range of r_G from -2 to $+2$ corresponds to $\mu \pm \sigma\sqrt{12}$) or $\mu \pm 3.46\sigma$. If a better approximation to the Gaussian function is required and calculation time is not a problem, $N = 12$ is particularly convenient because the resulting variance and standard deviation are unity.

A particularly elegant method for obtaining random numbers drawn from the Gaussian distribution was suggested by Box and Müller. This method makes use of the fact that, although the simple transformation method requires an integration of the Gaussian function, it is possible to find a function that generates the two-dimensional Gaussian distribution,

$$f(z_1, z_2) = \frac{1}{2\pi} \exp\left(-\frac{(z_1^2 + z_2^2)}{2}\right) = \frac{1}{\sqrt{2\pi}} \exp\left(-\frac{z_1^2}{2}\right) \times \frac{1}{\sqrt{2\pi}} \exp\left(-\frac{z_2^2}{2}\right)$$

(5.20)

From this equation, the authors obtained expressions that generate two Gaussian deviates, z_1 and z_2, from two uniform deviates, r_1 and r_2:

$$z_1 = \sqrt{-2\ln r_1} \, \cos 2\pi r_2$$
$$z_2 = \sqrt{-2\ln r_1} \, \sin 2\pi r_2$$

(5.21)

Example 5.4. A uniform 10-cm long rod has one end held at 0°C and the other at 100°C so that the temperature along the rod is expected to vary linearly from 0° to 100°C. Let us attempt to simulate data that would be obtained by measuring the temperature at regular intervals along the rod. We shall assume that the parent population is described by the equation

$$T = a_0 + b_0 x$$

(5.22)

with $a_0 = 0$°C and $b_0 = 10$°C/cm, and that 10 measurements are made at 1-cm intervals from $x = 0.5$ to $x = 9.5$ cm, with negligible uncertainties in x_i and uniform measuring uncertainties in T_i of $\sigma_T = 1.0$°C.

This is a common application of the Monte Carlo technique: simulation of the effects of measuring uncertainties by smearing data points. If a particular variable has a mean value T_i, with uncertainties σ_i and Gaussian uncertainties are assumed, then we obtain the smeared value of T_i from the relation

$$T_i' = T_i + \sigma_i r_i$$

(5.23)

TABLE 5.2
Simulated temperature versus position data
for a 10-cm rod held at $T = 0$°C at $x = 0.0$ cm
and at $T = 100$°C at $x = 10.0$ cm†

i	X_i (cm)	T_i (°C)
0.5	5.00	4.71
1.5	15.00	15.43
2.5	25.00	23.24
3.5	35.00	35.77
4.5	45.00	45.39
5.5	55.00	52.26
6.5	65.00	65.71
7.5	75.00	76.96
8.5	85.00	85.97
9.5	95.00	93.77

†A uniform temperature gradient was assumed. The uncertainty in the measurement of T was assumed to be $\sigma_T = 1.0$ °C.

where r_i is a random variable drawn from the standard Gaussian distribution with mean 0 and standard deviation 1. The calculation is equivalent to drawing the random variable T_i' directly from a Gaussian distribution with mean T_i and standard deviation σ_i.

Program 5.1. `Hot Rod` Illustrates use of routines to generate uniform and Gaussian deviates, as well as the Gaussian smearing technique. The program calculates `T` from Equation (5.22) and the smeared value `Tprime` from Equation (5.23) in a call to the routine `GaussSmear`. The program is listed in Appendix E.

Program 5.2. `MonteLib` A program unit that includes some general Monte Carlo procedures. The routines are listed in Appendix E.

`RandomDeviate` The Wichmann-Hill random number generator.

`RandomGaussDeviate` The Box-Müller method of Equations (5.20). The routine calls `RandomDeviate` twice to generate each pair of Gaussian deviates.

`SetRandomDeviateSeed` Routine to set the initial values of the three seeds for `RandomDeviate` and for `RandomGaussDeviate`.

`GetRandomDeviateSeed` Routine to return current values of the three random number generator seeds and return to the calling program.

`GaussSmear` Routine to add a Gaussian uncertainty to a variable.

`PoissonRecur` Calculates the Poisson probability function by a recurrence relation.

`PoissonDeviate` Draws a random number distributed according to the Poisson probability function.

The data generated by the program `HotRod` are shown in Table 5.2, with values of T_i for the parent population, predicted by Equation (5.22), and of T_i' for the sample population, calculated from Equation (5.23) for various values of x_i. Note that, as we should expect, the modified values of T are scattered about the values calculated from Equation (5.22).

Choice of a Method

Which of these methods for generating samples from the Gaussian probability distribution is the best? The answer depends on need and circumstance. For general use it is convenient to keep a version of the Box-Müller method in your program library. This routine produces a continuous range of samples limited only by the computer word size. For high precision work, however, we should be aware that subtle correlations between adjacent uniform deviates have been shown to distort the tails of the Gaussian distribution of these numbers. If highest speed is essential, then the transformation method with a precalculated table of the integral and some pointers for quick access to the table should be the choice. This method requires making decisions on the range and resolution of the generated variable and some extra programming to create and access the integral table, but the lookup method can be very fast. Finally, if you are stranded on a desert island with only your laptop computer and have an urgent need for random selections from a Gaussian distribution, the method of summing N random numbers is sufficiently simple that you should be able to write and debug the routine in a few minutes, provided you can remember that the magic number is $N = 12$ for a variance of 1.

Poisson Distribution

Poisson statistics are important in most Monte Carlo calculations, but they are usually implied rather than calculated explicitly. Nevertheless, we sometimes wish to generate data that are distributed according to the Poisson function, and application of the transformation method to the problem is particularly simple and instructive. To find an integer x drawn from the Poisson distribution with mean μ, a *Poisson deviate*, we generate a random variable r from the uniform distribution, replace the integral of Equation (5.10) by the sum

$$r = \sum_{x=0}^{x} P_P(x;\mu) = \sum_{x=0}^{x} \frac{\mu^x}{x!} e^{-\mu} \tag{5.24}$$

and solve Equation (5.24) for x.

Although the Poisson function does not have the convenient scaling properties of the Gaussian function, and thus different calculations are required for each value of the mean μ, very few calculations are actually needed because we are interested in this distribution only at small values of μ, say $\mu \le 16$, and only at integral values of the argument x. At larger values of μ, the Poisson distribution becomes indistinguishable from the Gaussian and it is generally more convenient to employ the Gaussian function in calculations.

> **Example 5.5.** An instructor is preparing an exercise on Poisson statistics for his class. He plans to provide each student with a simulated data set corresponding to 200 Geiger counter measurements of cosmic ray flux recorded in 10-s intervals with an assumed mean counting rate of 8.4 counts per interval. The data will correspond to the number of counts recorded in each 10-s interval.
>
> Students will be asked to make histograms of their individual data samples, find the means and standard deviations of the data, and compare their distributions with the predictions of Gaussian and Poisson probability functions.

For each student, a set of values of x is generated from Equation (5.24) with $\mu = 8.4$ and 200 different random numbers. The transformation method is used with a precalculated table of sums so that the value of x associated with each value of r can be selected by a simple search. To assure that each student's data set is independent, either all sets are generated in a single computer run or else the random number seeds are saved at the end of each run and used to start the next run.

Program 5.3. PoisDcay illustrates the procedure used to generate an individual data set of 200 random variables drawn from the Poisson probability distribution with mean $\mu = 8.4$. The generated data are displayed as a histogram, rather than as a table. The program is listed in Appendix E.

The main program initializes histogram routines in the program unit Hists (see Appendix D) with calls to HistInit and HistSetup, and calls the function PoissonDeviate with second argument init = TRUE to set up a table of sums of $P_P(i; \mu)$ from $i = 0$ to n indexed by n; that is, to form the array

$$S_n = \sum_{i=0}^{n} P_P(i; \mu) \quad \text{for } n = 1, 2, \ldots, n_{max} \tag{5.25}$$

so that

$$S_n = S_{n-1} + P_P(n; \mu) \quad \text{with } S_0 = P_P(0; \mu) = e^{-\mu} \tag{5.26}$$

where $n_{max} = 8\sqrt{\mu}$ is selected as a reasonable upper range for the Poisson curve.

For each event, the program calls PoissonDeviate with second argument init = FALSE to select a value from the table. The routine PoissonDeviate generates a random number r from the uniform distribution and searches the table beginning at S_0, to find the value of n for which $S_n \geq r$. The value of n at which this occurs is the desired random sample from the Poisson distribution. As the samples are generated they are entered in a histogram by calls to the routine Histogram.

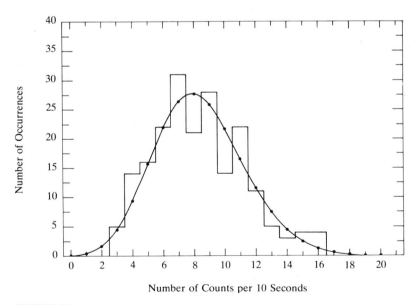

FIGURE 5.3
Histogram of 200 random variables generated by Program 5.3 from the Poisson distribution with mean $\mu = 8.4$.

TABLE 5.3
Poisson probability $P_p(i; \mu)$ and summed probability $S_i = \sum_{i=0}^{n} P_p(i; \mu)$ for $\mu = 8.4$†

n	$P_p(n; \mu)$	S_n	n	$P_p(n; \mu)$	S_n
0	0.0002248673	0.0002248673	16	0.0066035175	0.9940781736
1	0.0018888855	0.0021137528	17	0.0032629145	0.9973410882
2	0.0079333192	0.0100470720	18	0.0015226935	0.9988637816
3	0.0222132938	0.0322603658	19	0.0006731908	0.9995369724
4	0.0466479169	0.0789082827	20	0.0002827401	0.9998197126
5	0.0783685004	0.1572767830	21	0.0001130961	0.9999328086
6	0.1097159005	0.2669926835	22	0.0000431821	0.9999759908
7	0.1316590806	0.3986517641	23	0.0000157709	0.9999917616
8	0.1382420346	0.5368937988	24	0.0000055198	0.9999972814
9	0.1290258990	0.6659196977	25	0.0000018547	0.9999991361
10	0.1083817551	0.7743014529	26	0.0000005992	0.9999997353
11	0.0827642494	0.8570657023	27	0.0000001864	0.9999999217
12	0.0579349746	0.9150006768	28	0.0000000559	0.9999999776
13	0.0374349066	0.9524355835	29	0.0000000162	0.9999999938
14	0.0224609440	0.9748965275	30	0.0000000045	0.9999999983
15	0.0125781286	0.9874746561	31	0.0000000012	1.0000000000

†The summation was terminated arbitrarily at $n \simeq \mu + 8\sqrt{\mu} \simeq 31$, and $P_p(31; \mu)$ was set to 1.

A histogram of 200 variables drawn from the Poisson distribution by this program is shown in Figure 5.3 with the parent distribution represented as a solid curve (although it is, of course, not defined between integer values of the abscissa). The values of the Poisson function, calculated by the routine PoissonRecur, and the sums, calculated by the routine PoissonDeviate, for $\mu = 8.4$ and for x ranging from 0 to 16, are displayed in Table 5.3.

We note that with the precalculated table it is only necessary to increment a counter a few times and compare two real numbers to obtain each random variable, whereas, without the table, it would have been necessary to calculate the Poisson function several times for each generated sample, in addition to comparing the two real numbers.

Exponential Distribution

If the Monte Carlo problem includes the generation of unstable states, random numbers drawn from an exponential distribution will be needed. Here the transformation method is clearly the method of choice because the integral Equation (5.10) and resultant equation can be solved analytically.

Example 5.6. Consider an experiment to study the decay rate of a radioactive source with estimated mean life of τ seconds. The experiment involves collecting counts over successive time intervals Δt with a Geiger counter and scaler combination and plotting the number of counts in each interval against the mean interval time.

We wish to simulate this experiment with a Monte Carlo calculation. The normalized probability density function for obtaining a count at time t from an exponential distribution with mean life τ is given by

$$P_e(t;\tau) = \begin{cases} 0 & \text{for } t < 0 \\ \dfrac{e^{-t/\tau}}{\tau} & \text{for } t \geq 0 \end{cases} \tag{5.27}$$

We can obtain an expression for random samples t_i from this distribution by applying Equation (5.10) to obtain

$$t_i = -\tau \ln r_i \tag{5.28}$$

Thus, to obtain each value of t_i, we find a random number from the uniform distribution and calculate t_i from Equation (5.28).

Let us consider a second method of generating the histogram of the previous example, a method that is much more efficient, but that severely limits any later treatment of the data.

We can calculate the fraction of events that the parent distribution predicts would fall into each of the Δt wide histogram bins from the equation

$$\Delta N'(t) = \int_{t-e}^{t+e} \frac{e^{-t/\tau}}{\tau} \, dt = e^{-t/\tau}\big|_{t-e}^{t+e} \simeq \frac{\Delta t}{\tau} e^{-t/\tau} \tag{5.29}$$

where we have written $e = \Delta t/2$. The effect of the statistical errors is to smear each of these calculated values in a way consistent with the Poisson distribution with mean $\mu = \Delta N_i'$. For small values of $\Delta N_i'$ we find the smeared value ΔN_i directly from Equation (5.24):

$$r = \sum_{x=0}^{\Delta N} P_P(x; \Delta N') \tag{5.30}$$

For larger values of $\Delta N_i'$ calculation with the Poisson equation would be too tedious, but we can use Gaussian smearing as in Example 5.4 with $\sigma_i = \sqrt{\mu}$. Note that the Poisson equation *must* be used for bins with low statistics to assure a positive number of counts in each bin.

Although these two methods of generating a data set or histogram produce equivalent statistical results for Example 5.6, they differ in important details. The full Monte Carlo method required generating individual "events" that can be recorded and studied. For example, we could check the statistical behavior of the data by subdividing the sample into several smaller groups. We could also investigate the effect of decreasing as well as increasing the binning time intervals Δt. Finally, if we should wish to expand the study, perhaps to consider experimental geometry and detector efficiency, the full Monte Carlo method will allow that. The smearing method, on the other hand, produces only the 10 numbers, representing the counts in the 10 bins. Aside from merging the bins, we have no control over the data for future calculations. It is strictly a fast, "one-shot" procedure with a specific limited aim.

5.5   EFFICIENCY

Because the relative error in a result calculated by the Monte Carlo method is inversely proportional to the square root of the number of *successful* events generated, it is important, especially for a long calculation, to have the highest possible program efficiency. Rejected events do not improve the statistical accuracy and every effort should be made to reduce the time spent on calculations that lead to "misses" rather than "hits". There are several ways to improve generation efficiency:

1. **Don't be a purist.** The Monte Carlo method is basically a way of doing complicated multidimensional integrals. If you can save time by doing part of the problem by analytic methods, do so.
2. **Program carefully.** Do not repeat calculations if the results can be saved for later use.
3. If possible, test the low yield sections of the simulation early and cut out as soon as a "miss" occurs. Except for particular loss studies, it is usually not profitable to follow the calculation of an event that is known to end in failure.
4. Try to reduce the variance of the results by limiting ranges wherever possible. One application of this technique can be illustrated in Example 5.1, where the area of a circle of radius r_c is calculated by inscribing it within a square. Making the side of the square larger than the diameter of the circle would be wasteful and would increase the variance of the area determination.
5. When repeating a calculation to find the effects of varying a parameter, consider setting up the program in such a way that the identical sequence of random numbers is repeated throughout the calculation, except for calculations specifically associated with the change. This technique will not improve the variance of the *overall* calculation, but will reduce the variance of the *difference* of results from two calculations.
6. Inspect each probability function carefully before beginning a calculation and estimate the resolution and detail that will be required in the calculation. If a distribution has fine structure, try to determine whether or not such structure is of interest and must be preserved. If necessary, consider breaking the calculations into separate regions and varying the sampling sensitivity as appropriate for each region.
7. Be critical. Examine your generated variables to see that they fall within the expected ranges and follow expected distributions. In a large program, errors that affect the results in subtle ways may be buried within the program and be very difficult to detect. The only way to prevent problems is to make detailed checks at every stage of the program.

SUMMARY

Pseudorandom numbers: Numbers created by a computer algorithm such that successive numbers appear to be uncorrelated with previous numbers. They are referred to as *random numbers* or *random deviates*.

Uniform deviates: Pseudorandom numbers that are uniformly distributed between 0 and 1:

$$p(r) = \begin{cases} 1 & \text{for } 0 \leq r < 1 \\ 0 & \text{otherwise} \end{cases}$$

Normalized distribution: A distribution that is scaled so that its integral over a specified range is equal to unity.

Transformation integral: Transforms the variable r drawn randomly from the uniform distribution into a variable x drawn randomly from the distribution $P(x)$:

$$\int_{r=0}^{r} 1 \, dr = \int_{x=-\infty}^{x} P(x) \, dx$$

Rejection method: A method of generating random numbers drawn from particular distributions by rejecting those that fall outside the geometrical limits of the specified distribution.

Gaussian deviate: Random number drawn from a Gaussian distribution.

Quick Gaussian deviate: The sum of N random numbers is approximately Gaussian distributed with $\mu = N/2$ and $\sigma = \sqrt{N/12}$. Choose $N = 12$ and calculate $r_G = \Sigma r_i - N/2$ to obtain r_G drawn from the standard Gaussian distribution with $\mu = 0$ and $\sigma = 1$.

Box-Müller method for Gaussian deviates: Select r_1 and r_2 from the uniform distribution and calculate

$$z_1 = \sqrt{-2 \ln r_1} \, \cos 2\pi r_2 \quad \text{and} \quad z_2 = \sqrt{-2 \ln r_1} \, \sin 2\pi r_2$$

to obtain z_1 and z_2 drawn from the standard Gaussian distribution.

Data smearing: Method for adding random variations to calculations to simulate the effects of finite measuring errors, $T_i' = T_i + \sigma_i r_i$.

Random numbers from the exponential distribution: To obtain a random number t_i drawn from the exponential distribution, calculate $t_i = -\tau \ln r_i$ from a random deviate r_i.

EXERCISES

5.1. Write a computer program that incorporates the Wichmann and Hill pseudorandom number generator and use it to generate 100 random numbers beginning with seeds $s_1 = 13$, $s_2 = 117$, and $s_3 = 2019$. Make a histogram of the numbers and draw a line representing the expected number of events in each bin. Calculate χ^2 for the agreement between the expected and generated number of events and find the associated probability.

5.2. (*a*) Generate 1000 random numbers uniformly distributed between $-\pi$ and $+\pi$.

(*b*) Generate 1000 random numbers between $x = 0$ and 1, distributed according to the distribution function $P(x) = (5x + 3)$. Use the transformation method with an analytic integration.

(*c*) Find the mean and standard deviation of each distribution and compare them to the predicted values.

(*d*) Make a 20-bin histogram of each distribution and plot on each the predicted distribution.

(*e*) Calculate χ^2 to compare each generated distribution to its parent distribution.

5.3. Write a general routine to generate random integers drawn from the binomial distribution by the transformation method. Use the routine to generate 1000 events corresponding to the distribution of heads or tails when a coin is tossed 50 times. Plot your results and compare them to the direct prediction of Equation (2.4).

5.4. Write a Monte Carlo routine to simulate 200 rolls of a pair of dice and find the frequency of occurrences of each possible sum. Plot a histogram of the occurrences with statistical error bars and plot the prediction of the binomial distribution. Calculate χ^2 for the agreement between the prediction and the data, and find the χ^2 probability. Compare your results to the exact probability calculation of Exercise 2.3.

5.5. Make a histogram of 200 random numbers that follow the Gaussian distribution by finding the distribution of the sums of groups of 12 random variates drawn from the uniform distribution. Calculate the mean and standard deviation of the generated numbers and the uncertainty in the mean.

5.6. Generate 1000 random numbers between $x = -3$ and $+3$, distributed according to the Lorentzian distribution with mean $\mu = 0$ and half-width $\Gamma = 1.0$. Use the transformation method with a numerical integration and interpolation. (See Appendix A.) Make a 20-bin histogram of the generated numbers and plot the Lorentzian curve on the distribution. Calculate χ^2 to compare the generated distribution to the parent distribution.

5.7. Use the transformation method to produce a sequence of 200 random numbers x drawn from the distribution

$$P(x) = \sin x \quad \text{for } 0 \le x < \pi$$

$$= 0 \quad \text{elsewhere}$$

Make a histogram of the events and compare it to the expected distribution. Note that the calculation can be done analytically and requires an inverse trigonometric function.

5.8. Use the rejection method to generate 500 random deviates drawn from the distribution $y(x) = a_1 + a_2 x^2$, with $a_1 = 3.4$ and $a_2 = 12.1$. Find the mean and standard deviation of the generated numbers and compare them to the expected values.

5.9. Write a Monte Carlo program to generate 200 cubes with sides $a = 2.0 \pm 0.1$ cm, $b = 3.0 \pm 0.1$ cm, and $c = 4.0 \pm 0.2$ cm. Plot the distribution of the volumes of the cubes and find the mean volume, the standard deviation of the distribution, and the uncertainty in the mean. Compare the standard deviation of the distribution to the value predicted by the error propagation equation.

5.10. A *Pascal triangle* provides an interesting illustration of the relation between the binomial and Gaussian probability distributions. Assume an arrangement of pins in the form of a triangle as illustrated.

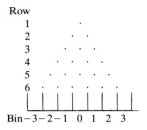

A ball, dropped into the device strikes the top pin and has a 50% probability of striking either of the two pins below it in the next row. The ball bounces down until it reaches the bottom where it is collected in one of the vertical bins.

(*a*) Find a general expression for the probability that a ball will land in a given bin after dropping through N rows of pins.

(*b*) Assume that 512 balls are dropped onto the top pin. Find the number of balls in each bottom bin for a device with three rows of pins above the bins. Repeat for devices with four, five and six rows of pins.

(*c*) Find the standard deviation of the distribution of balls for each example; that is, assume that the bin number is the independent variable so that $\bar{x} = 0$.

(*d*) Plot histograms of the distribution of the balls with Gaussian curves with the means and standard deviations determined in (*c*).

5.11. Write a Monte Carlo program to simulate the Pascal triangle device described in the previous exercise. Compare the results obtained by the two methods.

CHAPTER
6

LEAST-SQUARES FIT TO A STRAIGHT LINE

6.1 DEPENDENT AND INDEPENDENT VARIABLES

We often wish to determine one characteristic y of an experiment as a function of some other quantity x. That is, instead of making a number of measurements of a single quantity x, we make a series of N measurements of the pair (x_i, y_i), one for each of several values of the index i, which runs from 1 to N. Our object is to find a function $y = y(x)$ that describes the relation between these two measured variables. In this chapter we shall consider the problem of pairs of variables (x_i, y_i) that are linearly related to one another, and refer to data from two undergraduate laboratory experiments as examples. In the following chapters, we shall discuss methods of finding relationships that are not linear.

> **Example 6.1.** A student is studying electrical currents and potential differences. He has been provided with a 1-m nickel-silver wire mounted on a board, a lead-acid battery, and an analog voltmeter. He connects cells of the battery across the wire and measures the potential difference or voltage between the negative end and various positions along the wire. From examination of the meter, he estimates the uncertainty in each potential measurement to be 0.05 V. The uncertainty in the position of the probe is less than 1 mm and is considered to be negligible.

TABLE 6.1
Potential difference V as a function of position along a current-carrying nickel-silver wire†

Point number	Position x_i (cm)	Potential difference V_i (V)	x_i^2	$x_i V_i$	Fitted temperature $a + bx$
1	10.0	0.37	100	3.70	0.33
2	20.0	0.58	400	11.60	0.60
3	30.0	0.83	900	24.90	0.86
4	40.0	1.15	1,600	46.00	1.12
5	50.0	1.36	2,500	68.00	1.38
6	60.0	1.62	3,600	97.20	1.64
7	70.0	1.90	4,900	133.00	1.91
8	80.0	2.18	6,400	174.40	2.17
9	90.0	2.45	8,100	220.50	2.43
Sums	450.0	12.44	28,500	779.30	

$\Delta = N\Sigma x_i^2 - (\Sigma x_i)^2 = (9 \times 28,500) - (450)^2 = 54,000$

$a = (\Sigma x_i^2 \Sigma V_i - \Sigma x_i \Sigma x_i V_i)/\Delta = (28,500 \times 12.44 - 450.0 \times 779.30)/54,000$
$\quad = 0.0714$

$b = (N\Sigma x_i V_i - \Sigma x_i \Sigma V_i)/\Delta = (9 \times 779.30 - 450.0 \times 12.44)/54,000 = 0.0262$

$\sigma_a^2 \simeq \sigma_V^2 \Sigma x_i^2/\Delta = 0.05^2 \times 28,500/54,000 = 0.001319 \qquad \sigma_a \simeq 0.036 \qquad \sigma_a' = 0.019$

$\sigma_b^2 \simeq N\sigma_V^2/\Delta = 9 \times 0.05^2/54,000 = 0.417 \times 10^{-6} \qquad \sigma_b \simeq 0.00065 \qquad \sigma_b' = 0.00034$

†A uniform uncertainty in V of 0.05 V is assumed. A linear fit to the data, calculated by the method of determinants, gives $a = 0.07 \pm 0.04$ V and $b = 0.0262 \pm 0.0006$ V/cm, with $\chi^2 = 1.95$ for 7 degrees of freedom. The χ^2 probability for the fit is approximately 96%.

The data are listed in Table 6.1 and are plotted in Figure 6.1 to show the potential difference as a function of wire length x. The estimated common uncertainty in each measured voltage is indicated on the graph by the vertical error bars. The horizontal bars indicate only the position of the ordinates and do not correspond to uncertainties in the abscissae. From these measurements, we wish to find the linear function $y(x)$ (shown as a solid line) that describes the way in which the voltage V varies as a function of position x along the wire.

Example 6.2. In another experiment, a student is provided with a radioactive source enclosed in a small 8-mm-diameter plastic disk and a Geiger counter with a 1-cm-diameter end window. Her object is to investigate the $1/r^2$ law by recording Geiger counter measurements over a fixed period of time at various distances from the source between 20 and 100 cm. Because the counting rate is not expected to vary from measurement to measurement, except for statistical fluctuations, the student can record data long enough to obtain good statistics over the entire range of the experiment. She uses an automatic recording system and records counts for thirty 15-s intervals at each position. For analysis in this experiment, she sums the counts from the 30 measurements at each positions. The separate 15-s interval measurements at each position can be used in other statistical studies.

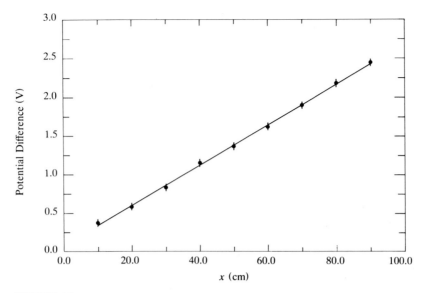

FIGURE 6.1
Potential difference as a function of position along a conducting wire (Example 6.1). The uniform uncertainties in the potential measurements are indicated by the vertical error bars. The straight line is the result of a least-squares fit to the data.

TABLE 6.2
Number of counts detected in $7\frac{1}{2}$-min intervals as a function of distance form the source†

i	Distance d (m)	$x = 1/d^2$	Counts C_i	Weight $(1/C_i^2)$ σ_{C_i}	w_i	$w_i x_i$	$w_i C_i$	$w_i x_i^2$	$w_i x_i C_i$	Fitted counts $a + bx_i$
1	20	25.00	901	30.0	0.00111	0.0278	1	0.694	25.0	887
2	25	16.00	652	25.5	0.00153	0.0254	1	0.393	16.0	610
3	30	11.11	443	21.0	0.00226	0.0251	1	0.279	11.1	461
4	35	8.16	339	18.4	0.00295	0.0241	1	0.197	8.2	370
5	40	6.25	283	16.8	0.00353	0.0221	1	0.138	6.3	311
6	45	4.94	281	16.8	0.00356	0.0176	1	0.087	4.9	271
7	50	4.00	240	15.5	0.00417	0.0167	1	0.067	4.0	242
8	60	2.78	220	14.8	0.00455	0.0126	1	0.035	2.8	205
9	75	1.78	180	13.4	0.00556	0.0099	1	0.018	1.8	174
10	100	1.00	154	12.4	0.00649	0.0065	1	0.007	1.0	150
Sums					0.03570	0.1868	10	1.912	81.0	

$\sigma_i = \sqrt{y_i} \qquad w_i = 1/\sigma_i^2 = 1/y_i$

$\Delta = \Sigma w_i \Sigma w_i x_i^2 - (\Sigma w_i x_i)^2 = 0.03570 \times 1.912 - (0.1868)^2 = 0.0334$

$a = [\Sigma w_i C_i \Sigma w_i x_i^2 - \Sigma w_i x_i \Sigma w_i x_i C_i]/\Delta = [10 \times 1.912 - 0.1868 \times 81.0]/\Delta$
$\quad = 119.5$

$b = [\Sigma w_i \Sigma w_i x_i C_i - \Sigma w_i x_i \Sigma w_i C_i]/\Delta = [0.03570 \times 81.0 - 0.1868 \times 10]/\Delta$
$\quad = 30.7$

$\sigma_a^2 \approx \Sigma w_i x_i^2/\Delta = 1.912/0.0334 = 57.3 \qquad \sigma_a \approx 7.6$
$\sigma_b^2 \approx \Sigma w_i/\Delta = 0.03570/0.0334 = 1.07 \qquad \sigma_b \approx 1.1$

†A linear fit to the data of the function $C = a + bx$ by the method of determinants gives $a = 119 \pm 8$ and $b = 31 \pm 1$, with $\chi^2 = 11.1$ for 8 degrees of freedom. The χ probability for the fit is about 20%.

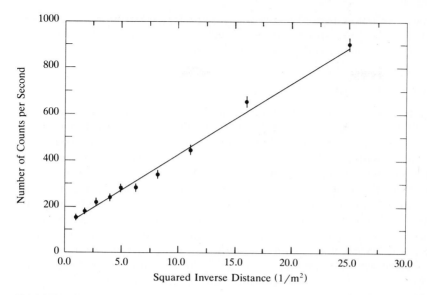

FIGURE 6.2
Number of counts in constant time intervals from a radioactive source as a function of the inverse distance from source to Geiger counter (Example 6.2). The vertical error bars indicate the statistical uncertainties in the counts. The straight line is the result of a least-squares fit to the data.

The data are listed in Table 6.2 and plotted against $x = 1/r^2$ in Figure 6.2. The vertical error bars on the data points represent the statistical uncertainties in the measured number of counts and are equal to the square root of the number of counts. The uncertainty in the measurement of the distance from the source to the counter was assumed to be negligible.

Linear Approximation

In both of these examples, the functional relationship between the dependent and independent variables can be approximated by a straight line of the form

$$y(x) = a + bx. \tag{6.1}$$

We shall consider in this chapter a method for determining the most probable values for the coefficients a and b.

We cannot fit a straight line to the data exactly in either example because it is impossible to draw a straight line through all the points. For a set of N arbitrary points, it is always possible to fit a polynomial of degree $N - 1$ exactly, but for our experiments, the coefficients of the higher-order terms would have questionable significance. We assume that the fluctuations of the individual points above and below the solid curves are caused by experimental uncertainties in the individual measurements. In Chapter 11 we shall develop a method for testing whether higher-order terms are significant.

Measuring Uncertainties

If we were to make a series of measurements of the dependent quantity y_i for one particular value x_i of the independent quantity, we would find that the measured values were distributed about a mean in the manner discussed in the previous chapter with a probability of $\sim 68\%$ that any single measurement of y_i be within 1 standard deviation of the mean. By making a number of measurements for each value of the independent quantity x_i, we could determine mean values $\bar{y}_i$ with any desired precision. Usually, however, we can make only one measurement y_i for each value of $x = x_i$, so that we must determine the value of y corresponding to that value of x with an uncertainty that is characterized by the standard deviation σ_i of the distribution of data for that point.

We shall assume for simplicity in all the following discussions that we can ascribe all the uncertainty in each measurement to the dependent variable. This is equivalent to assuming that the precision of the determination of x is considerably higher than that of y. This difference is illustrated in Figures 6.1 and 6.2 by the fact that the uncertainties are indicated by error bars for the dependent variables but not for the independent variables.

Our condition, that we neglect uncertainties in x and consider just the uncertainties in y, will be valid only if the uncertainties in y that would be produced by variations in x corresponding to the uncertainties in the measurement of x are much smaller than the uncertainties in the measurement of y. This is equivalent, in first order, to the requirement at each measured point that

$$\sigma_x \frac{dy}{dx} \ll \sigma_y$$

where dy/dx is the slope of the function $y = y(x)$.

We are not always justified in ascribing all uncertainties to the dependent parameter. Sometimes the uncertainties in the determination of both quantities x and y are nearly equal. But our fitting procedure will still be fairly accurate if we estimate the indirect contribution σ_{yI} from the uncertainty σ_x in x to the total uncertainty in y by the first-order relation

$$\sigma_{yI} = \sigma_x \frac{dy}{dx} \tag{6.2}$$

and combine this with the direct contribution σ_{yD}, which is the measuring uncertainty in y, to get

$$\sigma_y^2 = \sigma_{yI}^2 + \sigma_{yD}^2 \tag{6.3}$$

For both Examples 6.1 and 6.2 the condition would be reasonable because we predict a linear dependence of y with x. With the linear assumption, we treat the uncertainties in our data as if they were in the dependent variable only, while realizing that the corresponding fluctuations may have been originally derived from uncertainties in the determinations of both dependent and independent variables.

In those cases where the uncertainties in the determination of the independent quantity are considerably greater than those in the dependent quantity, it might be wise to interchange the definition of the two quantities.

6.2 METHOD OF LEAST SQUARES

Our data consist of pairs of measurements (x_i, y_i) of an independent variable x and a dependent variable y. We wish to find values of the parameters a and b that minimize the discrepancy between the measured values y_i and calculated values $y(x)$. We cannot determine the parameters exactly with only a finite number of observations, but can hope to extract the most probable estimates for the coefficients in the same way that we extracted the most probable estimate of the mean in Chapter 4.

Before proceeding, we must define our criteria for minimizing the discrepancy between the measured and predicted values y_i. For any arbitrary values of a and b, we can calculate the deviations Δy_i between each of the observed values y_i and the corresponding calculated or fitted values

$$\Delta y_i = y_i - y(x_i) = y_i - a - bx_i \qquad (6.4)$$

With well chosen coefficients, these deviations should be relatively small. However, the sum of these deviations is not a good measure of how well our calculated straight line approximates the data because large positive deviations can be balanced by negative ones to yield a small sum even when the fit of the function $y(x)$ to the data is bad. We might consider instead summing the absolute values of the deviations, but this leads to difficulties in obtaining an analytical solution. Instead we sum the squares of the deviations.

There in no correct unique method for optimizing the coefficients valid for all problems. There exists, however, a method that can be fairly well justified, that is simple and straightforward, and that is well established experimentally. This is the *method of least squares*, similar to the method discussed in Chapter 4, but extended to include more than one variable. It may be considered as a special case of the more general *method of maximum likelihood*.

Method of Maximum Likelihood

Our data consist of a sample of observations drawn from a parent distribution that determines the probability of making any particular observation. For the particular problem of an expected linear relationship between dependent and independent variables, we define parent coefficients a_0 and b_0 such that the actual relationship between y and x is given by

$$y_0(x) = a_0 + b_0 x \qquad (6.5)$$

We shall assume that each individual measured value of y_i is itself drawn from a Gaussian distribution with mean $y_0(x_i)$ and standard deviation σ_i. We should be aware that the Gaussian assumption may not always be exactly true. In Example 6.2 the $y_i = C_i$ were obtained in a counting experiment and therefore follow a Poisson distribution. However, for a sufficiently large number of counts y_i the distribution may be considered to be Gaussian. We shall discuss fitting with Poisson statistics in Section 6.6.

With the Gaussian assumption, the probability P_i for making the observed measurement y_i with standard deviation σ_i for the observations about the actual value $y_0(x_i)$ is

$$P_i = \frac{1}{\sigma_i \sqrt{2\pi}} \exp\left\{ -\frac{1}{2}\left[\frac{y_i - y_0(x_i)}{\sigma_i}\right]^2 \right\} \qquad (6.6)$$

The probability for making the observed set of measurements of the N values of y_i is the product of the probabilities for each observation:

$$P(a_0, b_0) = \Pi P_i = \Pi\left(\frac{1}{\sigma_i\sqrt{2\pi}}\right)\exp\left\{ -\frac{1}{2}\sum\left[\frac{y_i - y_0(x_i)}{\sigma_i}\right]^2 \right\} \qquad (6.7)$$

where the product Π is taken with i ranging from 1 to N and the product of the exponentials has been expressed as the exponential of the sum of the arguments. In these products and sums, the quantities $/1\sigma_i^2$ act as weighting factors.

Similarly, for any *estimated* values of the coefficients a and b, we can calculate the probability of obtaining the observed set of measurements

$$P(a, b) = \Pi\left(\frac{1}{\sigma_i\sqrt{2\pi}}\right)\exp\left\{ -\frac{1}{2}\sum\left[\frac{y_i - y(x_i)}{\sigma_i}\right]^2 \right\} \qquad (6.8)$$

with $y(x)$ defined by Equation (6.1) and evaluated at each of the values x_i.

We assume that the observed set of measurements is more likely to have come from the parent distribution of Equation (6.5) than from any other similar distribution with different coefficients and, therefore, the probability of Equation (6.7) is the maximum probability attainable with Equation (6.8). Thus, the maximum-likelihood estimates for a and b are those values that maximize the probability of Equation (6.8).

Because the first factor in the product of Equation (6.8) is a constant, independent of the values of a and b, maximizing the probability $P(a, b)$ is equivalent to minimizing the sum in the exponential. We define this sum to be our goodness-of-fit parameter χ^2:

$$\chi^2 \equiv \sum\left[\frac{y_i - y(x_i)}{\sigma_i}\right]^2 = \sum\left[\frac{1}{\sigma_i}(y_i - a - bx_i)\right]^2 \qquad (6.9)$$

We use the same symbol χ^2, defined earlier in Equation (4.32), because this is essentially the same definition in a different context.

Our method for finding the optimum fit to the data will be to find values of a and b that minimize this weighted sum of the squares of the deviations χ^2 and hence, to find the fit that produces the smallest sum of the squares or the *least-squares fit*. The magnitude of χ^2 is determined by four factors:

1. Fluctuations in the measured values of the variables y_i, which are random samples from a parent population with expectation values $y_0(x_i)$.
2. The values assigned to the uncertainties σ_i in the measured variables y_i. Incorrect assignment of the uncertainties σ_i will lead to incorrect values of χ^2.
3. The selection of the analytical function $y(x)$ as an approximation to the "true" function $y_0(x)$. It might be necessary to fit several different functions in order to find the appropriate function for a particular set of data.
4. The values of the parameters of the function $y(x)$. Our objective is to find the "best values" of these parameters.

6.3 MINIMIZING χ^2

To find the values of the coefficients a and b that yield the minimum value for χ^2, we set to zero the partial derivatives of χ^2 with respect to each of the parameters

$$\frac{\partial}{\partial a}\chi^2 = \frac{\partial}{\partial a} \sum \left[\frac{1}{\sigma_i^2}(y_i - a - bx)^2 \right]$$

$$= -2 \sum \left[\frac{1}{\sigma_i^2}(y_i - a - bx_i) \right] = 0$$

$$\frac{\partial}{\partial b}\chi^2 = \frac{\partial}{\partial b} \sum \left[\frac{1}{\sigma_i^2}(y_i - a - bx)^2 \right]$$

$$= -2 \sum \left[\frac{x_i}{\sigma_i^2}(y_i - a - bx_i) \right] = 0$$

$$(6.10)$$

These equations can be rearranged as a pair of linear simultaneous equations in the unknown parameters a and b:

$$\sum \frac{y_i}{\sigma_i^2} = a \sum \frac{1}{\sigma_i^2} + b \sum \frac{x_i}{\sigma_i^2}$$

$$\sum \frac{x_i y_i}{\sigma_i^2} = a \sum \frac{x_i}{\sigma_i^2} + b \sum \frac{x_i^2}{\sigma_i^2}$$

$$(6.11)$$

The solutions can be found in any one of a number of different ways, but, for generality we shall use the method of determinants. (See Appendix B.) The solutions are

$$a = \frac{1}{\Delta} \begin{vmatrix} \sum \dfrac{y_i}{\sigma_i^2} & \sum \dfrac{x_i}{\sigma_i^2} \\[2mm] \sum \dfrac{x_i y_i}{\sigma_i^2} & \sum \dfrac{x_i^2}{\sigma_i^2} \end{vmatrix} = \frac{1}{\Delta}\left(\sum \frac{x_i^2}{\sigma_i^2} \sum \frac{y_i}{\sigma_i^2} - \sum \frac{x_i}{\sigma_i^2} \sum \frac{x_i y_i}{\sigma_i^2} \right)$$

$$b = \frac{1}{\Delta} \begin{vmatrix} \sum \dfrac{1}{\sigma_i^2} & \sum \dfrac{y_i}{\sigma_i^2} \\[2mm] \sum \dfrac{x_i}{\sigma_i^2} & \sum \dfrac{x_i y_i}{\sigma_i^2} \end{vmatrix} = \frac{1}{\Delta}\left(\sum \frac{1}{\sigma_i^2} \sum \frac{x_i y_i}{\sigma_i^2} - \sum \frac{x_i}{\sigma_i^2} \sum \frac{y_i}{\sigma_i^2} \right) \quad (6.12)$$

$$\Delta = \begin{vmatrix} \sum \dfrac{1}{\sigma_i^2} & \sum \dfrac{x_i}{\sigma_i^2} \\[2mm] \sum \dfrac{x_i}{\sigma_i^2} & \sum \dfrac{x_i^2}{\sigma_i^2} \end{vmatrix} = \sum \frac{1}{\sigma_i^2} \sum \frac{x_i^2}{\sigma_i^2} - \left(\sum \frac{x_i}{\sigma_i^2} \right)^2$$

For the special case in which all the uncertainties are equal ($\sigma = \sigma_i$), they cancel and the solutions may be written

$$a = \frac{1}{\Delta'} \begin{vmatrix} \sum y_i & \sum x_i \\ \sum x_i y_i & \sum x_i^2 \end{vmatrix} = \frac{1}{\Delta'}\left(\sum x_i^2 \sum y_i - \sum x_i \sum x_i y_i \right)$$

$$b = \frac{1}{\Delta'} \begin{vmatrix} N & \sum y_i \\ \sum x_i & \sum x_i y_i \end{vmatrix} = \frac{1}{\Delta'}\left(N \sum x_i y_i - \sum x_i \sum y_i \right) \quad (6.13)$$

$$\Delta' = \begin{vmatrix} N & \sum x_i \\ \sum x_i & \sum x_i^2 \end{vmatrix} = N \sum x_i^2 - \left(\sum x_i \right)^2$$

Examples

For the data of Example 6.1 (Table 6.1), we assume that the uncertainties in the measured voltages V are all equal and that the uncertainties in x_i are negligible. We can therefore use Equations (6.13). We accumulate four sums $\sum x_i$, $\sum y_i = \sum T_i$, $\sum x_i^2$, and $\sum x_i y_i = \sum x_i T_i$ and combine them according to Equations (6.13) to find numerical values for a and b. The steps of the calculation are illustrated in Table 6.1, and the resulting fit is shown as a solid line on Figure 6.1.

Determination of the parameters a and b from Equation (6.12) is somewhat more tedious, because the uncertainties σ_i must be included. Table 6.2 shows steps in the calculation of the data of Example 6.2 with the uncertainties σ_i in the numbers of counts C_i determined by Poisson statistics so that $\sigma_i^2 = C_i$.

The values of a and b found in this calculation were used to calculate the straight line through the data points in Figure 6.2.

It is important to note that the value of C_i to be used in determining the uncertainty σ_i must be the actual number of events observed. If, for example, the student had decided to improve her statistics by collecting data at the larger distances over longer time periods Δt_i and to normalize all her data to a common time interval Δt_c,

$$C_i' = C_i \times \Delta T_c / \Delta T_i$$

then the statistical uncertainty in C' would be given by

$$\sigma_i' = \sqrt{C} \times \Delta T_c / \Delta T_i$$

Program 6.1. FitLine Solution of Equations (6.11) by the determinant method of Equation (6.12) is illustrated in the program TestDet, listed in Appendix E. The program uses routines in the common program units FitVars, FitUtil, and GenUtil that are shared with other fitting programs and are listed in Appendix E. The Pascal sample programs use single precision variables for simplicity, although double, or higher, precision is highly recommended, especially for fits with more than two parameters.

FitLine Main program. Calls utility routines CenterWrite and ReadChar in the program unit GenUtil to select input data and Output, and Plotit in program unit FitUtil to list output and display graphs. Variable definitions and data arrays are in the program unit FitVars. Graphical displays are produced by the routines in the program unit QuikPlot discussed in Appendix D.

LineFit Calculates the determinants, the parameters a_1 and a_2, and their uncertainties.

Calculate Calculates the function $y = a_1 + a_2 x$ and stores the result in the array yCalc.

CalcChi Calculates the value of χ^2. The probability associated with a particular value of χ^2 is calculated by a call to ChiPrb from the routine Output.

Program 6.1 uses Equation (6.12) for solving both Examples 6.1 and 6.2, although separate routines written for each problem would be more efficient. We could, for example, increase the speed for fitting the data of Example 6.1 by using Equations (6.13) rather than Equations (6.12), because the measurements of Example 6.1 have common errors. Considering the speed of today's computers, in general, any gain would be offset by the loss in programming simplicity.

For the data of Example 6.2, we could simplify the fitting routine by replacing the statistical errors sigY[i] by the explicit expression $\sqrt{y_i}$. However, in most calculations that involve statistical errors, there are also other errors to be considered, such as those arising from background subtractions, so the loss of generality would more than compensate for any increased efficiency in the calculations.

6.4   ERROR ESTIMATION

Common Uncertainties

If the standard deviations σ_i for the data points y_i are unknown and if we assume that they are all equal, $\sigma_i^2 = \sigma^2$, then we can estimate them from the data and the results of our fit. The requirement of equal errors may be satisfied if the uncertainties are instrumental and all the data are recorded with the same instrument and on the same scale, as was assumed in Example 6.1.

In Chapter 2 we obtained, for our best estimate of the variance of a data sample,

$$\sigma^2 \simeq s^2 \equiv \frac{1}{N - m} \Sigma (y_i - \bar{y})^2 \tag{6.14}$$

where $N - m$ is the number of degrees of freedom and is equal to the number of measurements minus the number of parameters determined from the fit. In Equation (6.14) we identify y_i with the measured value of the dependent variable, and for $\bar{y}$, the expected mean value of y_i, we use the value calculated from Equation (6.1) for each data point with the fitted parameters a and b. Thus, our estimate $\sigma_i = \sigma$ for the standard deviation of an individual measurement is

$$\sigma^2 \simeq s^2 = \frac{1}{N - 2} \Sigma (y_i - a - bx_i)^2 \tag{6.15}$$

By comparing Equation (6.15) with Equation (6.9), we see that it is just this common uncertainty that we have minimized in the least-squares fitting procedure. Thus, we can obtain the common error in our measurements of y from the fit, although at the expense of any information about the quality of the fit.

Variable Uncertainties

In general the uncertainties σ_i in the dependent variables y_i will not all be the same. If, for example, the quantity y represents the number of counts in a detector per unit time interval (as in Example 6.2), then the errors are statistical and the uncertainty in each measurement y_i is directly related to the magnitude of y (as discussed in Section 4.2), and the standard deviations σ_i associated with these measurements is

$$\sigma_i^2 = C_i \tag{6.16}$$

In principle, the value of y_i, which should be used in calculating the standard deviations σ_i by Equation (6.16), is the value $y_0(x_i)$ of the parent population. In practice we use the measured values that are only samples from that population. In the limit of an infinite number of determinations, the average of all the measurements would very closely approximate the parent value, but generally we cannot make more than one measurement of each value of x, much less an infinite number. We could approximate the parent value

$y_0(x_i)$ by using the calculated value $y(x)$ from our fit, but that would complicate the fitting procedure. We shall discuss this possibility further in the following section.

Contributions from instrumental and other uncertainties may modify the simple square root form of the statistical errors. For example, uncertainties in measuring the time interval during which the events of Example 6.2 were recorded might contribute, although statistical fluctuations generally dominate in counting experiments. Background subtractions are another source of uncertainty. In many counting experiments, there is a background under the data that may be removed by subtraction, or may be included in the fit. In Example 6.2, the cosmic and other backgrounds contribute to a counting rate even when the source is moved far away from the detector, as indicated by the nonzero intercept of the fitted line of Figure 6.2 on the C axis. If the student had chosen to record the radiation background counts C_b in a separate measurement and to subtract C_b from each of her measurements C_i to obtain

$$C_i' = C_i - C_b$$

then the uncertainty in C' would have been given by combining in quadrature the uncertainties in the two measurements:

$$\sigma_i'^2 = \sigma_i^2 + \sigma_b^2$$

χ^2 Probability

For those data for which we know the uncertainties σ_i in the measured values y_i we can calculate the value of χ^2 from Equation (6.9) and test the goodness of our fit. For our two-parameter fit to a straight line, the number of degrees of freedom will be $N - 2$. Then, for the data of Example 6.2, we should hope to obtain $\chi^2 \simeq 10 - 2 = 8$. The actual value, $\chi^2 = 11.1$, is listed in Table 6.2, along with the probability ($p = 20\%$). (See Table C.4.) We interpret this probability in the following way. Suppose that we have obtained a χ probability of $p\%$ for a certain set of data. Then, we should expect that, if we were to repeat the experiment many times, approximately $p\%$ of the experiments would yield χ^2 values as high as the one that we obtained or higher. This subject will be discussed further in Chapter 11.

In Example 6.1, we obtained a value of $\chi^2 = 1.95$ for 7 degrees of freedom, corresponding to a probability of about 96%. Although this probability may seem to be gratifyingly high, the very low value of χ^2 gives a strong indication that the common uncertainty in the data may have been overestimated and it might be wise to use the value of χ^2 to obtain a better estimate of the common uncertainty. From Equations (6.15) and (6.9), we obtain an expression for the revised common uncertainty σ_c' in terms of χ^2 and the original estimate, σ_c:

$$\sigma_c'^2 \simeq \sigma_i^2 \times \chi^2/(N - 2) \tag{6.17}$$

or, more generally

$$\sigma_c'^2 \simeq \sigma_i^2 \times \chi_\nu^2 \tag{6.18}$$

where $\chi_\nu^2 = \chi^2/\nu$ and ν is the number of degrees of freedom in the fit. Thus, for Example 6.1, we find $\sigma_c'^2 = 0.05^2 \times 1.95/(9-2) = 0.0007$, or $\sigma_c' = \sim 0.03$ V.

Uncertainties in the Parameters

In order to find the uncertainty in the estimation of the coefficients a and b in our fitting procedure, we use the error propagation method discussed in Chapter 3. Each of our data points y_i has been used in the determination of the parameters and each has contributed some fraction of its own uncertainty to the uncertainty in our final determination. Ignoring systematic errors, which would introduce correlations between uncertainties, the variance σ_z^2 of the parameter z is given by Equation (3.10) as the sum of the squares of the products of the standard deviations σ_i of the data points with the effects that the data points have on the determination of z:

$$\sigma_z^2 = \sum \left[\sigma_i^2 \left(\frac{\partial z}{\partial y_i} \right)^2 \right] \tag{6.19}$$

Thus, to determine the uncertainties in the parameters a and b, we take the partial derivatives of Equation (6.12):

$$
\begin{aligned}
\frac{\partial a}{\partial y_j} &= \frac{1}{\Delta} \left(\frac{1}{\sigma_j^2} \sum \frac{x_i^2}{\sigma_i^2} - \frac{x_j}{\sigma_j^2} \sum \frac{x_i}{\sigma_i^2} \right) \\
\frac{\partial b}{\partial y_j} &= \frac{1}{\Delta} \left(\frac{x_j}{\sigma_j^2} \sum \frac{1}{\sigma_i^2} - \frac{1}{\sigma_j^2} \sum \frac{x_i}{\sigma_i^2} \right)
\end{aligned}
\tag{6.20}
$$

We note that the derivatives are functions only of the variances and of the independent variables x_i. Combining these equations with the general expression of Equation (6.19) and squaring, we obtain for σ^2,

$$
\begin{aligned}
\sigma_a^2 &\simeq \sum_{j=1}^{N} \frac{\sigma_j^2}{\Delta^2} \left[\frac{1}{\sigma_j^4} \left(\sum \frac{x_i^2}{\sigma_i^2} \right)^2 - \frac{2x_j}{\sigma_j^4} \sum \frac{x_i^2}{\sigma_i^2} \sum \frac{x_i}{\sigma_i^2} + \frac{x_j^2}{\sigma_j^4} \left(\sum \frac{x_i}{\sigma_i^2} \right)^2 \right] \\
&= \frac{1}{\Delta} \left[\sum \frac{1}{\sigma_j^2} \left(\sum \frac{x_i^2}{\sigma_i^2} \right)^2 - 2 \sum \frac{x_j}{\sigma_j^2} \sum \frac{x_i^2}{\sigma_i^2} \sum \frac{x_i}{\sigma_i^2} + \sum \frac{x_j^2}{\sigma_j^2} \left(\sum \frac{x_i}{\sigma_i^2} \right)^2 \right] \\
&= \frac{1}{\Delta^2} \left(\sum \frac{x_i^2}{\sigma_i^2} \right) \left[\sum \frac{1}{\sigma_j^2} \sum \frac{x_i^2}{\sigma_i^2} - \left(\sum \frac{x_i}{\sigma_i^2} \right)^2 \right] \\
&= \frac{1}{\Delta} \sum \frac{x_i^2}{\sigma_i^2}
\end{aligned}
\tag{6.21}
$$

and for σ_b^2,

$$\sigma_b^2 \simeq \sum_{j=1}^{N} \frac{\sigma_j^2}{\Delta^2} \left[\frac{1}{\sigma_j^4} \left(\sum \frac{1}{\sigma_i^2} \right)^2 - \frac{2x_j}{\sigma_j^4} \sum \frac{1}{\sigma_i^2} \sum \frac{x_i}{\sigma_i^2} + \frac{1}{\sigma_j^4} \left(\sum \frac{x_i}{\sigma_i^2} \right)^2 \right]$$

$$= \frac{1}{\Delta^2} \left[\sum \frac{1}{\sigma_j^2} \left(\sum \frac{1}{\sigma_i^2} \right)^2 - 2 \sum \frac{x_j}{\sigma_j^2} \sum \frac{1}{\sigma_i^2} \sum \frac{x_i}{\sigma_i^2} + \sum \frac{1}{\sigma_j^2} \left(\sum \frac{x_i}{\sigma_i^2} \right)^2 \right]$$

$$= \frac{1}{\Delta^2} \left(\sum \frac{1}{\sigma_i^2} \right) \left[\sum \frac{1}{\sigma_j^2} \sum \frac{1}{\sigma_i^2} - \left(\sum \frac{x_i}{\sigma_i^2} \right)^2 \right]$$

$$= \frac{1}{\Delta} \sum \frac{1}{\sigma_i^2} \tag{6.22}$$

For the special case of common uncertainties in y_i, $\sigma_i = \sigma$, these equations reduce to

$$\sigma_a^2 = \frac{\sigma^2}{\Delta'} \Sigma x_i^2 \quad \text{and} \quad \sigma_b^2 = N \frac{\sigma^2}{\Delta'} \tag{6.23}$$

with σ given by Equation (6.15) and Δ' given by Equation (6.13).

The uncertainties in the parameters σ_a and σ_b, calculated from the original error estimates, are listed in Tables 6.1 and 6.2. For Example 6.1, revised uncertainties σ_a' and σ_b', based on the revised common data uncertainty calculated from Equation (6.18), are also listed.

6.5 SOME LIMITATIONS OF THE LEAST-SQUARES METHOD

When a curve is fitted by the least-squares method to a collection of statistical counting data, the data must first be *histogrammed*; that is, a histogram must be formed of the corrected data, either during or after data collection. In Example 6.2, the data were collected over intervals of time Δt, with the size of the interval chosen to assure that a reasonable number of counts would be collected in each time interval. For data that vary linearly with the independent variable, this treatment poses no special problems, but one could imagine a more complex problem in which fine details of the variation of the dependent variable y with the independent variable x are important. Such details might well be lost if the binning were too coarse. On the other hand, if the binning interval were too fine, there might not be enough counts in each bin to justify the Gaussian probability hypothesis. How does one choose the appropriate bin size for the data?

A handy rule of thumb when considering the Poisson distribution is to assume that *large enough* = 10. A comparison of the Gaussian and Poisson distributions for mean $\mu \simeq 10$ and standard deviation $\sigma = \sqrt{\mu}$ (see Figures 2.4 and 2.5) shows very little difference between the two distributions. We might expect this because the mean is more than 3 standard deviations away from the origin. Thus, we may be reasonably confident about the results of a fit if no histogram contains less than 10 counts and if we are not placing excessive reliance on the actual value of χ^2 obtained from the fit. If a bin does have fewer than the allowed minimum number of counts, it may be possible to merge that bin with an adjacent one. Note that there is no requirement that intervals on the abscissa be equal, although we must be careful in our choice of the appropriate value of x_i for the merged bin. We should also be aware that such mergers necessarily reduce the resolution of our data and may, when fitting functions more complicated than a straight line, obscure some interesting features.

In general, the choice of bin width will be a compromise between the need for sufficient statistics to maintain a small statistical error in the values of y_i and thus in the fitted parameters, and the need to preserve interesting structure in the data. When full details of any structure in the data must be preserved, it might be advisable to apply the maximum-likelihood method directly to the data, event by event, rather than to use the least-squares method with its necessary binning of the data. We shall return to this subject in Chapter 10.

There is also a question about our use of the experimental errors in the fitting process, rather than the errors predicted by our estimate of the parent distribution. For Example 6.2, this corresponds to our choosing $\sigma_i^2 = y_i$ rather than $\sigma_i^2 = y(x_i) = a + bx_i$. We shall consider the possibility of using errors from our estimate of the parent distribution, as well as the direct application of the Poisson probability function, in the following section.

Another important point to consider when fitting curves to data is the possibility of rounding errors, which can reduce the accuracy of the results. With manual calculations, it is important to avoid rounding the numbers until the very end of the calculation. With computers, problems may arise because of finite computer word length. This problem can be especially severe with matrix and determinant calculations, which often involve taking small differences between large numbers. Depending on the computer and the software, it may be necessary to use double-precision variables in the fitting routine.

6.6   ALTERNATE FITTING METHODS

In this section we shall attempt to solve the problem of fitting a straight line to a collection of data points by using errors determined from the estimated parent distribution rather than from the measurements, and by directly applying Poisson statistics, rather than Gaussian statistics. Because it is not possible to derive a set of independent linear equations for the parameters with these conditions, explicit expressions for the parameters a and b cannot be obtained. However, with fast computers, solving coupled, nonlinear equations is not difficult, although the clarity and elegance of the straightforward least-squares method can be lost.

Poisson Uncertainties

Let us consider a collection of purely statistical data that obey Poisson statistics (as in Example 6.2) so that the uncertainties can be expressed by Equation (6.16). We begin by substituting the approximation $\sigma_i^2 = y(x_i) = a + bx_i$ into the definition of χ^2 in Equation (6.9), which is based on Gaussian probability, and minimizing the value of χ^2 as in Equations (6.10). The result is a pair of simultaneous equations that can be solved for a and b:

$$N = \sum \frac{y_i^2}{(a + bx_i)^2}$$

$$\sum x_i = \sum \frac{x_i y_i^2}{(a + bx_i)^2} \tag{6.24}$$

Poisson Probability

Next, let us replace the Gaussian probability $P(a, b)$ of Equation (6.8) by the corresponding probability for observing y_i counts from a Poisson distribution with mean $\mu_i = y(x_i)$,

$$P(a, b) = \prod \left(\frac{[y(x_i)]^{y_i}}{y_i!} e^{-y(x_i)} \right) \tag{6.25}$$

and apply the method of maximum likelihood to this probability. It is easier and equivalent to maximize the natural logarithm of the probability with respect to each of the coefficients a and b:

$$\ln P(a, b) = \sum [y_i \ln y(x_i)] - \sum y(x_i) + \text{constant} \tag{6.26}$$

where the constant term is independent of the parameters a and b. The result of taking partial derivatives of Equation (6.26) is a pair of simultaneous equations similar to those of Equation (6.24),

$$N = \sum \frac{y_i}{a + bx_i}$$

$$\sum x_i = \sum \frac{x_i y_i}{a + bx_i} \tag{6.27}$$

but with less emphasis on fitting the larger values of y_i.

Neither the coupled simultaneous Equations (6.24) nor the Equations (6.27) can be solved directly for a and b, but each pair can be solved by an iterative method in which values of a and b are chosen and then adjusted until the two simultaneous equations are satisfied. (See Appendix A.3.)

TABLE 6.3
**Comparison of fits to a selection of statistical data from Example 6.2
for three different fitting methods†**

i	Inverse distance squared x_i	Number of counts C_i	(i) Standard	(ii) Gaussian $\sigma_i = y(x_i)$	(iii) Poisson $\sigma_i = y(x_i)$
1	25.00	44	32.1	15.4	15.0
2	16.00	18	21.1	14.1	13.7
3	11.11	17	15.1	12.9	12.5
4	8.16	6	11.5	11.6	11.2
5	6.25	8	9.2	10.3	9.9
6	4.94	9	7.6	9.0	8.7
7	4.00	9	6.4	7.8	7.4
8	2.78	11	4.9	6.5	6.1
9	1.78	3	3.7	5.2	4.9
10	1.00	3	2.7	3.9	3.6
Sums		128	114.3	96.7	93.0
	a		1.52	16.0	15.6
	b		1.22	−0.127	−0.127
	χ^2		13.7	11.6	10.3

†(i) Standard least-squares method with Gaussian statistics and experimental errors; (ii) Gaussian statistics and analytic errors; (iii) Poisson statistics and analytic errors. The analytic errors are expressed as $\sigma_i = a + bx_i$.

Example 6.3. Because we expect the methods discussed here to be equivalent to the standard method for large data samples, we selected a low statistics sample to emphasize the differences. We chose from the measurements of Example 6.2 only those events collected at each detector position during the first 15-s interval, a total of 128 events at 10 different positions. The results of (i) calculations by the standard method, (ii) calculations with Gaussian statistics and with errors given by $\sigma_i = y(x_i) = a + bx_i$, and (iii) calculations with Poisson statistics with errors as in method (ii) are listed in Table 6.3. We note that method (i) appears to underestimate the number of events in the sample, whereas method (ii) overestimates the number. Method (iii) with Poisson statistics and errors calculated as in method (ii) finds the exact number.

We can avoid questions of finite binning and the choice of statistics by making direct use of the maximum-likelihood method, treating the fitting function as a probability distribution. This method also allows detailed handling of problems in which the probability associated with individual measurements varies in a complex way from observation to observation. We shall pursue this subject further in Chapter 10.

In general, however, the simplicity of the least-squares method and the difficulty of solving the equations that result from other methods, particularly with more complicated fitting functions, leads us to choose the standard method of least squares for most problems. We make the following two assumptions to simplify the calculation:

1. The shapes of the individual Poisson distributions governing the fluctuations in the observed y_i are nearly Gaussian.
2. The uncertainties σ_i in the observations y_i may be obtained from the uncertainties in the data and may be approximated by $\sigma_i^2 \simeq y_i$ for statistical uncertainties.

SUMMARY

Linear function: $y(x) = a + bx$.
Chi-square:

$$\chi^2 = \sum \left[\frac{1}{\sigma_i} (y_i - a - bx_i) \right]^2$$

Least-squares fitting procedure: Minimize χ^2 with respect to each of the coefficients simultaneously.
Coefficients of least-squares fitting:

$$a = \frac{1}{\Delta} \begin{vmatrix} \sum \dfrac{y_i}{\sigma_i^2} & \sum \dfrac{x_i}{\sigma_i^2} \\ \sum \dfrac{x_i y_i}{\sigma_i^2} & \sum \dfrac{x_i^2}{\sigma_i^2} \end{vmatrix} = \frac{1}{\Delta} \left(\sum \frac{x_i^2}{\sigma_i^2} \sum \frac{y_i}{\sigma_i^2} - \sum \frac{x_i}{\sigma_i^2} \sum \frac{x_i y_i}{\sigma_i^2} \right)$$

$$b = \frac{1}{\Delta} \begin{vmatrix} \sum \dfrac{1}{\sigma_i^2} & \sum \dfrac{y_i}{\sigma_i^2} \\ \sum \dfrac{x_i}{\sigma_i^2} & \sum \dfrac{x_i y_i}{\sigma_i^2} \end{vmatrix} = \frac{1}{\Delta} \left(\sum \frac{1}{\sigma_i^2} \sum \frac{x_i y_i}{\sigma_i^2} - \sum \frac{x_i}{\sigma_i^2} \sum \frac{y_i}{\sigma_i^2} \right)$$

$$\Delta = \begin{vmatrix} \sum \dfrac{1}{\sigma_i^2} & \sum \dfrac{x_i}{\sigma_i^2} \\ \sum \dfrac{x_i}{\sigma_i^2} & \sum \dfrac{x_i^2}{\sigma_i^2} \end{vmatrix} = \sum \frac{1}{\sigma_i^2} \sum \frac{x_i^2}{\sigma_i^2} - \left(\sum \frac{x_i}{\sigma_i^2} \right)^2$$

Estimated uniform variance s^2:

$$\sigma^2 \simeq \sigma^2 = \frac{1}{N - 2} \sum (y_i - \bar{y})^2$$

Statistical fluctuations:

$$\sigma_i^2 \simeq y_i \quad \text{(raw data counts)}$$

Uncertainties in coefficients:

$$\sigma_a^2 = \frac{1}{\Delta} \sum \frac{x_i^2}{\sigma_i^2} \qquad \sigma_b^2 = \frac{1}{\Delta} \sum \frac{1}{\sigma_i^2}$$

EXERCISES

6.1. Fit the data of Example 6.2 as if all the data had equal uncertainties $\sigma_i = \bar{\sigma} = 20$, where $\bar{\sigma}$ is the average of the given values of σ. Note that the fitted parameters are independent of the value of σ, but the values of χ^2, σ_a, and σ_b are not.

6.2. Derive Equation (6.23) from Equations (6.21) and (6.22).

6.3. Show that Equation (6.12) reduces to Equation (6.13) if $\sigma_i = \sigma$.

6.4. Derive a formula for making a linear fit to data with an intercept at the origin so that $y = bx$. Apply your method to fit a straight line through the origin to the following coordinate pairs. Assume uniform uncertainties $\sigma_i = 1.5$ in y_i. Find χ^2 for the fit and the uncertainty in b.

x_i	2	4	6	8	10	12	14	16	18	20	22	24
y_i	5.3	14.4	20.7	30.1	35.0	41.3	52.7	55.7	63.0	72.1	80.5	87.9

6.5. A student hangs masses on a spring and measures the spring's extension as a function of the applied force in order to find the spring constant k. Her measurements are:

Mass (kg)	200	300	400	500	600	700	800	900
Extension (cm)	5.1	5.5	5.9	6.8	7.4	7.5	8.6	9.4

There is an uncertainty of 0.2 in each measurement of the extension. The uncertainty in the masses is negligible. For a perfect spring, the extension ΔL of the spring will be related to the applied force by the relation $k \, \Delta L = F$, where for this problem $F = mg$, and $\Delta L = L - L_0$, and L_0 is the unstretched length of the spring. Use these data and the method of least squares to find the spring constant k, the unstretched length of the spring L_0, and their uncertainties. Find χ^2 for the fit and the associated probability.

6.6. Outline a procedure for solving the simultaneous Equations (6.27). Refer to Appendix A.

CHAPTER
7

LEAST-SQUARES
FIT
TO A
POLYNOMIAL

7.1   DETERMINANT SOLUTION

So far we have discussed fitting a straight line to a group of data points. However, suppose our data (x_i, y_i) were not consistent with a straight line fit. We might construct a more complex function with extra parameters and try varying the parameters of this function to fit the data more closely. A very useful function for such a fit is a power-series polynomial

$$y(x) = a_1 + a_2 x + a_3 x^2 + a_4 x^3 + \cdots + a_m x^{m-1} \tag{7.1}$$

where the dependent variable y is expressed as a sum of power series of the independent variable x with coefficients a_1, a_2, a_3, a_4, and so forth.

For problems in which the fitting function is linear in the parameters, the method of least squares is readily extended to any number of terms m, limited only by our ability to solve m linear equations in m unknowns and by the precision with which calculations can be made. We can rewrite Equation (7.1) as

$$y(x) = \sum_{k=1}^{m} a_k x^{k-1} \tag{7.2}$$

where the index k runs from 1 to m. In fact, we can generalize the method even

115

further by writing Equation (7.2) as

$$y(x) = \sum_{k=1}^{m} a_k f_k(x) \tag{7.3}$$

where the functions $f_k(x)$ could be the powers of x as in Equation (7.2), $f_1(x) = 1$, $f_2(x) = x$, $f_3(x) = x^2$, and so forth, or they could be other functions of x as long as they *do not involve the parameters* a_1, a_2, a_3, and so forth.

With this definition, the probability function of Equation (6.8) can be written as

$$P(a_1, a_2, \ldots, a_m) = \prod \left(\frac{1}{\sigma_i \sqrt{2\pi}} \right) \exp \left\{ -\frac{1}{2} \sum \frac{1}{\sigma_i^2} \left[y_i - \sum_{k=1}^{m} a_k f_k(x_i) \right]^2 \right\} \tag{7.4}$$

and Equation (6.9) for χ^2 becomes

$$\chi^2 = \sum \left[\frac{1}{\sigma_i} \left[y_i - \sum_{k=1}^{m} a_k f_k(x_i) \right] \right]^2 \tag{7.5}$$

The method of least squares requires that we minimize χ^2, our measure of the goodness of fit to the data, with respect to the coefficients a_1, a_2, a_3, and so forth. The minimum is determined by taking partial derivatives with respect to each parameter in the expression for χ^2 of Equation (7.5), and setting them to zero:

$$\frac{\partial}{\partial a_l} \chi^2 = \frac{\partial}{\partial a_l} \sum \left[\frac{1}{\sigma_i} \left[y_i - \sum_{k=1}^{m} a_k f_k(x_i) \right] \right]^2$$

$$= -2 \sum \left\{ \frac{f_l(x_i)}{\sigma_i^2} \left[y_i - \sum_{k=1}^{m} a_k f_k(x_i) \right] \right\} = 0 \tag{7.6}$$

Thus, we obtain a set of m coupled linear equations for the m parameters a_l, with the index l running from 1 to m:

$$\sum y_i \frac{f_l(x_i)}{\sigma_i^2} = \sum_{k=1}^{m} \left\{ a_k \sum \left[\frac{1}{\sigma_i^2} f_l(x_i) f_k(x_i) \right] \right\}$$

or $$\sum y_i \frac{f_1(x_i)}{\sigma_i^2} = \sum \frac{f_1(x_i)}{\sigma_i^2} [a_1 f_1(x_i) + a_2 f_2(x_i) + a_3 f_3(x_i) \cdots]$$

$$\sum y_i \frac{f_2(x_i)}{\sigma_i^2} = \sum \frac{f_2(x_i)}{\sigma_i^2} [a_1 f_1(x_i) + a_2 f_2(x_i) + a_3 f_3(x_i) \cdots] \tag{7.7}$$

$$\sum y_i \frac{f_3(x_i)}{\sigma_i^2} = \sum \frac{f_3(x_i)}{\sigma_i^2} [a_1 f_1(x_i) + a_2 f_2(x_i) + a_3 f_3(x_i) \cdots]$$

and so forth.

The solutions can be found by the method of determinants, as in Chapter 6. We shall display the full solution for the particular case of $m = 3$:

$$
a_1 = \frac{1}{\Delta}
\begin{vmatrix}
\sum y_i \dfrac{f_1(x_i)}{\sigma_i^2} & \sum \dfrac{f_1(x_i)f_2(x_i)}{\sigma_i^2} & \sum \dfrac{f_1(x_i)f_3(x_i)}{\sigma_i^2} \\[2ex]
\sum y_i \dfrac{f_2(x_i)}{\sigma_i^2} & \sum \dfrac{f_2(x_i)f_2(x_i)}{\sigma_i^2} & \sum \dfrac{f_2(x_i)f_3(x_i)}{\sigma_i^2} \\[2ex]
\sum y_i \dfrac{f_3(x_i)}{\sigma_i^2} & \sum \dfrac{f_3(x_i)f_2(x_i)}{\sigma_i^2} & \sum \dfrac{f_3(x_i)f_3(x_i)}{\sigma_i^2}
\end{vmatrix}
$$

$$
a_2 = \frac{1}{\Delta}
\begin{vmatrix}
\sum \dfrac{f_1(x_i)f_1(x_i)}{\sigma_i^2} & \sum y_i \dfrac{f_1(x_i)}{\sigma_i^2} & \sum \dfrac{f_1(x_i)f_3(x_i)}{\sigma_i^2} \\[2ex]
\sum \dfrac{f_2(x_i)f_1(x_i)}{\sigma_i^2} & \sum y_i \dfrac{f_2(x_i)}{\sigma_i^2} & \sum \dfrac{f_2(x_i)f_3(x_i)}{\sigma_i^2} \\[2ex]
\sum \dfrac{f_3(x_i)f_1(x_i)}{\sigma_i^2} & \sum y_i \dfrac{f_3(x_i)}{\sigma_i^2} & \sum \dfrac{f_3(x_i)f_3(x_i)}{\sigma_i^2}
\end{vmatrix} \tag{7.8}
$$

$$
a_3 = \frac{1}{\Delta}
\begin{vmatrix}
\sum \dfrac{f_1(x_i)f_1(x_i)}{\sigma_i^2} & \sum \dfrac{f_1(x_i)f_2(x_i)}{\sigma_i^2} & \sum y_i \dfrac{f_1(x_i)}{\sigma_i^2} \\[2ex]
\sum \dfrac{f_2(x_i)f_1(x_i)}{\sigma_i^2} & \sum \dfrac{f_2(x_i)f_2(x_i)}{\sigma_i^2} & \sum y_i \dfrac{f_2(x_i)}{\sigma_i^2} \\[2ex]
\sum \dfrac{f_3(x_i)f_1(x_i)}{\sigma_i^2} & \sum \dfrac{f_3(x_i)f_2(x_i)}{\sigma_i^2} & \sum y_i \dfrac{f_3(x_i)}{\sigma_i^2}
\end{vmatrix}
$$

with
$$
\Delta =
\begin{vmatrix}
\sum \dfrac{f_1(x_i)f_1(x_i)}{\sigma_i^2} & \sum \dfrac{f_1(x_i)f_2(x_i)}{\sigma_i^2} & \sum \dfrac{f_1(x_i)f_3(x_i)}{\sigma_i^2} \\[2ex]
\sum \dfrac{f_2(x_i)f_1(x_i)}{\sigma_i^2} & \sum \dfrac{f_2(x_i)f_2(x_i)}{\sigma_i^2} & \sum \dfrac{f_2(x_i)f_3(x_i)}{\sigma_i^2} \\[2ex]
\sum \dfrac{f_3(x_i)f_1(x_i)}{\sigma_i^2} & \sum \dfrac{f_3(x_i)f_2(x_i)}{\sigma_i^2} & \sum \dfrac{f_3(x_i)f_3(x_i)}{\sigma_i^2}
\end{vmatrix}
$$

We note that, as in the straight-line fits in Chapter 6, the denominator Δ is a function only of the independent variable x and the uncertainties σ_i in the dependent variable, and is not a function of the dependent variable y_i itself. For the special case of a quadratic power series in x, $y(x_i) = a_1 + a_2 x_i + a_3 x_i^2$, we have $f_1(x_i) = 1$, $f_2(x_i) = x_i$, and $f_3(x_i) = x^2$, so that Equations (7.8) become

$$a_1 = \frac{1}{\Delta} \begin{vmatrix} \sum y_i \frac{1}{\sigma_i^2} & \sum \frac{x_i}{\sigma_i^2} & \sum \frac{x_i^2}{\sigma_i^2} \\ \sum y_i \frac{x_i}{\sigma_i^2} & \sum \frac{x_i^2}{\sigma_i^2} & \sum \frac{x_i^3}{\sigma_i^2} \\ \sum y_i \frac{x_i^2}{\sigma_i^2} & \sum \frac{x_i^3}{\sigma_i^2} & \sum \frac{x_i^4}{\sigma_i^2} \end{vmatrix}$$

$$a_2 = \frac{1}{\Delta} \begin{vmatrix} \sum \frac{1}{\sigma_i^2} & \sum y_i \frac{1}{\sigma_i^2} & \sum \frac{x_i^2}{\sigma_i^2} \\ \sum \frac{x_i}{\sigma_i^2} & \sum y_i \frac{x_i}{\sigma_i^2} & \sum \frac{x_i^3}{\sigma_i^2} \\ \sum \frac{x_i^2}{\sigma_i^2} & \sum y_i \frac{x_i^2}{\sigma_i^2} & \sum \frac{x_i^4}{\sigma_i^2} \end{vmatrix} \qquad (7.9)$$

$$a_3 = \frac{1}{\Delta} \begin{vmatrix} \sum \frac{1}{\sigma_i^2} & \sum \frac{x_i}{\sigma_i^2} & \sum y_i \frac{1}{\sigma_i^2} \\ \sum \frac{x_i}{\sigma_i^2} & \sum \frac{x_i^2}{\sigma_i^2} & \sum y_i \frac{x_i}{\sigma_i^2} \\ \sum \frac{x_i^2}{\sigma_i^2} & \sum \frac{x_i^3}{\sigma_i^2} & \sum y_i \frac{x_i^2}{\sigma_i^2} \end{vmatrix}$$

with

$$\Delta = \begin{vmatrix} \sum \frac{1}{\sigma_i^2} & \sum \frac{x_i}{\sigma_i^2} & \sum \frac{x_i^2}{\sigma_i^2} \\ \sum \frac{x_i}{\sigma_i^2} & \sum \frac{x_i^2}{\sigma_i^2} & \sum \frac{x_i^3}{\sigma_i^2} \\ \sum \frac{x_i^2}{\sigma_i^2} & \sum \frac{x_i^3}{\sigma_i^2} & \sum \frac{x_i^4}{\sigma_i^2} \end{vmatrix}$$

Example 7.1. A student plans to use a thermocouple to monitor temperatures and must first calibrate it against a thermometer. The thermocouple consists of a junction of a copper wire and a constantan wire. In order to measure the junction voltage with high precision, she connects the sample junction in series with a reference junction that is held at 0°C in an ice water bath. The data, therefore, will be valid only for calibrating the relative variation of the junction voltage with temperature. The absolute voltage must be determined in a separate experiment by measuring it at one specific temperature.

TABLE 7.1
**Experimental data for the determination of the relative output voltage V
of a thermocouple junction as a function of temperature T
of the junction†**

Trial i	Temperature T (°C)	Measured voltage V (mV)	Calculated voltage $V(T)$ (mV)
1	0.	−0.849	−0.918
2	5.	−0.738	−0.728
3	10.	−0.537	−0.536
4	15.	−0.354	−0.341
5	20.	−0.196	−0.143
6	25.	−0.019	0.058
7	30.	0.262	0.261
8	35.	0.413	0.467
9	40.	0.734	0.676
10	45.	0.882	0.888
11	50.	1.258	1.102
12	55.	1.305	1.319
13	60.	1.541	1.539
14	65.	1.768	1.761
15	70.	1.935	1.987
16	75.	2.147	2.215
17	80.	2.456	2.446
18	85.	2.676	2.679
19	90.	2.994	2.915
20	95.	3.200	3.155
21	100.	3.318	3.396

$a_1 = -0.918 \pm 0.030$
$a_2 = 0.0377 \pm 0.0013$
$a_3 = 0.000055 \pm 0.000013$

†The common uncertainty in the voltage measurement is assumed to be 0.05 V. The value of χ^2
for the fit was $\chi^2 = 26.6$ for 18 degrees of freedom, with a probability of 8.8%. Parameters
obtained from the fit are listed at the bottom of the table.

The student measures the difference in output voltage between the two
junctions for a temperature variation in the sample junction from 0 to 100°C in
steps of 5°C. The measurements are made on the 3-mV scale of the voltmeter, and
fluctuations of the needle indicate that the uncertainties in the measurements are
approximately 0.05 mV for all readings.

Data from the experiment are listed in Table 7.1 and are plotted in Figure
7.1. To a first approximation, the variation of V with T is linear, but close
inspection of the graph reveals a slight curvature. Theoretically, we expect a
good fit to these data with a quadratic curve of the form $V = a_1 + a_2 T + a_3 T^2$.
The parameters for the fit to the data of Example 7.1 have been obtained
by evaluating the sums and determinants of Equations (7.9). For a second-

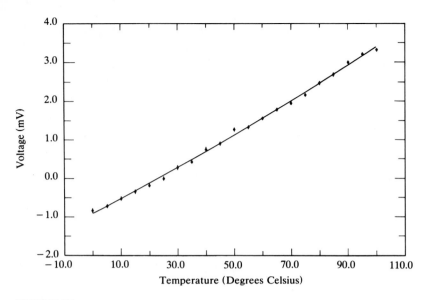

FIGURE 7.1
Thermocouple voltage versus temperature (Example 7.1). The curved line was calculated by fitting to the data the second-degree polynomial $V = a_1 + a_2T + a_3T^2$ by the least-squares method. Uniform uncertainties were assumed.

degree polynomial with 21 data points, Equation (7.5) becomes

$$\chi^2 \equiv \sum_{i=1}^{21} \frac{1}{\sigma_i^2}\left[y_i - a_1 - a_2 x_i - a_3 x_i^2\right]^2 \tag{7.10}$$

The values of χ^2 and the parameters a_1, a_2, and a_3 determined from the fit are listed in Table 7.1, as are the calculated values of $V(T_i) = y(x_i)$. The calculated values of V are also represented by the solid line on the graph of Figure 7.1. We obtain $\chi^2 = 26.6$ for this fit, or $\chi_\nu^2 = \chi^2/\nu = 1.5$, where the number of degrees of freedom ν is related to the number of events N and the number of free parameters m by $\nu = N - m$. The probability for obtaining χ^2 this high or higher can be determined from the χ^2-probability distribution (see Table C.4) and is about 8.8%, indicating a reasonable fit to the data.

As an alternative to calculating χ^2 from the fit, we could extend Equation (6.15) to three parameters and calculate the average uncertainty in the temperature readings to obtain

$$\sigma^2 \simeq s^2 = \frac{1}{N-m}\sum_{i=1}^{21}\left[\left[y_i - \left(a_1 + a_2 x_i + a_3 x_i^2\right)\right]\right]^2 = \frac{\chi^2}{N-m} \tag{7.11}$$

which is just the value of the uncertainty that would make $\chi_\nu^2 = 1$. For example

7.1, we obtain for an estimate of the variance,

$$\sigma'^2 = \sigma^2 \times \chi^2/(N - n) = 0.05 \times 26.6/18 = 0.06°C$$

suggesting, perhaps, that the student slightly underestimated the uncertainty in her measurements of V.

7.2 MATRIX SOLUTION

The techniques of least-squares fitting fall under the general name of regression analysis. Because we have been considering only problems in which the fitting function

$$y(x_i) = \sum_{k=1}^{m} a_k f_k(x_i) \tag{7.12}$$

is linear in the *parameters* a_k, we are considering only linear regression or multiple linear regression, usually shortened to multiple regression. In Chapter 8 we shall deal with techniques for handling problems with fitting functions that are not linear in the parameters.

Matrix Equations

We have not yet determined the uncertainties in the three coefficients we obtained when we fitted the second-order equation to the data of Example 7.1. We could fit the uncertainties by extending the method used for the linear fits of Examples 6.1 and 6.2. However, the algebra becomes even more tedious as the number of terms in the fitted equation increases, and in fact, our method only yielded estimates of the variances σ_k^2 and not of the covariances σ_{kl}^2, which are often important for fitted parameters. Rather than pursue the determinant method, we shall discuss immediately the more elegant and general matrix method for solving the multiple regression problem. Some of the properties of matrices are discussed in Appendix B.

Equations (7.7) can be expressed in matrix form as the equivalence between a row matrix β and the product of a row matrix a with a symmetric matrix α, all of order m:

$$\beta = a\alpha \tag{7.13}$$

The elements of the row matrix β are defined by

$$\beta_k \equiv \sum \left[\frac{1}{\sigma_i^2} y_i f_k(x_i) \right] \tag{7.14}$$

those of the symmetric matrix α by

$$\alpha_{lk} \equiv \sum \left[\frac{1}{\sigma_i^2} f_l(x_i) f_k(x_i) \right] \tag{7.15}$$

and the elements of the row matrix a are the parameters of the fit. For $m = 3$, the matrices may be written as

$$\boldsymbol{\beta} = \begin{bmatrix} \beta_1 & \beta_2 & \beta_3 \end{bmatrix} \qquad \mathbf{a} = \begin{bmatrix} a_1 & a_2 & a_3 \end{bmatrix} \qquad (7.16)$$

and

$$\boldsymbol{\alpha} = \begin{bmatrix} \alpha_{11} & \alpha_{12} & \alpha_{13} \\ \alpha_{21} & \alpha_{22} & \alpha_{23} \\ \alpha_{31} & \alpha_{32} & \alpha_{33} \end{bmatrix} \qquad (7.17)$$

To solve for the parameter matrix **a** we multiply both sides of Equation (7.13) on the right by the inverse $\boldsymbol{\epsilon}$ of the matrix $\boldsymbol{\alpha}$, defined such that $\boldsymbol{\alpha}\boldsymbol{\epsilon} = \boldsymbol{\alpha}\boldsymbol{\alpha}^{-1} = \mathbf{1}$, the unity matrix. We obtain

$$\boldsymbol{\beta}\boldsymbol{\epsilon} = \mathbf{a}\boldsymbol{\alpha}\boldsymbol{\epsilon} = \mathbf{a} \qquad (7.18)$$

which gives

$$\mathbf{a} = \boldsymbol{\beta}\boldsymbol{\epsilon} = \boldsymbol{\beta}\boldsymbol{\alpha}^{-1} \qquad (7.19)$$

Equation (7.19) can also be expressed as

$$a_l = \sum_{k=1}^{m} (\beta_k \epsilon_{kl}) = \sum_{k=1}^{m} \left\{ \epsilon_{kl} \sum \left[\frac{1}{\sigma_i^2} y_i f_k(x_i) \right] \right\} \qquad (7.20)$$

where the β_k's are given by Equation (7.16).

The solution of Equation (7.19) requires that the matrix $\boldsymbol{\alpha}$ be inverted. This generally is not a simple procedure, except for matrices of very low order, but computer routines are readily available. The inversion of a matrix is discussed in Appendix B.

The symmetric matrix $\boldsymbol{\alpha}$ is called the *curvature matrix* because of its relationship to the curvature of the χ^2 function in parameter space. The relationship becomes apparent when we take the second derivatives of χ^2 with respect to the parameters. From Equation (7.6), we have for the partial derivative of χ^2 with respect to any arbitrary parameter a_l,

$$\frac{\partial \chi^2}{\partial a_l} = -2 \sum \left\{ \frac{f_l(x_i)}{\sigma_i^2} \left[y_i - \sum_{k=1}^{m} a_k f_k(x_i) \right] \right\} \qquad (7.21)$$

and the second cross-partial derivative with respect to two such parameters is

$$\frac{\partial^2 \chi^2}{\partial a_l \partial a_k} = 2 \sum \left[\frac{1}{\sigma_i^2} f_l(x_i) f_k(x_i) \right] = 2\alpha_{lk} \qquad (7.22)$$

Estimation of Errors

The variance of $\sigma_{a_l}^2$ for the uncertainty in the determination of any parameter a_l is the sum of the variances of each of the data points σ_i multiplied by the square of the effect that each data point has on the determination of the parameter a_l [see Equation (6.19)]. Similarly, the covariance of two parameters

a_j and a_l is given by

$$\sigma_{a_j a_l}^2 = \sum \left[\sigma_i^2 \frac{\partial a_j}{\partial y_i} \frac{\partial a_l}{\partial y_i} \right] \tag{7.23}$$

(which also gives the variance for $j = l$), where we have assumed that there are no correlations between uncertainties in the measured variables y_i. Taking the derivatives in Equation (7.23) of a_l with respect to y_i we obtain

$$\frac{\partial a_l}{\partial y_i} = \sum_{k=1}^{m} \left[\epsilon_{lk} \frac{1}{\sigma_i^2} f_k(x_i) \right] \tag{7.24}$$

and, substituting into Equation (7.23), we obtain for the weighted sum of the squares of the derivatives,

$$\sigma_{a_j a_l}^2 = \sum \left\{ \sigma_i^2 \sum_{k=1}^{m} \left[\epsilon_{jk} \frac{1}{\sigma_i} f_k(x_i) \right] \sum_{p=1}^{m} \left[\epsilon_{lp} \frac{1}{\sigma_i} f_p(x_i) \right] \right\}$$

$$= \sum_{k=1}^{m} \left\{ \epsilon_{jk} \sum_{p=1}^{m} \left[\epsilon_{lp} \sum \left(\frac{1}{\sigma_i^2} f_p(x_i) f_k(x_i) \right) \right] \right\}$$

$$= \sum_{k=1}^{m} \left\{ \epsilon_{jk} \sum_{p=1}^{m} \left[\epsilon_{lp} \cdot \alpha_{pk} \right] \right\}$$

$$= \sum_{k=1}^{m} \left[\epsilon_{kj} \cdot \mathbf{1}_{lk} \right] = \epsilon_{jl} \tag{7.25}$$

where we have switched the order of the sums over the dummy indices i, k, and l and have used the fact that because the curvature matrix $\boldsymbol{\alpha}$ is symmetric, its inverse $\boldsymbol{\epsilon}$ must also be symmetric, so that $\epsilon_{kj} = \epsilon_{jk}$. The elements of the unity matrix, which result from the summed products of the elements of $\boldsymbol{\alpha}$ with its inverse $\boldsymbol{\epsilon}$, are represented by $\mathbf{1}_{lk}$.

The inverse matrix $\boldsymbol{\epsilon} \equiv \boldsymbol{\alpha}^{-1}$ is called the error matrix or the covariance matrix because its elements are the variances and covariances of the fitted parameters $\sigma_{a_j a_l} = \epsilon_{jl}$.

Example 7.2. The matrix method is illustrated by a straight-line fit $V = a_1 + a_2 T$ to a selection of data from Example 7.1. To show clearly each step of the calculation, we have selected just six points spaced at 25° intervals between 0 and 100° and have assumed a common uncertainty in the dependent variable $\sigma_V = 0.05$ mV. The data are listed in the columns 2 and 3 of Table 7.2a.

We begin by calculating each of the fitting functions $f_1 = 1$ and $f_2 = x$ at each value of the independent variable T. These are listed in columns 4 and 5 of Table 7.2a. For each measured value of x, the values of β_k, the elements of the column matrix $\boldsymbol{\beta}$, and of α_{lk}, the elements of the symmetric matrix $\boldsymbol{\alpha}$, are calculated according to Equations (7.14) and (7.15). The individual terms in the

TABLE 7.2
Matrix solution for linear fit to data of Example 2†
(a) Data and components of matrix elements

i	T	V	$f_1(x_i)$	$f_2(x_i)$	β_1'	β_2'	α_{11}'	α_{12}'	α_{22}'	V_{fit}
1	0	−0.849	1	0	−339.6	0	400	0	0	−0.947
2	20	−0.196	1	20	−78.4	−1,458	400	8,000	160,000	−0.101
3	40	0.734	1	40	293.6	11,744	400	16,000	640,000	0.745
4	60	1.541	1	60	616.4	36,984	400	24,000	1,440,000	1.590
5	80	2.456	1	80	982.4	78,592	400	32,000	2,560,000	2.436
6	100	3.318	1	100	1327.6	132,720	400	40,000	4,000,000	3.281
					2802.0	258,472	2400	120,000	8,00,000	

(b) Matrices

$$\alpha = \begin{bmatrix} 2,400 & 120,000 \\ 120,000 & 8,800,000 \end{bmatrix} \qquad \epsilon = \begin{bmatrix} 1.310 \times 10^{-03} & -1.786 \times 10^{-05} \\ -1.786 \times 10^{-05} & 3.571 \times 10^{-07} \end{bmatrix}$$

$$\beta = \begin{bmatrix} 2,802 & 258,472 \end{bmatrix} \qquad a = \begin{bmatrix} -0.947 & 0.0423 \end{bmatrix}$$

†The uniform uncertainty in V was assumed to be 0.05 mV as in Example 1. The columns labelled β_1' and α_{11}', etc., correspond to the individual contributions by each measured coordinate pair to the summed values of β and α. The value of χ^2 for the fit was 9.1 for 4 degrees of freedom corresponding to a probability of 5.5%.

calculation of β_1 and β_2 are listed in columns 6 and 7 of Table 7.2a and the individual terms in the calculation of α_{lk} are listed in columns 8 through 10. (We assume symmetry in α.) The resulting matrices are displayed in Table 7.2b.

The symmetric matrix α is inverted to obtain the variance matrix ϵ with elements ϵ_{kl}, shown in Table 7.2b, and the product matrix of the fitted parameters $\alpha = \beta\epsilon$ is calculated and displayed in Table 7.2b. The calculated values of the fitted variable V for each value of the independent variable T are listed in the last column of Table 7.2a.

Program 7.1. MultRegr Multiple regression problems are usually solved with the help of computer programs. The program MultRegr calls a set of routines for fitting any function that is linear in the parameters $a_1, a_2, \ldots, a_m$ to a set of N data points. Branches in the program on the global character variable PAE permit selection of the fitting function for each example in this chapter, with PAE = ´P´ for the power series in x of Example 7.1. The program uses several program units in addition to some that were referred to in Chapter 6. All routines are listed in Appendix E.

MakeAB7 Routines to set up the arrays alpha and beta, corresponding to the matrices α and β. These routines call the function Funct to calculate the individual terms in the fitting function.

Matrix The routines MatInv, to invert a matrix, and LinearBySquare, to find the product of a linear and a square matrix. Matrix manipulation is discussed in Appendix B.

TABLE 7.3
Error matrix from a fit by the matrix method to the data of Table 7.1†

$$
\begin{bmatrix}
8.907 \times 10^{-04} & -3.473 \times 10^{-05} & 2.823 \times 10^{-07} \\
-3.473 \times 10^{-05} & 1.913 \times 10^{-06} & -1.783 \times 10^{-08} \\
2.823 \times 10^{-07} & -1.783 \times 10^{-08} & 1.783 \times 10^{-10}
\end{bmatrix}
$$

†The table gives the variances and covariances of the fitted parameters. The values of the parameters and of χ^2 are listed in Table 7.1.

FitFunc7 Fitting function and χ^2 calculation. In general, every problem requires its own "FitFuncs" routine. For Example 7.1, the individual terms of the power series in x, which are required for the matrix fitting method, are calculated by the function PowerFunc selected through a branch on the variable PAE in the function Funct.

When we use the matrix method to fit a polynomial function to a data sample, the resulting parameters must be identical to those calculated by the determinant method, but we also obtain the full error matrix. The error matrix obtained by fitting of a second-degree polynomial to the complete data sample of Example 7.1 is listed in Table 7.3.

The error matrix can be used to estimate the uncertainty in a calculated result, including the effects of the correlations of the errors. As an example, let us suppose that we wish to find the predicted value of the voltage V and its uncertainty for a temperature of exactly 80°C. We should calculate

$$
V = a_1 + a_2 T + a_3 T^2 \tag{7.26}
$$

using the parameters determined by the fit to the data. The uncertainty in the calculated value of V, which results from the uncertainty in the parameters, is given by Equation (3.13),

$$
s^2 = \left(\frac{\partial V}{\partial a_1}\right)^2 \sigma_1^2 + \left(\frac{\partial V}{\partial a_2}\right)^2 \sigma_2^2 + \left(\frac{\partial V}{\partial a_3}\right)^2 \sigma_3^2
$$

$$
+ 2\left(\frac{\partial V}{\partial a_1}\frac{\partial V}{\partial a_2}\right)^2 \sigma_{12}^2 + 2\left(\frac{\partial V}{\partial a_1}\frac{\partial V}{\partial a_3}\right)^2 \sigma_{13}^2 + 2\left(\frac{\partial V}{\partial a_2}\frac{\partial V}{\partial a_3}\right)^2 \sigma_{23}^2
$$

$$
= 1 \cdot \epsilon_{11} + T^2 \cdot \epsilon_{22} + T^4 \cdot \epsilon_{33} + 2\left(T \cdot \epsilon_{12} + T^2 \cdot \epsilon_{13} + T^3 \cdot \epsilon_{23}\right) \tag{7.27}
$$

where ϵ_{12} and so on are the covariant terms in the symmetric error matrix. If we used only the diagonal terms in the error matrix, our result would be $V = (2.45 \pm 0.14)$ V. However, the off-diagonal terms are mainly negative, and including them reduces the uncertainty by almost a factor of 10 to 0.015, so that we should quote $V = (2.45 \pm 0.02)$ V.

7.3 INDEPENDENT PARAMETERS

Suppose we take the data of Example 6.1 or Example 6.2 and fit to them the quadratic polynomial function $y = a_1 + a_2 x + a_3 x^2$ as we did for Example 7.1. We should expect to find a rather small and possibly meaningless result for the coefficient a_3 of the quadratic term, but, because a_3 was not set equal to zero by definition, as in the analysis of Chapter 6, we might also find that the values of a_1 and a_2 have changed, sometimes considerably, from the values obtained in the linear fit. In general, the polynomial fitting procedure that we have considered will yield values for the coefficients that depend on the degree of the polynomial fitted to the data.

This interdependence arises from the fact that we have specified our coordinate system without regard to the region of parameter space from which our data points are extracted. The value of a_1 represents the intercept on the ordinal axis, the coefficient a_2 represents the slope at this same point, and other coefficients represent higher orders of curvature at this same intercept point. if the data are not clustered about this intercept point, its location might be highly dependent on the polynomial used to fit the data.

We might be able to extract more meaningful information about the data if we were to determine instead coefficients a_1', a_2', a_3', and so forth, which represent the average value, the average slope, the average curvature, and so forth, of the data. Such coefficients would be independent of our choice of coordinate system and would represent physical characteristics of the data that are independent of the degree of the fitted polynomial.

Orthogonal Polynomials

We should like to fit the data to a function that is similar to that of Equation (7.1) but that yields the desired independent of the coefficients. The appropriate function to use is the sum of orthogonal polynomials,[1] which has the form

$$y(x) = a_1 + a_2(x - \beta) + a_3(x - \gamma_1)(x - \gamma_2)$$
$$+ a_4(x - \delta_1)(x - \delta_2)(x - \delta_3) + \cdots \qquad (7.28)$$

Following the development of Section 7.1, we must minimize χ^2 to determine the coefficients a_1, a_2, a_3, a_4, and so on, with the further criterion that the addition of higher-order terms to the polynomial will not affect the evaluation of lower-order terms. This criterion will be used to determine the extra parameters β, γ_1, γ_2, and so on.

---

[1] Any polynomial such as that of Equation (7.1) can be rewritten as a sum of orthogonal polynomials

$$y = a + \sum_{j=1}^{n} [b_j f_j(x_i)]$$

with the orthogonal property that $\Sigma[f_j(x_i)f_k(x_i)] = 0$ for $j \neq k$.

The goodness-of-fit parameter χ^2 is defined as

$$\chi^2 \equiv \sum \left[\frac{\Delta y_i}{\sigma_i}\right]^2 = \sum \left[\frac{1}{\sigma_i^2}[y_i - y(x_i)]^2\right] \tag{7.29}$$

Setting the derivatives of χ^2 with respect to each of the m coefficients a_1, a_2, and so forth to 0 yields m simultaneous equations

$$\sum y_i = Na_1 + a_2\sum(x_i - \beta) + a_3\sum(x_i - \gamma_1)(x_i - \gamma_2)$$
$$+ a_4\sum(x_i - \delta_1)(x_i - \delta_2)(x_i - \delta_2) + \cdots \tag{7.30}$$

$$\sum x_i y_i = a_1\sum x_i + a_2\sum x_i(x_i - \beta) + a_3\sum x_i(x_i - \gamma_1)(x_i - \gamma_2)$$
$$+ a_3\sum x_i(x_i - \delta_1)(x_i - \delta_2)(x_i - \delta_2) + \cdots \tag{7.31}$$

$$\sum x_i^2 y_i = a_1\sum x_i^2 + a_2\sum x_i^2(x_i - \beta) + a_3\sum x_i^2(x_i - \gamma_1)(x_i - \gamma_2)$$
$$+ a_4\sum x_i^2(x_i - \delta_1)(x_i - \delta_2)(x_i - \delta_2) + \cdots \tag{7.32}$$

$$\sum x_i^3 y_i = a_1\sum x_i^3 + a_2\sum x_i^3(x_i - \beta) + a_3\sum x_i^3(x_i - \gamma_1)(x_i - \gamma_2)$$
$$+ a_4\sum x_i^3(x_i - \delta_1)(x_i - \delta_2)(x_i - \delta_2) + \cdots \tag{7.33}$$

where we have omitted a factor of σ_i^2 in the denominator for clarity.

Parameters

Let us examine Equation (7.30). If we restrict ourselves to a zeroth-degree polynomial, that is, to only one coefficient a_1, all the other coefficients are equal to 0 by definition. The coefficient a_1, therefore, is specified completely by the first term on the right-hand side of Equation (7.30):

$$a_1 = \frac{1}{N}\sum y_i = \bar{y} \tag{7.34}$$

If we restrict ourselves to a first-degree polynomial, the coefficient a_2 of the second term of Equation (7.30) is not 0. However, if a_1 is to be independent of the value of a_2, the second term itself must be 0. Hence, the requirement that

$$\sum(x_i - \beta) = 0$$

leads to the value for β,

$$\beta = \frac{1}{N}\sum x_i = \bar{x} \tag{7.35}$$

and a_2 can be determined directly from Equation (7.31) by substituting the values of a_1 and β with higher-order coefficients (a_3, a_4 etc.) set to 0.

Similarly, if we consider a quadratic function, the third term of Equation (7.30) must be 0 even when the coefficient a_3 is not 0. This constraint leads to a quadratic equation in γ_1 and γ_2 that is not sufficient to specify either parameter. We have the additional constraint, however, that the coefficient a_2 must be specified completely by Equations (7.30) and (7.31). Thus, the third term in Equation (7.31) must be 0 regardless of the value of the coefficient a_2, and we have two simultaneous quadratic equations for the parameters γ_1 and γ_2,

$$\sum (x_i - \gamma_1)(x_i - \gamma_2) = 0 \quad \text{and} \quad \sum x_i(x_i - \gamma_1)(x_i - \gamma_2) = 0 \quad (7.36)$$

Similarly, the coefficient a_3 must be determined completely by Equation (7.32) (and the predetermined values of a_1 and a_2), and this constraint yields three simultaneous equations for the parameters δ_1, δ_2, and δ_3:

$$\sum (x_i - \delta_1)(x_i - \delta_2)(x_i - \delta_3) = 0$$

$$\sum x_i(x_i - \delta_1)(x_i - \delta_2)(x_i - \delta_3) = 0 \qquad (7.37)$$

$$\sum x_i^2(x_i - \delta_1)(x_i - \delta_2)(x_i - \delta_3) = 0$$

The extrapolation to higher orders is straightforward.

Coefficients

Once the parameters β, γ, δ, and so on are determined by the constraint equations, the coefficients a_1, a_2 and so on can be found from the resulting $n + 1$ simultaneous equations. The value for the first coefficient a_1 is specified completely by minimizing χ^2 with respect to a_1 in Equation (7.30) and is given in Equation (7.34). The value of the second coefficient a_2 is determined by minimizing χ^2 with respect to both a_1 and a_2 in Equations (7.30) and (7.31) and substituting the value of a_1 from Equation (7.34) into Equation (7.31). Similarly, the value of a_3 can be determined from Equation (7.32) after substituting the values of a_1 and a_2 determined from Equations (7.30) and (7.31). Each succeeding equation yields a value for the next higher-order coefficient.

Note that the value determined for any coefficient is thus independent of the value specified for any higher-order coefficient, but is not independent of the value of lower-order coefficients. The coefficients are given by

$$a_1 = \frac{1}{N} \sum y_i = \bar{y}$$

$$a_2 = \frac{\sum y_i(x_i - \beta)}{\sum (x_i - \beta)^2}$$

$$a_3 = \frac{\sum y_i(x_i - \gamma_1)(x_i - \gamma_2)}{\sum [(x_i - \gamma_1)(x_i - \gamma_2)]^2} \qquad (7.38)$$

$$a_4 = \frac{\sum y_i(x_i - \delta_1)(x_i - \delta_2)(x_i - \delta_3)}{\sum [(x_i - \delta_1)(x_i - \delta_2)(x_i - \delta_3)]^2}$$

and so forth.

Simplification

For the general case of arbitrarily chosen data points (x_i, y_i), this procedure is cumbersome even with computer techniques because it requires the solution of coupled, nonlinear equations. There is, however, a special type of data for which the calculations can be considerably simplified, namely, data that meet the following two criteria: (i) the independent variables x_i are equally spaced and (ii) the uncertainties are constant, $\sigma_i = \sigma$, and can therefore be ignored.

Consider the experiments of Examples 6.1 (measurement of temperature versus position) and 7.1 (voltage versus temperature). Those data satisfy the required conditions and, therefore, we could use a simplified method of independent parameters to obtain a fit. The resulting values of the coefficients for these particular experiments might not have any great physical significance (that is, $a_1 = \overline{T}$ the average temperature of the data points in Example 6.1 is not a particularly useful number), but by using this technique of fitting orthogonal polynomials we could try fitting higher-degree polynomials without changing the values of the coefficients already calculated for a straight-line or quadratic fit. The experiment of Example 6.2 (the decay of a radioactive state) fulfills only the first of the two criteria, because the x data points are equally spaced but the uncertainties are statistical, so that we cannot ignore the factor of σ_i^2 that belongs in the denominators of the fitting Equations (7.30) through (7.33).

For an experiment similar to that of Example 7.1, where we have made N measurements of equally spaced values of the independent variable x ranging from x_1 to x_N in steps of Δ,

$$\Delta = x_{i+1} - x_i$$

and the uncertainties are due to instrumental errors with a common standard deviation $\sigma_i = \sigma$, Equations (7.35) through (7.37) reduce to

$$\beta = \frac{1}{N}\sum x_i = \bar{x} = \frac{1}{2}(x_i + x_n)$$

$$\gamma = \beta \pm \sqrt{\frac{1}{N}\sum(x_i - \beta)^2} = \beta \pm \Delta\sqrt{\frac{1}{12}(N^2 - 1)} \qquad (7.39)$$

$$\delta = \beta, \beta \pm \sqrt{\frac{\sum\left[x_i(x_i - \beta)^3\right]}{\sum\left[x_i(x_i - \beta)\right]}} = \beta, \beta \pm \Delta\sqrt{\frac{1}{20}(3N^2 - 7)}$$

A more comprehensive list of parameters for orthogonal polynomials can be found in Anderson and Houseman.

Table 7.4 shows coefficients a_1, a_2, a_3, and a_4 as well as the values of χ^2 and the χ^2 probability obtained when we fit the data of Example 7.1, by the standard least-squares method and by the independent parameter method of Equation (7.39). We have made separate fits with first-, second-, and third-degree polynomials ($m = 2$, 3, and 4). As expected, adding extra terms does not change the values of the lower-order coefficients obtained by the independent parameter method and therefore we display the parameters only once in Table 7.4.

TABLE 7.4
Values of χ^2 and parameters obtained by fitting the data of Example 7.1 by the standard least-squares method and by the method of independent parameters, as a function of the number of parameters m of the fit†

m	Standard least squares			Independent parameters
	2	3	4	
χ^2	43.5 (0.12%)	26.6 (8.8%)	24.9 (9.4%)	
α_1	-1.01 ± 0.02	(-0.92 ± 0.03)	(-0.89 ± 0.03)	1.15
α_2	$(4.31 \pm 0.04)10^{-2}$	$(3.8 \pm 0.1)10^{-2}$	$(3.4 \pm 0.3)10^{-2}$	4.31×10^{-2}
a_3		$(5.5 \pm 1.3)10^{-5}$	$(1.5 \pm 0.8)10^{-4}$	5.49×10^{-5}
a_4			$(-6.5 \pm 5.1)10^{-7}$	6.51×10^{-7}

†The values of χ^2 are the same for both methods. The numbers in parentheses correspond to the χ^2 probability for the fit with 21-m degrees of freedom.

There is a marked improvement in χ^2 in going from the two-parameter (linear) fits to three-parameter (quadratic) fits. Unless a theoretical reason dictates that our data should follow a cubic distribution, there is no justification in making a four-parameter (cubic) fit to these data, because the value of χ^2 for $m = 3$ is satisfactory (26.6 for 18 degrees of freedom, corresponding to $P = 8.8\%$), and adding more terms does not improve the fit. If a cubic function had been predicted by theoretical considerations, we should be obligated to say that our data are not sensitive to the presence of a cubic term.

Legendre Polynomials

Although the method of fitting to orthogonal polynomials outlined in the previous section can be tedious, there are predefined sets of orthogonal polynomials that are often useful in fitting data. One important set is the *Legendre polynomials*

$$y(x) = a_0 P_0(x) + a_1 P_1(x) + \cdots = \sum_{L=0}^{m-1} [a_L P_L(x)] \qquad (7.40)$$

where $x = \cos\theta$ and the terms $P_L(x)$ in the function are given by

$$P_0(x) = 1 \qquad P_2(x) = 1/2(3x^2 - 1)$$
$$P_1(x) = x \qquad P_3(x) = 1/2(5x^3 - 3x) \qquad (7.41)$$

and higher-order terms can be determined from the recurrence relation

$$P_L(x) = \frac{1}{L}[(2L - 1)xP_{L-1}(x) - (L - 1)P_{L-2}(x)] \qquad (7.42)$$

TABLE 7.5
Angular distribution of gamma rays emitted in the reaction $^{13}C(p,\gamma)^{14}N$ produced by incident protons at $E_p = 4.5$ MeV†

θ (degrees)	$x = \cos\theta$	C_i counts	σ_{C_i}	Y_i all terms	Y_i even term
0	1.000	1400	37.4	1365.8	1361.3
10	0.985	1386	37.2	1325.2	1321.1
20	0.940	1130	33.6	1217.0	1213.9
30	0.866	1045	32.3	1075.8	1074.5
40	0.766	971	31.2	943.5	944.4
50	0.643	862	29.4	852.5	855.6
60	0.500	819	28.6	813.9	818.6
70	0.342	808	28.4	816.9	821.9
80	0.174	862	29.4	836.5	840.2
90	0.000	829	28.2	848.6	849.6
100	−0.174	824	28.7	842.8	840.2
110	−0.342	839	29.0	827.5	821.9
120	−0.500	819	28.6	825.4	818.6
130	−0.643	901	30.0	861.0	855.6
140	−0.766	925	30.4	945.7	944.4
150	−0.866	1044	32.3	1069.8	1074.5
160	−0.940	1224	35.0	1202.9	1213.9

†The calculated numbers of counts were obtained from least-squares Legendre polynomials fits to the data of the form $Y_i(x) = \sum_{L=1}^{5} a_L P_{L-1}(x_i)$, for separate fits with all terms and with even terms only.

Legendre polynomials are orthogonal when averaged over all values of $x = \cos\theta$:

$$\int_{-1}^{1} [P_L(x)P_M(x)]\,dx = \begin{cases} 2/(2L+1) & \text{for } L = M \\ 0 & \text{for } L \neq M \end{cases} \quad (7.43)$$

Example 7.3. Let us consider an experiment in which ^{13}C is bombarded by 4.5-MeV protons. In the subsequent reaction, some of the protons are captured by the ^{13}C nucleus, which then decays by gamma emission, producing gamma rays with energies up to 11 MeV. A measurement of the angular distribution of the emitted gamma rays gives information about the angular momentum states of the energy levels in the residual nucleus ^{14}N.

Table 7.5 lists simulated data for this experiment. Gamma ray counts were recorded at 17 angles from 0 to 116°. Columns 1 through 4 list the angles at which the measurements were made, the cosine of the angle ($x = \cos\theta$), the measured number of counts (C_i), and the uncertainties σ_{C_i} in the counts. The uncertainties are assumed to be purely statistical. These data are plotted in Figure 7.2 as a function of the angle θ. There appears to be symmetry around $\theta = 90°$, and consideration of the reaction process predicts that the data should

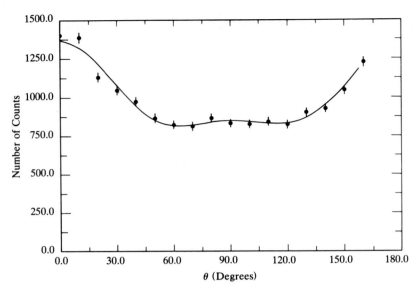

FIGURE 7.2
Angular distribution of gamma rays emitted from the simulated reaction $^{13}C(p, \gamma)^{14}N$ produced by incident protons at $E_p = 4.5$ MeV (Example 7.3). The calculated curve represents a fit to the data of a series of even Legendre polynomials up to $L = 4$. Statistical uncertainties were assumed.

be described by a fourth-order Legendre polynomial with only even terms:

$$C = a_0 P_0(x) + a_2 P_2(x) + a_4 P_4(x) \quad \text{with } x = \cos\theta \qquad (7.44)$$

Let us apply the matrix method of least squares of Section 7.2 to this problem to fit the series of Legendre polynomials of Equation (7.41) to these data. We shall first fit a fourth-order Legendre polynomial that includes both odd and even orders. The fitting function is of the form

$$y(x) = a_1 f_1(x) + a_2 f_2(x) + \cdots + a_m f_m(x) \qquad (7.45)$$

which is linear in the fitting parameters, the coefficients a_i.

> **Computer fits.** Routines used for fitting a series of Legendre polynomials to these data are included in Program 7.1. The procedure LegPoly in the program unit FitFunc7 calculates the terms of the Legendre polynomials through tenth order. The procedure is selected through a branch on the global variable PAE in the function Funct with PAE = `A` for all terms to order $n = m - 1$, or PAE = `E` to fit with just the even terms. Note that the index k of the term in the fitting function, in general, does not correspond to the order L of the Legendre polynomial.

The efficiency of the calculation, (and therefore the speed of the linear regression calculation) could be improved in a number of ways. The simplest

TABLE 7.6
Coefficients and χ^2 from least-squares fit to Legendre polynomial series

	χ^2	a_1	a_2	a_3	a_4	a_5
All terms	17.2 (14%)	937.4 ± 7.6	0.7 ± 12.8	259 ± 14	10 ± 17	158 ± 18
Even terms	17.6 (22%)	938.1 ± 7.5		261 ± 14		161 ± 16

change would be to calculate the functions once at each value of the independent variable, and store the calculated values in an array.

Parameters obtained by fitting a series in Legendre polynomials for terms up to $L = 4$ are listed in Table 7.6. Separate fits were made with all terms and with only the even terms in the series. As expected, the coefficients of terms involving odd orders are comparable to their uncertainties and negligible compared to those involving even polynomials. The full error matrix for the fit with even terms is listed in Table 7.7.

In view of the strong theoretical argument that only even Legendre polynomials are required for this reaction, it would be appropriate to fit a series that includes only the even terms. The parameters obtained in this fit are also displayed in Table 7.6, and the numbers of counts calculated from these parameters are listed. The function calculated with even terms is illustrated as a curve on the data of Figure 7.2.

Because we are fitting with orthogonal functions, we might have expected to obtain identical values of the coefficient a_1 from both fits. (We expect the higher-order even coefficients to change because the presence or absence of lower-order coefficients must affect the higher coefficients.) The fact that there is some dependence of a_1 on higher-order terms is a result of the fact that a given experiment does not sample uniformly the entire range of the Legendre polynomial, so the orthogonality relation Equation (7.43) is not satisfied by a finite data set. This is in contrast to the situation in the previous section, where we set up orthogonal functions based on the data themselves. Nevertheless, it is generally good practice to use orthogonal fitting functions whenever possible to minimize both the correlations between coefficients and the dependence of higher coefficients on the presence of lower ones.

The values of χ^2 and the χ^2 probability for the two fits are also given in Table 7.6. We note that χ^2 for the three-parameter fit is necessarily higher than that for the five-parameter fit, but χ^2 per degree of freedom is smaller and the χ^2 probability is higher.

TABLE 7.7
Error matrix for a least-squares fit to even Legendre polynomials

$$\begin{bmatrix} 5.624E + 01 & -5.256E + 00 & -6.272E + 00 \\ -5.256E + 00 & 1.865E + 02 & -2.690E + 01 \\ -6.272E + 00 & -2.690E + 01 & 2.798E + 02 \end{bmatrix}$$

7.4   NONLINEAR FUNCTIONS

In all the procedures developed so far we have assumed that the fitting function was linear in the coefficients. By that we mean that the function can be expressed as a sum of separate terms each multiplied by a single coefficient. How can we fit data with a function that is not linear in the coefficients? For example, suppose we have measured the distribution of decay times of an unstable state and that the distribution can be represented by the normalized function $P(t) = (1/\tau)e^{-t/\tau}$, where τ is the mean lifetime of the state. Can we find the parameter τ by the least-squares method? The method of least squares does not yield a straightforward analytical solution for such functions. In Chapter 8 we shall investigate methods of searching parameter space for values of the coefficients that will minimize the goodness-of-fit criterion χ^2. Here we shall consider approximate solutions to such problems using linear-regression techniques.

Linearization

It is possible to transform some functions into linear functions. For example, if we were to fit an exponential decay problem of the form

$$y = ae^{-bx} \tag{7.46}$$

where a and b are the unknown parameters, it would seem reasonable to take logarithms of both sides and to fit the resulting straight line equation

$$\ln y = \ln a - bx \tag{7.47}$$

The method of least squares minimizes the value of χ^2 with respect to each of the coefficients $\ln a$ and $\ln b$ where χ^2 is given by

$$\chi^2 = \sum \left\{ \frac{1}{\sigma_i'^2} [\ln y_i + \ln a - bx_i]^2 \right\} \tag{7.48}$$

where we must use weighted uncertainties σ_i' instead of σ_i to account for the transformation of the dependent variable:

$$\sigma_i' = \frac{d(\ln y_i)}{dy} \sigma_i = \frac{1}{y_i} \sigma_i \tag{7.49}$$

The importance of weighting the uncertainties is illustrated in Figure 7.3, which shows the function of Equation (7.46) graphed both on a linear and on a logarithmic scale. (For plotting, we use logarithms to base 10 rather than natural logarithms.) The uncertainties are given by $\sigma_i = \sqrt{y_i}$ and therefore increase with increasing y_i. However, on the logarithmic scale, they appear to decrease with increasing y_i and are very large for very small $\ln y_i$. If we were to ignore this effect in fitting Equation (7.47), we would overemphasize the uncertainties for small values of y_i.

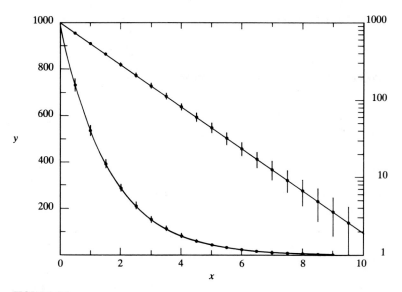

FIGURE 7.3
Graph of the function $y = ae^{-bx}$ calculated on a linear and a logarithmic scale. The error bars are given by $\sigma_i = \sqrt{y_i}$. The curved line corresponds to the linear scale on the left, and the straight line to the logarithmic scale on the right.

In general, if we fit the function $f(y)$ rather than y, the uncertainties σ_i in the measured quantities must be modified by

$$\sigma_i' = \frac{df(y)}{dy_i}\sigma_i \tag{7.50}$$

Errors in the Parameters

If we modify the fitting function so that instead of fitting the data points y_i with the coefficient $a, b, \ldots$, we fit modified data points $y_i' = f(y_i)$ with coefficients $a', b', \ldots$, then our estimates of the errors in the coefficients will pertain to the uncertainties in the modified coefficients $a', b', \ldots$, rather than to the desired coefficients $a, b, \ldots$. If the relationship between the two sets of coefficients is defined to be

$$a' = f_a(a) \qquad b' = f_b(b) \tag{7.51}$$

then the correspondence between the uncertainties $\sigma_a', \sigma_b', \ldots$ in the modified coefficients and the uncertainties $\sigma_a, \sigma_b, \ldots$ in the desired coefficients is obtained in a manner similar to that for σ_i' and σ_i in Equation (7.50):

$$\sigma_a' = \frac{df_a(a)}{da}\sigma_a \qquad \sigma_b' = \frac{df_b(b)}{db}\sigma_b \tag{7.52}$$

Thus, if the modified coefficient is $a' = \ln a$, the estimated error in a is determined from the estimated error in a', according to Equation (7.52) with $f_a = \ln a$:

$$\sigma'_a = \frac{d(\ln a)}{da} \sigma_a = \frac{\sigma_a}{a} \qquad (7.53)$$

Values of χ^2 for testing the goodness of fit should be determined from the original uncertainties of the data σ_i and from the unmodified equation, although Equation (7.48) should give approximately equivalent results when weighted with the modified uncertainties σ'_i.

In Example 6.2, we considered an experiment to check the decrease in the number of counts C as a function of distance r from a radiative source. We expected a relation of the form

$$C(r) = b/r^2 \qquad (7.54)$$

and therefore changed the independent variable to $x = 1/r^2$ and fitted a straight line to the C versus x data. Because uncertainties were assigned only to the dependent variable C, the fit was not distorted by that transformation.

Suppose, instead, that our objective had been to determine the exponent a in the expression for C:

$$C(r) = br^{-a} \qquad (7.55)$$

Taking logarithms of both sides, we obtain the linear equation

$$\ln(C) = \ln(b) - a \ln r$$

or
$$C_i = b' - ar' \qquad (7.56)$$

with $C' = \ln C$, $r' = \ln r$, and $b' = \ln b$. The uncertainties σ' in C' would be given by Equation (7.49) as

$$\sigma' = \sigma/C$$

and we could find the exponent a by fitting a straight line to Equation (7.56) using these weighted uncertainties.

Although the method of taking logarithms of an exponential or a power function to produce a function that is linear in the parameters may be convenient for quick estimates, with fast computers it is generally better to solve such problems by one of the approximation methods developed for fitting nonlinear functions. These methods will be explored in Chapter 8.

SUMMARY

Linear function: Function that is linear in its parameter a_k:

$$y(x) = \sum_{k=1}^{m} a_k f_k(x)$$

Least squares fit to a function that is linear in its parameters:

$$\Delta = \begin{vmatrix} \sum \dfrac{f_1(x_i)f_1(x_i)}{\sigma_i^2} & \sum \dfrac{f_1(x_i)f_2(x_i)}{\sigma_i^2} & \sum \dfrac{f_1(x_i)f_3(x_i)}{\sigma_i^2} & \cdots \\[3mm] \sum \dfrac{f_2(x_i)f_1(x_i)}{\sigma_i^2} & \sum \dfrac{f_2(x_i)f_2(x_i)}{\sigma_i^2} & \sum \dfrac{f_2(x_i)f_3(x_i)}{\sigma_i^2} & \cdots \\[3mm] \sum \dfrac{f_3(x_i)f_1(x_i)}{\sigma_i^2} & \sum \dfrac{f_3(x_i)f_2(x_i)}{\sigma_i^2} & \sum \dfrac{f_3(x_i)f_3(x_i)}{\sigma_i^2} & \cdots \\[3mm] \vdots & \vdots & \vdots & \vdots \end{vmatrix}$$

$$a_1 = \frac{1}{\Delta} \begin{vmatrix} \sum y_i \dfrac{f_1(x_i)}{\sigma_i^2} & \sum \dfrac{f_1(x_i)f_2(x_i)}{\sigma_i^2} & \sum \dfrac{f_1(x_i)f_3(x_i)}{\sigma_i^2} & \cdots \\[3mm] \sum y_i \dfrac{f_2(x_i)}{\sigma_i^2} & \sum \dfrac{f_2(x_i)f_2(x_i)}{\sigma_i^2} & \sum \dfrac{f_2(x_i)f_3(x_i)}{\sigma_i^2} & \cdots \\[3mm] \sum y_i \dfrac{f_3(x_i)}{\sigma_i^2} & \sum \dfrac{f_3(x_i)f_2(x_i)}{\sigma_i^2} & \sum \dfrac{f_3(x_i)f_3(x_i)}{\sigma_i^2} & \cdots \\[3mm] \vdots & \vdots & \vdots & \vdots \end{vmatrix}$$

For the jth coefficient, a_j is found by replacing the jth column in the expression for Δ with the first column in the expression for a_1.
Chi square:

$$\chi^2 = \sum_{i=1}^{N} \left[\frac{1}{\sigma} [y_i - y(x_i)] \right]^2 = \sum_{i=1}^{N} \left[\frac{1}{\sigma} \left[y_i - \sum_{k=1}^{m} a_k f_k \right] \right]^2$$

Sample variance σ^2:

$$\sigma^2 \simeq s^2 = \frac{1}{N-m} \sum_{i=1}^{N} [y_i - y(x_i)]^2$$

Matrix solution: $\boldsymbol{\alpha} = \boldsymbol{\beta}\boldsymbol{\epsilon} = \boldsymbol{\beta}\boldsymbol{\alpha}^{-1}$ where a is a linear matrix of the coefficients and

$$\beta_k \equiv \sum \left[\frac{1}{\sigma_i^2} y_i f_k(x_i) \right]$$

$$\alpha_{lk} \equiv \sum \left[\frac{1}{\sigma_i^2} f_l(x_i) f_k(x_i) \right]$$

Error or variance matrix: The diagonal elements of the square matrix $\boldsymbol{\epsilon} = \boldsymbol{\alpha}^{-1}$ are the variances of the parameters a_k and the off-diagonal elements are the covariances:

$$\sigma_{a_l}^2 = \epsilon_{ll} \qquad \sigma_{a_{lk}}^2 = \epsilon_{lk}$$

Orthogonal polynomials:

$$y(x) = a_1 + a_2(x - \beta) + a_3(x - \gamma_1)(x - \gamma_2)$$
$$+ a_4(x - \delta_1)(x - \delta_2)(x - \delta_3) + \cdots$$

$$a_1 = \bar{y} \qquad\qquad a_2 = \frac{\Sigma y_i(x_i - \beta)}{\Sigma(x_i - \beta)^2}$$

$$a_3 = \frac{\Sigma y_i(x_i - \gamma_1)(x_i - \gamma_2)}{\Sigma[(x_i - \gamma_1)(x_i - \gamma_2)]^2} \qquad a_4 = \frac{\Sigma y_i(x_i - \delta_1)(x_i - \delta_2)(x_i - \delta_3)}{\Sigma[(x_i - \delta_1)(x_i - \delta_2)(x_i - \delta_3)]^2}$$

For equally spaced values of x, $x_{i+1} - x_i = \Delta$,

$$\beta = \tfrac{1}{2}(x_i + x_N) \qquad \gamma = \beta \pm \Delta\sqrt{\tfrac{1}{12}(N^2 - 1)}$$

$$\delta = \beta, \beta \pm \Delta\sqrt{\tfrac{1}{20}(3N^2 - 7)}$$

Legendre polynomials:

$$y(x) = \sum_{L=1}^{m-1} [a_L P_L(x)]$$

$$P_0(x) = 1 \qquad P_1(x) = x$$

$$P_L(x) = \frac{1}{L}[(2L - 1)xP_{L-1}(x) - (L - 1)P_{L-2}(x)] \quad \text{(recursion relation)}$$

Nonlinear functions:
   If $y_i' = f(y_i)$, then

$$\sigma_i' = \frac{df(y)}{dy_i}\sigma_i$$

and if $a' = f_a(a)$ and $b' = f_b(b)$, then

$$\sigma_a' = \frac{df_a(a)}{da}\sigma_a \qquad \sigma_b' = \frac{df_b(b)}{db}\sigma_b$$

EXERCISES

7.1. Show by direct calculation using the data of Example 7.2 listed in Table 7.2 that $\alpha\epsilon = 1$ where **1** is the unity matrix.

7.2. The tabulated data represent the lower bin limit x and the bin contents y of a histogram of data that fall into two peaks.

i	1	2	3	4	5	6	7	8	9	10
x_i	50	60	70	80	90	100	110	120	130	140
y_i	5	7	11	13	21	43	30	16	15	10

i	11	12	13	14	15	16	17	18	19	20
x_i	150	160	170	180	190	200	210	220	230	240
y_i	13	42	90	75	29	13	8	4	6	3

Use the method of least squares to find the amplitudes a_1 and a_2 and their uncertainties by fitting to the data the function

$$y(x) = a_1 L(x; \mu_1, \Gamma_1) + a_2 L(x; \mu_2, \Gamma_2)$$

with $\mu_1 = 102.1$, $\Gamma_1 = 30$, $\mu_2 = 177.9$, and $\Gamma_2 = 20$. The function $L(x; \mu, \Gamma)$ is the Lorentzian function of Equation (2.32). Assume statistical uncertainties ($\sigma_i = \sqrt{y_i}$). Find χ^2 for the fit and the full error matrix.

7.3. From the parameters listed in Table 7.6 for the fit of even terms to the data of Example 7.3, determine the predicted value of the cross section for $\theta = 90°$ and its uncertainty. Calculate the uncertainty from the diagonal errors, listed in Table 7.6 and from the full error matrix listed in Table 7.7 and compare the two results.

7.4. Fit fourth-degree power series polynomials instead of Legendre polynomials to the data of Example 7.3. Let $x = \cos\theta$ and fit a polynomial with all terms to x^4 and another polynomial with only the even terms. Compare your results to those obtained from the fit to Legendre polynomials displayed in Table 7.6.

7.5. Derive the expression for γ_1 and γ_2 of Equation (7.36).

7.6. Derive an expression for $P_4(\cos\theta)$. [See Equation (7.42).]

7.7. Show by direct integration that $P_0(x)$, $P_1(x)$, and $P_2(x)$ are orthogonal and obey Equation (7.43).

7.8. In an experiment to measure the angular distribution of elastically scattered particles, a beam of particles strikes a liquid hydrogen target and counts are recorded at selected angles to the direction of the incident beam. Measurements are made both with the target filled with liquid hydrogen (*full target*) and with an empty target (*empty target*). The empty-target measurements were made with one-half the number of incident particles used for the full-target signal. By subtracting the suitably scaled empty-target signal from the full-target signal, the angular distribution of scattering on pure hydrogen can be determined.

Assume that the following data were obtained in such an experiment. Uncertainties in the numbers of counts are statistical.

$\cos\theta$ (lower limit)	-1.0	-0.8	-0.6	-0.4	-0.2	0.0	0.2	0.4	0.6	0.8
$\cos\theta$ (upper limit)	-0.8	-0.6	-0.4	-0.2	0.0	0.2	0.4	0.6	0.8	1.0
Counts, full target	184	128	99	49	53	55	70	81	136	216
Counts, empty target	5	4	4	1	3	1	4	9	8	7

(*a*) Scale the empty-target data to the same number of incident antiprotons used in recording the full-target data and make a subtraction to obtain the number of interactions on the hydrogen. Pay particular attention to the uncertainties in the difference.

(b) Use the least-squares method to fit the function

$$y(x) = a_1 P_0(x) + a_2 P_1(x) + a_3 P_2(x)$$

to the subtracted data, to obtain the coefficients a_1, a_2, and a_3, where the functions $P_L(x)$ are the Legendre polynomials defined in Equation (7.41).

7.9. Follow the procedure outlined in Section 7.4 to find the exponent a in Equation (7.55), using the data of Example 6.2 (Table 6.2).

7.10. A 1-m-long plastic plate with rulings at 10-cm intervals is dropped through a photogate to measure the acceleration of gravity g in an undergraduate laboratory experiment. The time is recorded as each ruling passes through the gate. The passage of the first ruling starts the timer. Data from such an experiment are tabulated. The recorded time is related to the distance that the ruler has fallen by $y = y_0 - v_0 t - 1/2 g t^2$. Note that neither the initial height y_0 nor the initial speed v_0 are known.

Ruling #	0	1	2	3	4	5	6	7	8	9	10
Time (s)	0.000	0.079	0.132	0.174	0.212	0.244	0.271	0.301	0.325	0.349	0.373

Use the least-squares method with a second-degree polynomial to find g and its uncertainty. Measure y from the photogate so that you can set $y = 0$ when ruling #0 passes the gate, $y = 1$ when ruling #1 passes, and so forth. Choose t as the independent and y as the dependent variable. Assume a uniform uncertainty in t of 0.001 s and a negligible uncertainty in y. Because the uncertainty is in the independent variable, it must be transformed to the dependent variable by the method discussed in Section 6.1. This will require initial estimates of g and v_0. After the fit has been made you may wish to repeat the fit using estimates of g and v_0 from the previous fit to improve the results.

CHAPTER
8

LEAST-SQUARES
FIT TO AN
ARBITRARY
FUNCTION

8.1 Nonlinear Fitting

The methods of least squares and multiple regression developed in the previous
chapters are restricted to fitting functions that are linear in the parameters as in
Equation (7-3):

$$y(x) = \sum_{j=1}^{m} \left[a_j f_j(x) \right] \tag{8.1}$$

This limitation is imposed by the fact that, in general, minimizing χ^2 can yield a
set of coupled equations that are linear in the m unknown parameters only if
the fitting functions $y(x)$ are themselves linear in the parameters. We shall
distinguish between the two types of problems by referring to *linear fitting* for
problems that involve equations that are linear in the parameters, such as those
discussed in Chapters 6 and 7, and *nonlinear fitting* for those problems that are
nonlinear in the parameters.

> **Example 8.1.** In a popular undergraduate physics laboratory experiment, a real
> silver 25 cent piece is irradiated with thermal neutrons to create two short-lived
> isotopes of silver, $_{47}\text{Ag}^{108}$ and $_{47}\text{Ag}^{110}$, that subsequently decay by beta emission.
> Students count the emitted beta particles in 15-s intervals for about 4 min to
> obtain a decay curve. Data collected from such an experiment are listed in Table
> 8.1 and plotted on a semilogarithmic graph in Figure 8.1. The data are reported at

141

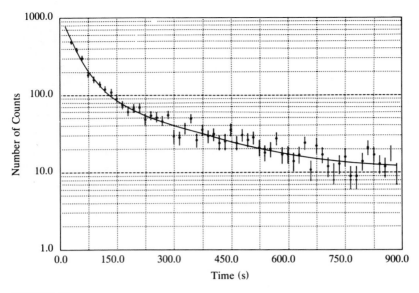

FIGURE 8.1
Number of counts detected from the decay of two excited states of silver as a function of time (Example 8.1). Time is reported at the end of each interval. Statistical uncertainties are assumed. The curve was obtained by a nonlinear least-squares fit of Equation (8.2) to the data.

the end of each 15-s interval, just as they were recorded by a scaler. The data points do not fall on a straight line because the probability function that describes the process is the sum of two exponential functions plus a constant background. We can represent the decay by the fitting function

$$y(x_i) = a_1 + a_2 e^{-t/a_4} + a_3 e^{-t/a_5} \tag{8.2}$$

where the parameter a_1 corresponds to the background radiation and a_2 and a_3 correspond to the amplitudes of the two excited states with mean lives a_4 and a_5, respectively. Clearly, Equation (8.2) is not linear in the parameters a_4 and a_5, although it is linear in the parameters a_1, a_2, and a_3.

We can use a graphical analysis method to find the two mean lifetimes by plotting the data on semilogarithmic paper after first subtracting from each data point the constant background contribution, which has been measured separately. (Note that the background counts have not been subtracted in Figure 8.1.) We then consider two regions of the plot: region a, at small values of T (e.g., $T < 120$ s) in which the short-lived state dominates the plot, and region b, at large values of T (e.g., $T > 200$ s) in which only the long-lived state contributes to the data. We can estimate the mean lifetime of the long-lived state by finding the slope of our best estimate of the straight line that passes through the data points in region b. From this result we can estimate the contribution of the long-lived component to region a and subtract that contribution from each of the data points, and thus make a new plot of the number of

TABLE 8.1
Geiger counter data from an irradiated silver piece, recorded in 15-s intervals†

Point number	Time	Measured counts	Calculated counts	Point number	Time	Measured counts	Calculated counts
1	15	775	748.3	31	465	24	24.0
2	30	479	519.8	32	480	30	23.0
3	45	380	370.4	33	495	26	22.1
4	60	302	272.0	34	510	28	21.3
5	75	185	206.7	35	525	21	20.5
6	90	157	162.7	36	540	18	19.8
7	105	137	132.5	37	555	20	19.2
8	120	119	111.5	38	570	27	18.5
9	135	110	96.3	39	585	17	18.0
10	150	89	85.0	40	600	17	17.4
11	165	74	76.5	41	615	14	16.9
12	180	61	69.7	42	630	17	16.5
13	195	66	64.2	43	645	24	16.0
14	210	68	59.5	44	660	11	15.6
15	225	48	55.5	45	675	22	15.2
16	240	54	51.9	46	690	17	14.9
17	255	51	48.8	47	705	12	14.6
18	270	46	45.9	48	720	10	14.3
19	285	55	43.3	49	735	13	14.0
20	300	29	40.9	50	750	16	13.8
21	315	28	38.7	51	765	9	13.5
22	330	37	36.7	52	780	9	13.3
23	345	49	34.8	53	795	14	13.1
24	360	26	33.1	54	810	21	12.9
25	375	35	31.5	55	825	17	12.7
26	390	29	30.0	56	840	13	12.6
27	405	31	28.6	57	855	12	12.4
28	420	24	27.3	58	870	18	12.3
29	435	25	26.1	59	885	10	12.1
30	450	35	25.0				

†The time is reported at the end of each interval. The calculated number of counts was found by method 4.

counts in region *a*, which we attribute to the short-lived state alone. The slope of the line through the corrected points gives us the mean lifetime of the short-lived state. Linear regression techniques discussed in Section 7.6 could be used to find the slope of the graph in each region.

Because the analytic methods of least-squares fitting cannot be used for nonlinear fitting problems, we must consider approximation methods and make searches of parameter space. In the following sections we shall discuss four nonlinear fitting methods: a simple *grid-search* method in which we simply calculate χ^2 at trial values of the parameters, and search for the lowest value, a *gradient-search* method that uses the slope of the function to improve the efficiency of the search, and two semianalytic methods that make use of the matrix method developed in Chapter 7, with a linear approximation to the nonlinear functions. As examples, we shall determine the parameters $(a_1 \cdots a_5)$ by fitting Equation (8.2) to the data of Example 8.1 using each of the four methods. The curve on Figure 8.1 is the result of such a fit.

Method of Least Squares

We can generalize the probability function, or *likelihood function*, of Equation (6.7) to any number of parameters,

$$P(a_1, a_2, \ldots, a_m) = \prod \left[\frac{1}{\sigma_i \sqrt{2\pi}} \right] \exp \left\{ -\frac{1}{2} \sum \left[\frac{y_i - y(x_i)}{\sigma_i} \right]^2 \right\} \quad (8.3)$$

and, as in the previous chapters, maximize the likelihood with respect to the parameters by minimizing the exponent, or the goodness-of-fit parameter χ^2:

$$\chi^2 \equiv \sum \left\{ \frac{1}{\sigma_i^2} [y_i - y(x_i)]^2 \right\} \quad (8.4)$$

where x_i and y_i are the measured variables, σ_i is the uncertainty in y_i, and $y(x_i)$ are values of the function calculated at x_i. According to the method of least squares, the optimum values of the parameters a_j are obtained by minimizing χ^2 simultaneously with respect to each parameter,

$$\frac{\partial \chi^2}{\partial a_j} = \frac{\partial}{\partial a_j} \sum \left\{ \frac{1}{\sigma_i^2} [y_i - y(x_i)]^2 \right\} = 0$$

$$= -2 \sum \left\{ \frac{1}{\sigma_i^2} [y_i - y(x_i)] \frac{\partial y(x_i)}{\partial a_j} \right\} \quad (8.5)$$

Taking partial derivatives of χ^2 with respect to each of the m parameters a_j will yield m coupled equations in the m unknown parameters a_j as in Section 7.1. If these equations are not linear in all the parameters, we must, in general, treat χ^2 as a continuous function of the m parameters, describing a hypersurface in an m-dimensional space, as expressed by Equation (8.4), and search that space for the appropriate minimum value of χ^2. Figure 8.2 illustrates such a hyperspace for a function of two parameters. Alternatively, we may apply to the m equations obtained from Equations (8.5) approximation methods developed for finding roots of coupled, nonlinear equations. A combination of both methods is often used.

Variation of χ^2 near a Minimum

For a sufficiently large event sample, the likelihood function becomes a Gaussian function of each parameter centered on those values a_j' that minimize χ^2:

$$P(a_j) = Ae^{-(a_j - a_j')^2 / 2\sigma_j^2} \quad (8.6)$$

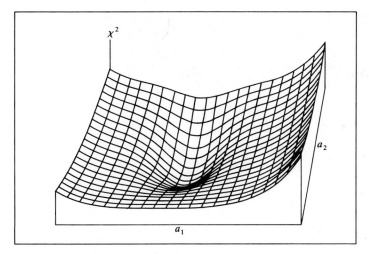

FIGURE 8.2
Chi-square hypersurface as a function of two parameters.

where A is a function of the other parameters, but not of a_j. Comparing Equation (8.3) for the likelihood function with Equation (8.4) for χ^2, we observe that we can express χ^2 as

$$\chi^2 = -2\ln[P(a_1, a_2, \ldots, a_m)] + 2\sum \ln(\sigma_i\sqrt{2\pi}) \tag{8.7}$$

Then, from Equation (8.6), we can write

$$\chi^2 = \frac{(a_j - a'_j)^2}{\sigma_j^2} + C \tag{8.8}$$

to show the variation of χ^2 with any single parameter a_j in the vicinity of a minimum with respect to that parameter. The constant C is a function of the uncertainties σ_i and the parameters a_k for $k \neq j$. Thus χ^2 varies as the square of distance from a minimum, and an increase of 1 standard deviation (σ) in the parameter from the value a'_j at the minimum increases χ^2 by 1. For a more general proof, see Arndt and MacGregor, appendix II.

We can see that this result is consistent with that obtained from a second-order Taylor expansion of χ^2 about the values α'_j, where the values of χ^2 and its derivatives at $a = a'$ are written as χ_0^2, $\partial\chi_0^2/\partial a_j$, and so forth:

$$\chi^2 \simeq \chi_0^2 + \sum_{j=1}^{m}\left\{\frac{\partial\chi_0^2}{\partial a_j}(a_j - a'_j)\right\} + \frac{1}{2}\sum_{k=1}^{m}\sum_{j=1}^{m}\left\{\frac{\partial^2\chi_0^2}{\partial a_k\,\partial a_j}(a_k - a'_k)(a_j - a'_j)\right\} \tag{8.9}$$

Because the condition for minimizing χ^2 is that the first partial derivative with respect to each parameter vanish (i.e., $\partial \chi^2 / \partial a_j = 0$), we can expect that near a local minimum in any parameter a_j, χ^2 will be a quadratic function of that parameter.

We can obtain another useful relation from Equation (8.8) by taking the second derivative of χ^2 with respect to the parameter a_j to obtain

$$\frac{\partial^2 \chi^2}{\partial a_j^2} = \frac{2}{\sigma_j^2} \tag{8.10}$$

We obtain the following expression for the uncertainty in the parameter in terms of the curvature of the χ^2 function in the region of the minimum:

$$\sigma_j^2 = 2 \left(\frac{\partial^2 \chi^2}{\partial a_j^2} \right)^{-1} \tag{8.11}$$

We note that for uncorrelated parameters, Equation (8.11) is equivalent to Equation (7.22) with Equation (7.25) for obtaining the uncertainties from the curvature matrix.

We can also use the quadratic relation to find the approximate location of a χ^2 minimum by considering the equation of a parabola that passes through three points that straddle the minimum, and solving for the value of the parameter at the minimum, as illustrated in Figure 8.3. If we have calculated three values of χ^2, $\chi_1^2 = \chi^2(a_{j1})$, $\chi_2^2 = \chi^2(a_{j2})$, and $\chi_3^2 = \chi^2(a_{j3})$, where $a_{j2} =$

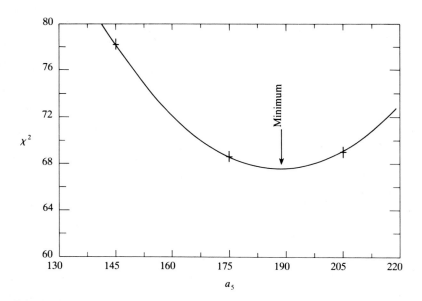

FIGURE 8.3
Plot of χ^2 versus a single parameter a in the region of a local minimum. The location of the minimum is calculated by fitting a parabola through the three indicated data points.

$a_{j1} + \Delta a_j$ and $a_{j3} = a_{j2} + \Delta a_j$, then the value a'_j of the parameter at the minimum of the parabola is given by

$$a'_j = a_{j3} - \Delta a_j \left[\frac{\chi_3^2 - \chi_2^2}{\chi_1^2 - 2\chi_2^2 + \chi_3^2} + \frac{1}{2} \right] \qquad (8.12)$$

In addition, we can estimate the errors in the fitting parameters a_j by varying each parameter about the minimum to increase χ^2 by 1 from the minimum value. The variation σ_j in the parameter a_j, which will increase χ^2 by 1 from its value at the minimum of the parabola, is given by

$$\sigma_j = \Delta a_j \sqrt{2(\chi_1^2 - 2\chi_2^2 + \chi_3^2)^{-1}} \qquad (8.13)$$

Alternatively, we can attempt to calculate the second derivative of χ^2 at the minimum and find the standard deviation from Equation (8.11).

However, if the covariant terms in the error matrix are important, it may be necessary to obtain the full error matrix, as in Chapter 7, by taking cross-partial derivatives of χ^2 in Equation (8.11).

8.2 SEARCHING PARAMETER SPACE

The method of least squares consists of determining the values of the parameters a_j of the function $y(x)$ that yield a minimum for the function χ^2 given in Equation (8.4). For nonlinear fitting problems, there are several ways of finding this minimum value. In Sections 8.3 and 8.4 we discuss approximation methods for finding solutions to the m coupled nonlinear equations in m unknowns that result from the minimization procedure of Equation (8.5).

Starting Values and Local Minima

Fitting nonlinear functions to data samples sometimes seems to be more of an art than a science. In part, this is in the nature of the approximation process, where the speed of convergence toward a solution may depend upon the choice of the method for finding solutions, the choice of starting values for the parameters, and possibly the choice of the step size. To use any of these methods, we must first determine starting values, estimates to be used by the fitting routine for initial calculations of the function and of chi square. For the pure search methods we must also define step sizes, the initial variations of the parameters. Neither starting values nor step sizes, of course, are needed in linear fitting.

Another problem in nonlinear fitting is that of multiple solutions or local minima. For an arbitrary function there may be more than one minimum of the χ^2 function within a reasonable range of values for the parameters, and thus, more than one set of solutions of the m coupled equations. An unfortunate choice of starting point may "drive" the solution toward a local minimum rather than to the absolute minimum that we seek. Before attempting a nonlinear least-squares fit, therefore, it is useful to search the parameter space to locate the main minima and identify the desired range of parameters over which to refine the search.

The first step is to find starting values for the parameters. A convenient approach, for which a computer graphics program is very useful, is to make plots of the data with curves calculated from trial values of the parameters. By visual inspection, one can often determine acceptable starting values with little or no further calculations. A basic requirement is that the area under the plotted curve be approximately the same as that under the data.

Another approach is to map the parameter space and search for values of the parameters that approximately minimize χ^2. In the simplest brute-force mapping procedure, the permissible range of each parameter a_j is divided into p equal increments Δa_j so that the m-parameter space is divided into $\prod_{j=1}^{m} p_j$ hypercubes. The value of χ^2 is then evaluated at the vertices of each hypercube. This procedure yields a coarse map of the behavior of χ^2 as a function of all the parameters a_j. At the vertex for which χ^2 has its lowest value, the size of the grid can be reduced to obtain more precise values of the parameters. For a simple two- or three-parameter fit, the parameters obtained by this procedure may be sufficiently precise that no further searching is required. For more than three parameters, the mapping is rather tedious and displaying the grid map is difficult.

A variation on the regular lattice method is a Monte Carlo search of the m-dimensional space. Trial values of the parameters are generated randomly from uniform distributions of the parameters, selected within predefined ranges, and a value of χ^2 determined for each trial. The general Monte Carlo method has been discussed in Chapter 5.

A more sophisticated method of locating the various minima of the χ^2 hypersurface involves traversing the surface from minimum to minimum by the path of lowest value in χ^2, as a river follows a ravine in travelling from lake to ocean. Starting at a point in the m-dimensional space, the search traverses the length of the local minimum, then continues in the same general direction but in a direction that minimizes the new values of χ^2. When a new local minimum is discovered, the search repeats the process until all local minimum have been located in the specified region of the space.

For relatively straightforward fitting problems, it should be sufficient to plot the data, make a reasonable estimate of the parameters to be used as starting values in the search procedure, and perform the fit by one or more of the methods described in the following sections. As a precaution, one should vary the starting values of the parameters to test whether or not the various fits converge to the same values of the parameters, within the expected uncertainties. If the dimensionality of the space is low enough, a grid of starting points may be used. For higher dimensionality, a Monte Carlo method may be used to select random starting points.

Bounds on the Parameters

From a particular set of starting values for the parameters, the search may converge toward solutions that are physically unreasonable. In Example 8.1 negative values for the parameters are not acceptable, and the current trial value of one of the parameters a_2, or a_3, may limit the possibility of determining values of the others. For example, if a_2 becomes very small or 0, a_4 cannot be determined at all. If it is not possible to find starting values for the parameters that prevent the search from wandering into these illegal regions, it may be necessary to place limits on them in the search procedure to keep them within physically allowable ranges. Simple *if then* statements in the routines may be sufficient. Care should be taken that the final value of any parameter is not at one of these artificially imposed limits.

Selection and Adjustment of Step Sizes

There are no hard and fast rules for selecting step sizes for the search methods. Clearly the steps will be different for different parameters and should be related to the slope of the χ^2 function. Very small step sizes result in slow convergence, whereas step sizes that are too large will overshoot the local minima and require constant readjustment to bracket the valleys. In the sample routines in Section 8.7, we choose initial step sizes to be proportional to the starting values of the parameters and readjust them if necessary after each local minimum is found. In the simple grid-search calculation, we adjust the step sizes to be those values that increase χ^2 by approximately 2 from its value at the local minimum.

Condition for Convergence

A change in χ^2 per degree of freedom (χ^2/dof) of less than about 1% from one trial set of parameters to the next is probably not significant. However, because of the problems of local minima and very flat valleys in the parameter space, it may not be sufficient to set an arbitrary condition for convergence, start a search, and let it run to completion. If the starting parameters are not chosen very carefully, the search may stop in a flat valley with an inappropriately large value of χ^2. If this happens, there are several possible ways to proceed. We can choose different starting values and retry the fit, as suggested in the previous sections, or we can set tighter convergence requirements (e.g., $\Delta\chi^2/\text{dof} < 0.1\%$) and rerun the search in the hope that the program will escape from the valley and reach the appropriate minimum. A convenient approach for small problems is to observe the process of the search and to cut it off manually when it appears that a stable minimum has been found. If a suitable minimum cannot be found, then different starting values should be tried. When fitting curves to several similar samples of data, we may find it satisfactory to establish suitable starting parameters, step sizes, and a cutoff criterion for the first set, and employ an automatic method to the remaining sets.

8.3 GRID-SEARCH METHOD

If the variation of χ^2 with each parameter a_j is not very sensitive to the values of the other parameters, then the optimum parameter values can be obtained most simply by minimizing χ^2 with respect to each of the parameters separately. This is the *grid-search* method. The procedure is simply to select starting values of the parameters, find the value of one of the parameters that minimizes χ^2 with respect to that parameter, set the parameter to that value, and repeat the procedure for each parameter in turn. The entire process is then repeated until a stable χ^2 minimum is obtained.

> **Grid search.** The procedure for a grid search may be summarized as follows:
>
> 1. Select starting values a_j and step or increment sizes Δa_j for each parameter and calculate χ^2 with the starting parameters.
> 2. Increment one parameter a_j by $\pm \Delta a_j$ and calculate χ^2, where the sign is chosen so that χ^2 decreases.
> 3. Repeat step 2 until χ^2 stops decreasing and begins to increase. The increase in χ^2 indicates that the search has crossed a ravine and started up the other side.
> 4. Use the last three values of a_j (which bracket the minimum) and the associated values of χ^2 to determine the minimum of the parabola, which passes through the three points as illustrated in Figure 8.3. [See Equation (8.12).]
> 5. Repeat to minimize χ^2 with respect to each parameter in turn.
> 6. Continue to repeat the procedure until the last iteration yields a predefined negligibly small decrease in χ^2.

The main advantage of the grid-search method is its simplicity. With successive iterations of the search, the absolute minimum of the χ^2 function in parameter space can be located to any desired precision.

The main disadvantage is that, if the variations of χ^2 with the parameters are strongly correlated, then the approach to the minimum may be very slow. Consider, for example, the contour plot of χ^2 as a function of two parameters in Figure 8.4. The χ^2 contours are generally approximately elliptical near the minimum. The degree of correlation of the parameters is indicated by the tilt of the ellipse. If two parameters are not correlated, so that the variation of χ^2 with each parameter is independent of the variation with the other, then the axes of the ellipse will be parallel to the coordinate axes. Thus, if a grid search is initiated near one end of a tilted ellipse, the search may follow a zigzag path is indicated by the solid line in Figure 8.4 and the search will be very inefficient. Nevertheless, the simplicity of the calculations involved in a grid search often compensates for this inefficiency.

Computer Routines

Sample routines illustrate each of the four types of nonlinear fitting procedures described in this chapter. Although technically only one program is used, with

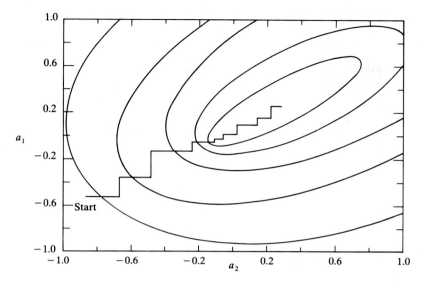

FIGURE 8.4
Contour plot of χ^2 as a function of two highly correlated variables. The zigzag line represents the search path approach to a local minimum by the grid-search method.

four different subprograms, we shall refer to four *programs* in the following sections. Several routines are common to all examples in this chapter, in addition to the program units FitVars, Matrix, FitUtil, GenUtil, and QuikPlot, previously referred to in Chapters 6 and 7. All routines are listed in Appendix E.

   NonLinFt The common calling routine to test the four different fitting methods. Program branches call the appropriate fitting and output routines. Variables are defined in the program unit FitVars and data input and output and graphical displays are handled in the unit FitUtil, as discussed in Chapters 6 and 7.

   FitFun8 Calculates the fitting function for Example 8.1. The function yFunction calculates y from Equation (8.2) and the procedure CalculateY stores the calculated values in the array yCalc. The exponential function of Example 8.1 is calculated for absolute values of the argument, which effectively limits parameters a_4 and a_5 to positive values. CalcChiSq calculates the value of χ^2 and the functions SigParab and SigMatrx determine the diagonal errors by χ^2 variation and from the diagonal terms in the error matrix, respectively.

   Program 8.1. Gridls The grid-search method is illustrated in the procedure Gridls, in the program unit GridSear listed in Appendix E.

Step sizes (deltaA) are set initially in the routine FetchParameters to be a fraction of the starting values of the parameters. (The step sizes must not be scaled to the parameters throughout the calculation, however, lest they become 0 if a parameter is 0, which would halt the search in that parameter.)

The main search routine, Grids, is entered with the value of χ^2 (chiSqr) as argument. In a loop over each of the m parameters in turn, the value of the parameter is varied until χ^2 has passed through a local minimum in the parameter. The three most recent values of χ^2 that bracket the minimum are stored in the variables chiSq1, chiSq2, and chiSq3. The best estimate of the parameter at this stage of the calculation is determined from the minimum of the parabola that passes through the three points. The step size (deltaA[j]) is then adjusted to be that value that increases χ^2 by 2 from its value at the local minimum.

One pass through Gridls represent a single zigzag along the path of Figure 8.4. The search is repeated until χ^2 does not change by more than the present level, chiCut.

A call to the function SigParab in the unit FitUtil at the end of the search returns an estimate of the uncertainty in each parameters in turn from a calculation of the independent variation needed to increase χ^2 by 1 from its minimum value.

Table 8.2 shows values of χ^2 and the parameters a_1 through a_5 for several stages of the calculation at the beginning, middle, and end of the search. The search is rather slow and tedious, but eventually a satisfactory solution is found. Note that the calculated uncertainties correspond to the diagonal terms in the error matrix for uncorrelated parameters. If correlations are considered to be

TABLE 8.2
Two exponentials plus constant background: grid-search method†

Trial	χ^2	a_1	a_2	a_3	a_4	a_5
0	406.6	10.0	900.0	80.0	27.0	225.0
1	143.0	14.5	1332.3	106.8	27.7	207.2
2	96.9	12.6	1233.9	127.9	28.2	198.4
3	79.4	11.6	1155.1	140.2	28.8	192.2
4	72.9	11.2	1100.3	147.0	29.3	189.2
⋮						
16	66.7	11.3	963.5	148.8	32.3	185.3
17	66.7	11.3	962.5	148.2	32.4	185.8
⋮						
39	66.3	10.9	959.3	139.1	33.3	195.4
40	66.2	10.8	959.2	138.9	33.3	195.7
Uncertainties		0.6	28.3	4.5	0.8	5.0

χ^2/dof = 1.23; probability = 12.1%; time = 83 s

†Stages in the fit to counts from the decay of excited states of silver. The values of χ^2 and the parameters are listed at the beginning, middle, and end of the search. The uncertainties in the parameters correspond to a change of 1 in χ^2 from its value at the end of the search. The time required to find the solution is listed in the last row of the table.

important, the matrix inversion methods discussed in the following sections could be used to find a better approximation to the errors.

8.4   GRADIENT-SEARCH METHOD

The search could be improved if the zigzagging direction of travel in Figure 8.4 were replaced by a more direct vector toward the appropriate minimum. In the *gradient-search* method of least squares, all the parameters a_j are incremented simultaneously, with relative magnitudes adjusted so that the resultant direction of travel in parameter space is along the gradient (or direction of maximum variation) of χ^2.

The gradient $\nabla \chi^2$ is a vector that points in the direction in which χ^2 increases most rapidly and has components in parameter space equal to the rate of change of χ^2 along each axis:

$$\nabla \chi^2 = \sum_{j=1}^{n} \left[\frac{\partial \chi^2}{\partial a_j} \hat{a}_j \right] \tag{8.14}$$

where $\hat{a}_j$ indicates a unit vector in the direction of the a coordinate axis. In order to determine the gradient, we estimate the partial derivatives numerically as discussed in Appendix A:

$$(\nabla \chi^2)_j = \frac{\partial \chi^2}{\partial a_j} \simeq \frac{\chi^2(a_j + f \Delta a_j) - x(a_j)}{f \Delta a_j} \tag{8.15}$$

where f is fraction of the step size Δa_j by which a_j is changed in order to determine the derivative.

The gradient has both magnitude and dimensions and, if the dimensions of the various parameters a_j are not all the same (which is usually the case), the components of the gradient do not even have the same dimensions. Let us define dimensionless parameters b_j by rescaling each of the parameters a_j to a size that characterizes the variation of χ^2 with a_j rather roughly. We shall use the step sizes Δa_j as the scaling constants, so that

$$b_j = \frac{a_j}{\Delta a_j} \tag{8.16}$$

The derivative with respect to b_j then becomes

$$\frac{\partial \chi^2}{\partial b_j} = \frac{\partial \chi^2}{\partial a_j} \Delta a_j \tag{8.17}$$

which may be calculated numerically as

$$\frac{\partial \chi^2}{\partial b_j} \simeq \frac{\chi^2(a_j + f \Delta a_j) - \chi^2(a_j)}{f \Delta a_j} \Delta a_j = \frac{\chi^2(a_j + f \Delta a_j) - \chi^2(a_j)}{f} \tag{8.18}$$

We can then define a dimensionless gradient γ, with unit magnitude and components

$$\gamma_j = \frac{\partial \chi^2 / \partial b_j}{\sqrt{\sum_{j=1}^{m} \left(\partial \chi^2 / \partial b_j \right)^2}} \tag{8.19}$$

In the numerical calculation of Equation (8.18), the quantities Δa_j and f occur only in the argument of χ^2 and not as scale factors.

The direction that the gradient-search method follows is the *direction of steepest descent*, which is opposite of the gradient $\boldsymbol{\gamma}$. The search begins by incrementing all parameters simultaneously by an amount Δa_j, with relative value given by the corresponding component γ_j of the dimensionless gradient and absolute magnitude given by the size constant Δa_j:

$$\delta a_j = - \gamma_j \Delta a_j \tag{8.20}$$

The minus sign ensures that the value of χ^2 decrease. The size constant Δa_j of Equation (8.20) is the same as that of Equation (8.16).

There are several possible methods of continuing the gradient search after a first step. The most straightforward is to recompute the gradient after each change in the parameters. One disadvantage of this method is that it is difficult to approach the bottom of the minimum asymptotically because the gradient tends to 0 at the minimum. Another disadvantage is that recomputation of the gradient at each step for small step sizes results in an inefficient search, but the use of larger step sizes makes location of the minimum less precise.

A reasonable variation on the method is to search along one direction of the original gradient in small steps, calculating only the value of χ^2 until χ^2 begins to rise again. At this point, the gradient is recomputed and the search continues in the new direction. Whenever the search straddles a minimum, a parabolic interpretation of χ^2 is used to improve the determination of the minimum.

A more sophisticated approach would be to use second partial derivatives of χ^2 to determine changes in the gradient along the search path:

$$\left. \frac{\partial \chi^2}{\partial a_j} \right|_{a_j + \delta a_j} \simeq \left. \frac{\partial \chi^2}{\partial a_j} \right|_{a_j} + \sum_{k=1}^{m} \left(\frac{\partial^2 \chi^2}{\partial a_j \, \partial a_k} \delta a_k \right) \tag{8.21}$$

If the search is already fairly near the minimum, this method does decrease the number of steps needed, but at the expense of more elaborate computation. If the search is not near enough to the minimum, this method can actually increase the number of steps required when first-order perturbations on the gradient are not valid.

The efficiency of the gradient search decreases markedly as the search approaches a minimum because the evaluation of the derivative according to the method of Equation (8.18) involves taking differences between nearly equal numbers. In fact, at the minimum of χ^2, these differences should vanish. For this reason, one of the methods discussed in the following sections may be used to locate the actual minimum once the gradient search has approached it fairly closely.

Program 8.2. Gradls The gradient-search method is illustrated in the procedure Gradls in program unit GradSear. On each entry to the main search routine, Gradls, the components of the gradient grad[j] are calculated numerically from Equation (8.18) in the procedure CalcGrad. The argument fract of this routine, corresponding to the variable f of Equation (8.18), determines the fraction of the step size (DeltaA) used in the numerical calculation of the partial derivative. Each parameter a[j] is then changed by the amount stepDown*deltaA[j]*grad[j], where stepDown is a scaling factor that is set initially in the main program and readjusted after each stage to the size needed to locate the minimum.

The initial value of delta determines to some extent the execution speed of each pass through the routine Gradls, and the value of chiCut determines when the search will stop. Because of the small gradient near the x^2 minimum, it may take many steps to reach a reasonable value of x^2, and the cutoff, chiCut, may have to be set at a very low value. In those cases, the terminating condition, which requires small changes in x^2 from one pass to the next, can be satisfied with excessively large values of x^2 and the search comes to a premature halt. User intervention can be provided as an alternate method of stopping the search.

At the conclusion of the search, the uncertainties in the parameters are estimated in the function SigParab as in the routine Gradls.

Table 8.3 shows values of x^2 and the parameters a_1 through a_5 for several stages of the calculation at the beginning, middle, and end of the search. For Example 8.1, the gradient search is considerably faster than the grid-search

TABLE 8.3
Two exponentials plus constant background: gradient-search method†

Trial	x^2	a_1	a_2	a_3	a_4	a_5
0	406.6	10.0	900.0	80.0	27.0	225.0
1	82.3	10.6	1061.0	94.0	34.4	254.2
2	72.6	9.8	984.0	98.8	36.8	237.4
3	69.8	9.9	966.9	100.9	36.8	244.6
4	69.3	9.8	953.7	101.6	36.7	242.1
⋮						
19	66.6	8.9	952.2	114.7	35.5	233.6
20	66.5	8.9	954.8	114.9	35.6	233.9
Uncertainties		0.6	26.5	3.8	0.8	7.0

$x^2/\text{dof} = 1.23$; probability $= 11.8\%$; time $= 36.2$ s

†Stages in the fit to counts from the decay of excited states of silver. The values of x^2 and the parameters are listed at the beginning, middle, and end of the search. The uncertainties in the parameters corresponding to a change of 1 in x^2 from its value at the end of the search. The time required to find the solution is listed in the last row of the table.

approach because all the parameters are varied together at each step. However, the gradient-search method has one disadvantage that is not illustrated. If the starting values of the parameters are too far from the final values, the grid search has a good chance of plodding along until it reaches the correct solution. The gradient search, on the other hand, may tend to get bogged down in local minima that correspond to long, flat valleys in the parameter space.

8.5   EXPANSION METHODS

Instead of searching the χ^2 hypersurface to map the variation of χ^2 with parameters, we should be able to find an approximate analytical functions that describes the χ^2 hypersurface and use this function to locate the minimum, with methods developed for linear least-squares fitting. The approximations will introduce errors into the calculated values of the parameters, but successive iterations of the analytical method should approach the χ^2 minimum with increasing accuracy. The main advantage of such an approach is that the number of points on the χ^2 hypersurface at which computations must be made will be fewer than for a grid or gradient search. This advantage is somewhat offset by the fact that the computations at each point are considerably more complicated. However, the analytical solution essentially chooses its own step size and, thus, the user is spared the problem of trying to optimize the step size for speed and precision.

Parabolic Expansion of χ^2

In Equation (8.9) we expanded χ^2 to second order in the parameters about a local minimum χ_0^2 where $a_j = a'_j$:

$$\chi^2 \simeq \chi_0^2 + \sum_{j=1}^{m} \left\{ \frac{\partial \chi_0^2}{\partial a_j} \delta a_j \right\} + \frac{1}{2} \sum_{k=1}^{m} \sum_{j=1}^{m} \left\{ \frac{\partial^2 \chi_0^2}{\partial a_j \partial a_k} \delta a_j \, \delta a_k \right\} \qquad (8.22)$$

which is equivalent to approximating the χ^2 hypersurface by a parabolic surface. Here we define $\delta a_j \equiv a_j - a'_j$, and χ_0^2 is given by

$$\chi_0^2 = \sum \left\{ \frac{1}{\sigma_i^2} [y_i - y'(x_i)]^2 \right\} \qquad (8.23)$$

where $y'(x_i)$ is the value of the function when $\delta a_j = 0$.

Applying the method of least squares, we minimize χ^2 as expressed in Equation (8.22) with respect to the *increments* (δa_j) in the parameters, and solve for the optimum values of these increments to obtain

$$\frac{\partial \chi^2}{\partial (\delta a_k)} = \frac{\partial \chi_0^2}{\partial a_k} + \sum_{j=1}^{m} \left\{ \frac{\partial^2 \chi_0^2}{\partial a_k \partial a_j} \delta a_j \right\} = 0 \qquad k = 1, m \qquad (8.24)$$

The result is a set of m linear equations in δa_j that we can write as

$$\beta_k - \sum_{j=1}^{m} \left(\delta a_j \, \alpha_{jk} \right) = 0 \qquad k = 1, m \qquad (8.25)$$

with
$$\beta_k \equiv -\frac{1}{2} \frac{\partial \chi_0^2}{\partial a_k} \quad \text{and} \quad \alpha_{jk} \equiv \frac{1}{2} \frac{\partial^2 \chi_0^2}{\partial a_j \, \partial a_k} \qquad (8.26)$$

The factors $\pm \frac{1}{2}$ are included for agreement with the conventional definitions of these quantities.

As in Chapter 7, we can treat Equation (8.25) as a matrix equation:

$$\boldsymbol{\beta} = \delta \mathbf{a} \, \boldsymbol{\alpha} \qquad (8.27)$$

where $\boldsymbol{\beta}$ and $\delta \mathbf{a}$ are row matrices and $\boldsymbol{\alpha}$ is a symmetric matrix of order m. We shall find that $\boldsymbol{\alpha}$ is the *curvature matrix* discussed in Section 7.2, so named because it measures the curvature of the χ^2 hypersurface.

Method of Computation

The solution of Equation (8.27) can be obtained by matrix inversion as in Section 7.2:

$$\delta \mathbf{a} = \boldsymbol{\beta} \boldsymbol{\epsilon} \qquad \delta a_k = \sum_{j=1}^{m} \left(\epsilon_{kj} \beta_j \right) \qquad (8.28)$$

where the error matrix $\boldsymbol{\epsilon} = \boldsymbol{\alpha}^{-1}$ is the inverse of the curvature matrix.

If the parameters are independent of one another, that is, if the variation of χ^2 with respect to each parameter is independent of the values of the other parameters, then the cross-partial derivatives α_{jk} $(j \neq k)$ will be 0 in the limit of a very large data sample and the matrix $\boldsymbol{\alpha}$ will be diagonal. The inverse matrix $\boldsymbol{\epsilon}$ will also be diagonal and Equation (8.27) will degenerate into m separate equations:

$$\delta a_j \simeq \frac{\beta_j}{\alpha_{jj}} = -\frac{\partial x_0^2}{\partial a_j} \div \frac{\partial^2 x_0^2}{\partial a_j^2} \qquad (8.29)$$

Computation of the matrix elements by Equation (8.26) requires knowledge of the first and second derivatives of χ^2 evaluated at the current values of the parameters. Analytic forms of the derivatives are generally quickest to compute, but may be difficult or cumbersome to derive. If it is not convenient or possible to provide analytic forms of the derivatives, then they can be computed by the method of finite differences (see Appendix A). In the following expressions, we use forward differences for efficient calculations. The intervals Δa_j should be chosen to be large enough to avoid roundoff errors but small enough to furnish reasonably accurate values of the derivatives near the minimum:

$$\frac{\partial \chi_0^2}{\partial a_j} \simeq \frac{\chi_0^2(a_j + \Delta a_j, a_k) - \chi_0^2(a_j, a_k)}{\Delta a_j}$$

$$\frac{\partial^2 \chi_0^2}{\partial^2 a_j} \simeq 4\left[\frac{\chi_0^2(a_j, a_k) - 2\chi_0^2(a_j + \delta a_j/2, a_k) + \chi_0^2(a_j + \Delta a_j, a_k)}{(\Delta a_j)^2}\right]$$

(8.30)

$$\frac{\partial \chi_0^2}{\partial a_j \partial a_k} \simeq \left[\chi_0^2(a_j, a_k)\right.$$

$$-\chi_0^2(a_j + \Delta a_j, a_k) - \chi_0^2(a_j, a_k + \Delta a_k)$$

$$\left.+\chi_0^2(a_j + \Delta a_j, a_k + \Delta a_k)\right]/\left[\Delta a_j \Delta a_k\right]$$

In actual practice, calculations are faster and, in general, more accurate if the elements of the matrix α are determined from the first-order expansion (to be discussed in the following text), which involves only first derivatives of $y(x)$ with respect to the parameters, rather than the second derivatives of χ^2 as expressed in Equation (8.30).

Fitting Procedure

Within the limits of the approximation of the χ^2 hypersurface by a parabolic extrapolation, we can solve Equation (8.27) directly to yield parameter increments δa_j such that χ^2 should be minimized for $a'_j + \delta a_j$. If the starting point is close enough to the minimum so that higher-order terms in the expansion can be neglected, this becomes an accurate and precise method. But if the starting point is not near enough, the parabolic approximation of the χ^2 hypersurface is not valid and the results will be in error. In fact, if the starting point is so far from the minimum that the curvature of χ^2 is negative, the solution will tend toward a maximum rather than a minimum. During computation, therefore, the diagonal elements α_{jj} of the matrix α must be set positive whether they are or not. The resulting magnitude for δa_j will be incorrect, but the sign may be correct.

Expansion of the Fitting Function

An alternative to expanding the χ^2 function to develop an analytic description for the hypersurface is to expand the fitting function $y(x)$ in the parameters a_j and to use the method of linear least squares to determine the optimum value for the parameter increments δa_j. If we carry out the derivation rigorously and drop higher-order terms, we should achieve the same result as for the expansion of χ^2 to first and second order.

First-Order Expansion

Let us expand the fitting function $y(x)$ in a Taylor series about the point a'_j, to first order in the *parameter increments* $\delta a_j = a_j - a'_j$:

$$y(x) \simeq y'(x) + \sum_{j=1}^{m} \left[\frac{\partial y'(x)}{\partial a_j} \delta a_j \right] \tag{8.31}$$

where $y'(x)$ is the value of the fitting function when the parameters have starting point values a'_j and the derivatives are evaluated at the starting point. The result is a linear function in the parameter increments δa_j to which we can apply the method of linear least squares developed in Chapter 7.

In this approximation, χ^2 can be expressed explicitly as a function of the parameter increments δa_j:

$$\chi^2 = \sum \left(\frac{1}{\sigma_i^2} \left\{ y_i - y'(x_i) - \sum_{j=1}^{m} \left[\frac{\partial y'(x_i)}{\partial a_j} \delta a_j \right] \right\}^2 \right) \tag{8.32}$$

Following the method of least squares, we minimize χ^2 with respect to each of the parameter increments δa_j by setting the derivatives equal to 0:

$$\frac{\partial \chi^2}{\partial \delta a_k} = -2 \sum \left(\frac{1}{\sigma_i^2} \left\{ y_i - y'(x_i) - \sum_{j=1}^{m} \left[\frac{\partial y'(x_i)}{\partial a_j} \delta a_j \right] \right\} \frac{\partial y'(x_i)}{\partial a_k} \right) = 0 \tag{8.33}$$

As before, this yields the set of m simultaneous Equations (8.25), which can be expressed as the matrix Equation (8.27):

$$\boldsymbol{\beta} = \delta \mathbf{a}\, \boldsymbol{\alpha} \tag{8.34}$$

where β_k is defined as in Equation (8.26) and a_{ij} is given by

$$\alpha_{jk} \simeq \sum \left[\frac{1}{\sigma_i^2} \frac{\partial y'(x_i)}{\partial a_j} \frac{\partial y'(x_i)}{\partial a_k} \right] \tag{8.35}$$

Second-Order Expansion

Suppose we make a Taylor expansion of the fitting function $y(x)$ to second order in the parameter increments δa_j:

$$y(x) \simeq y'(x) + \sum_{j=1}^{m} \left[\frac{\partial y'(x)}{\partial a_j} \delta a_j \right] + \frac{1}{2} \sum_{j=1}^{m} \sum_{k=1}^{m} \left[\frac{\partial^2 y'(x)}{\partial a_j \partial a_k} da_j\, da_k \right] \tag{8.36}$$

If we include the last term of Equation (8.36) in the expression for χ^2 of Equation (8.32) and again minimize χ^2 by setting to 0 the derivatives with respect to the increments δa_j, we again obtain Equation (8.25), this time with

$$\beta_k \equiv \sum \left\{ \frac{1}{\sigma_i^2} [y_i - y'(x_i)] \frac{\partial y'(x_i)}{\partial a_k} \right\} = -\frac{1}{2} \frac{\partial \chi_0^2}{\partial a_k}$$

$$\alpha_{jk} \equiv \sum \frac{1}{\sigma_i^2} \left\{ \frac{\partial y'(x_i)}{\partial a_j} \frac{\partial y'(x_i)}{\partial a_k} - [y_i - y'(x_i)] \frac{\partial^2 y'(x_i)}{\partial a_j \partial a_k} \right\}$$

$$= \frac{1}{2} \frac{\partial^2 \chi_0^2}{\partial a_j \partial a_k} \tag{8.37}$$

The resulting definitions for β_k and α_{ij} are identical to those of Equation (8.26) obtained by expanding the χ^2 function, and the χ^2-expansion method is therefore equivalent to a second-order expansion of the fitting function.

Let us compare Equations (8.37) with the analogous Equations (7.14) and (7.15) for linear least-squares fitting. The definitions of α_{jk} in Equations (8.37) and (7.15) are equivalent in the linear approximation, and thus α corresponds to the curvature matrix. The definition of β_k in Equation (8.37) is equivalent, in the linear approximation, to the definition of β_k in Equations (7.14) except for the substitution of $y_i - y'(x_i)$ for y_i. We can justify this substitution by noting that the solutions of Equation (8.34) are the parameter increments δa_j, whereas those of Equation (7.14) are the parameters themselves. In essence, we are applying linear least-squares methods to fit the parameter increments to difference data Δy_i between the actual data and the starting values of the fitting $y'(x_i)$:

$$\Delta y_i = y_i - y'(x_i) \tag{8.38}$$

Thus, the expression given in Equation (8.35) for α_{jk} is a first-order approximation to the curvature matrix that is given to second order in Equation (8.37). For linear functions, the second-order term vanishes. It is convenient to use the first-order approximation for fitting nonlinear functions and thus avoid the necessity of calculating the second derivatives in Equation (8.37). We note that this procedure can be somewhat justified on the grounds that, in the vicinity of the χ^2 minimum, we should expect the factor of $y_i - y'(x_i)$ in the expression for α of Equation (8.37) to be close to 0 so that the first term in the expression will dominate.[1]

Program 8.3. ChiFit The method of nonlinear least-squares fitting discussed in this section is illustrated in the routine ChiFit in program unit ExpndFit, listed in Appendix E. The main program NonLinFt calls routines to initialize variables and parameters as in Programs 8.1 and 8.2. Parameter variations deltaA[j] are set to appropriate values for calculating numerical derivatives in the routine

---

[1]See W. H. Press et al., page 523.

TABLE 8.4
Two exponentials plus constant background: χ^2 expansion method†

Trial	χ^2	a_1	a_2	a_3	a_4	a_5
0	406.6	10.0	900.0	80.0	27.0	225.0
1	86.2	11.1	933.8	140.4	33.8	170.5
2	66.6	10.8	861.2	128.9	33.9	201.7
3	66.1	10.4	958.2	131.2	34.0	205.4
Uncertainties		1.8	49.9	21.7	2.5	30.5

$\chi^2/\mathrm{dof} = 1.22$; probability $= 12.4\%$; time $= 9.1$ s

†All stages in the fit to counts from the decay of excited states of silver. The uncertainties in the parameters correspond to the square roots of the diagonal terms in the error matrix.

FetchParameters in the program unit FitUtil. The following program units are used by the fitting methods of Programs 8.3 and 8.4.

MakeAB8 Routines MakeBeta and MakeAlpha create the arrays of elements of the β matrix and α matrix. The first-order approximation of Equation (8.35) was used to calculate the components of the curvature matrix α_{jk}. This is equivalent to neglecting terms in the second derivatives of the fitting function $y(x)$ in the expression for α_{jk} in Equation (8.37).

NumDeriv Derivatives are calculated numerically by the functions dXiSq_da, d2XiSq_da2, and d2XiSq_dajk, in the unit. To avoid repetitive calculations, the values of the derivatives at each value of x and for the variation of each of the m parameters are calculated once for each trial and stored in arrays. Analytic expressions for the derivatives could be substituted directly for the functions dXiSq_da, d2XiSq_da2, and d2XiSq_dajk to increase the speed and accuracy of the calculation.

At the conclusion of the search, the inverse ϵ of the final value of the curvature matrix α is treated as the error matrix, and the errors in the parameters are obtained from the square roots of the diagonal terms by calls to the function SigMatrix in the unit FitFunc8.

Table 8.4 shows values of χ^2 and the parameters a_1 through a_5 for several stages of the calculation at the beginning, middle, and end of the search.

8.6 THE MARQUARDT METHOD

Convergence

One disadvantage inherent in the analytical methods of expanding either the fitting function $y(x)$ or χ^2 is that, although they converge quite rapidly to the point of minimum χ^2 from points nearby, they cannot be relied on to approach the minimum with any accuracy from a point outside the region where the χ^2 hypersurface is approximately parabolic. In particular, if the curvature of the χ^2

hypersurface is used, as in Equation (8.37) or (8.26), the analytical solution is clearly unreliable whenever the curvature becomes negative. Symptomatic of this problem is the need to set positive the diagonal elements α_{jj} of the matrix α so that all curvatures are treated as if they are positive.

In contrast, the gradient search of Section 8.4 is ideally suited for approaching the minimum from far away, but does not converge rapidly near the minimum. Therefore, we need an algorithm that behaves like a gradient search for the first portion of a search and behaves more like an analytical solution as the search converges. In fact, it can be shown (see Marquardt) that the path directions for gradient and analytical searches are nearly perpendicular to each other, and that the optimum direction is somewhere between these two vectors.

One advantage of combining these two methods into one algorithm is that the simpler first-order expansion of the analytical method will certainly suffice because the expansion need only be valid in the intermediate neighborhood of the minimum. Thus, we can neglect the second derivatives of Equation (8.37) and use the approximation of Equation (8.35) to find the curvature matrix α.

Gradient-Expansion Algorithm

A convenient algorithm (see Marquardt), which combines the best features of the gradient search with the method of linearizing the fitting function, can be obtained by increasing the diagonal terms of the curvature matrix α by a factor $1 + \lambda$ that controls the interpolation of the algorithm between the two extremes. Equation (8.34) becomes

$$\boldsymbol{\beta} = \delta a\, \boldsymbol{\alpha}' \quad \text{with} \quad \alpha'_{jk} = \begin{cases} \alpha_{jk}(1 + \lambda) & \text{for } j = k \\ \alpha_{jk} & \text{for } j \neq k \end{cases} \tag{8.39}$$

If λ is very small, Equations (8.39) are similar to the solution of Equation (8.34) developed from the Taylor expansion. If λ is very large, the diagonal terms of the curvature matrix dominate and the matrix equations degenerates into m separate equations

$$\beta_j \simeq \lambda\, \delta a_j\, \alpha_{jj} \tag{8.40}$$

which yield the vector increment $\delta\mathbf{a}$ in the same direction as the vector $\boldsymbol{\beta}$ of Equation (8.37) (or opposite to the gradient of χ^2).

The solution for the parameter increments δa_j follows from Equations (8.39) after matrix inversion

$$\delta a_j = \sum_{k=1}^{m} (\beta_k \epsilon'_{jk}) \tag{8.41}$$

where the β_k are given by Equation (8.37) and the matrix $\boldsymbol{\epsilon}'$ is the inverse of the matrix $\boldsymbol{\alpha}'$ with elements given by Equations (8.39).

The initial value of the constant factor λ should be chosen small enough to take advantage of the analytical solution, but large enough so that χ^2 decreases.

Because this algorithm approaches the gradient-search method with small steps for large λ, there should exist a value of λ such that $\chi^2(a + \delta a) < \chi^2(a)$. The recipe given by Marquardt is:

1. Compute $\chi^2(a)$.
2. Start initially with $\lambda = 0.001$.
3. Compute δa and $\chi^2(a + \delta a)$ with this choice of λ.
4. If $\chi^2(a + \delta a) > \chi^2(a)$, increase λ by a factor of 10 and repeat step 3.
5. If $\chi^2(a + \delta a) < x^2(a)$, decrease λ by a factor of 10, consider $a' = a + \delta a$ to be the new starting point, and return to step 3, substituting a' for a.

For each iteration it may be necessary to recompute the parameter increments δa_j from Equation (8.41), and the elements α_{jk} and β_j of the matrices, several times to optimize λ. As the solution approaches the minimum, the value of λ will decrease and the program should locate the minimum with a few iterations. A lower limit may be set for the value λ, but in practice this limit will seldom be reached.

Program 8.4. `Marquardt` The method of nonlinear least-squares fitting discussed in this section is illustrated in the routine `Marquardt`, listed in Appendix E. The procedure is identical to that of Program 8.3 except for the use of the variable `Lambda` by the `Marquardt` routine to adjust diagonal elements α_{jj} of the matrix α. The routine modifies the diagonal elements of α according ot Equation (8.39).

Table 8.5 shows values of χ^2 and the parameters a through a_5 for several stages of the calculation at the beginning, middle, and end of the search. Table 8.6 shows the error matrix from the fit.

TABLE 8.5
Two exponentials plus constant background: Marquardt method†

Trial	χ^2	a_1	a_2	a_3	a_4	a_5
0	406.6	10.0	900.0	80.0	27.0	225.0
1	82.9	11.0	933.5	139.3	33.9	173.9
2	66.4	10.8	960.1	130.6	33.8	201.2
3	66.1	10.4	958.3	131.4	33.9	205.0
Uncertainties		1.8	49.9	21.7	2.5	30.5

$\chi^2/\text{dof} = 1.22$; probability $= 12.4\%$; time $= 9.1$ s

†All stages in the fit to counts from the decay of excited states of silver. The uncertainties in the parameters correspond to the square roots of the diagonal terms in the error matrix.

TABLE 8.6
Elements of the error matrix (Marquardt method)†

l / k	1	2	3	4	5
1	3.38	− 3.69	27.98	− 2.34	− 49.24
2	− 3.69	2492.26	81.89	− 69.21	− 3.90
3	27.98	81.89	468.99	− 44.22	− 615.44
4	− 2.34	− 69.21	− 44.22	6.39	53.80
5	− 49.24	− 3.90	− 615.44	53.80	929.45

†Error matrix from a fit to the radioactive silver data. The diagonal terms are the variances σ_k^2 and the off-diagonal terms are the covariances σ_{kl}^2 of the parameters a_k.

8.7 COMMENTS ON THE FITS

Although the Marquardt method is the most complex of the four fitting routines, it is also the clear winner for finding fits most directly and efficiently. It has the strong advantage of being reasonably insensitive to the starting values of the parameters, although in the peak-over-background example in Chapter 9, it does have difficulty when the starting parameters of the function for the peak are outside reasonable ranges. The Marquardt method also has the advantage over the grid- and gradient-search methods of providing an estimate of the full error matrix and better calculation of the diagonal errors.

The routines of Programs 8.3 and 8.4 were tested with both numerical and analytical derivatives. Typical search paths with numerical derivatives are shown in Tables 8.4 and 8.5. For the sample problem with the assumed starting conditions, the minimum χ^2 was found in only a few steps by either method with essentially no time difference. Both methods are reasonably insensitive to starting values of parameters in which the fit is linear, but can be sensitive to starting values of the nonlinear parameters. Program 8.4 had remarkable success over a broad range of starting values, whereas Program 8.3 required better definition of the starting values of the parameters and generally required many more iterations.

The uncertainties in the parameters for these fits were calculated from the diagonal terms in the error matrices and are, in general, considerably larger than the uncertainties obtained in the grid- and gradient-search methods. Because the latter errors were obtained by finding the change in each parameter to produce as change of χ^2 of 1 from the minimum values, without reoptimizing the fit, there is a strong suggestion that correlations among the parameters play an important role in fitting Figure 8.1. This point of view is supported by examination of the error matrix from the method 4 fit (Table 8.6), which shows large off-diagonal elements.

With poorly selected starting values, the searches may terminate in local minima with unacceptably high values of χ^2 and, therefore, with unacceptable final values for the parameters. Termination in the sample programs is controlled simply by considering the reduction in χ^2 from one iteration to the next and stopping at a preselected difference. With this method, it is essential to check the results carefully to be sure that the absolute minimum has indeed been found.

SUMMARY

Nonlinear function: One that cannot be expressed as a sum of terms with the parameters appearing only as coefficients of the terms.

Minimum of χ^2:

$$a'_j = a_{j3} - \Delta a_j \left[\frac{\chi_3^2 - \chi_2^2}{\chi_1^2 - 2\chi_2^2 + \chi_3^2} + \frac{1}{2} \right]$$

Estimate of standard deviation from $\Delta x = 1$:

$$\sigma_j = \Delta a_j \sqrt{2(\chi_1^2 - 2\chi_2^2 + \chi_3^2)^{-1}}$$

Grid search: Vary each parameter in turn, minimizing χ^2 with respect to each parameter independently. Many successive iterations are required to locate the minimum of χ^2 unless the parameters are independent; that is, the variation of χ^2 with respect to one parameter is independent of the values of the other parameters.

Gradient search: Vary all the parameters simultaneously, adjusting relative magnitudes of the variations so that the direction of propagation in parameter space is along the direction of steepest descent of χ^2.

Direction of steepest descent: Opposite the gradient $\Delta \chi^2$:

$$(\nabla \chi^2)_j = \frac{\partial \chi^2}{\partial a_j} \simeq \frac{\chi^2(a_j + f\Delta a_j) - x(a_j)}{f\Delta a_j}$$

$$\delta a_j = \frac{-\left((\partial \chi^2/\partial a_j)\,\Delta a_j^2\right)}{\sqrt{\sum_{j=1}^{m}\left((\partial \chi^2/\partial a_j)\,\Delta a_j\right)^2}}$$

Parabolic expansion of χ^2:

$$\boldsymbol{\delta\alpha} = \boldsymbol{\beta\epsilon} \qquad \delta a_k = \sum_{j=1}^{m} (\epsilon_{kj}\beta_j)$$

with $\qquad \beta_k \equiv -\dfrac{1}{2}\dfrac{\partial \chi^2}{\partial a_k} \quad$ and $\quad \alpha_{jk} \equiv \dfrac{1}{2}\dfrac{\partial^2 \chi^2}{\partial a_j \partial a_k}$

Linearization of the fitting function:

$$\beta_k \equiv \sum \left\{ \frac{1}{\sigma_i^2} [y_i - y(x_i)] \frac{\partial y(x_i)}{\partial a_k} \right\} = -\frac{1}{2} \frac{\partial \chi^2}{\partial a_k}$$

$$\alpha_{jk} \equiv \sum \frac{1}{\sigma_i^2} \left\{ \frac{\partial y(x_i)}{\partial a_j} \frac{\partial y(x_i)}{\partial a_k} - [y_i - y(x_i)] \frac{\partial^2 y(x_i)}{\partial a_j \partial a_k} \right\}$$

$$= \frac{1}{2} \frac{\partial^2 \chi^2}{\partial a_j \partial a_k}$$

Gradient-expansion algorithm—the Marquardt method: Make λ just large enough to insure that χ^2 decreases:

$$\alpha'_{jk} = \begin{cases} \alpha_{jk}(1 + \lambda) & \text{for } j = k \\ \alpha_{jk} & \text{for } j \neq k \end{cases}$$

$$\alpha_{jk} \simeq \sum \left[\frac{1}{\sigma_i^2} \frac{\partial y(x_i)}{\partial a_j} \frac{\partial y(x_i)}{\partial a_k} \right] \qquad \beta_k = -\frac{1}{2} \frac{\partial \chi^2}{\partial a_k}$$

$$\delta a_j = \sum^m (\beta_k \epsilon'_k)$$

Uncertainty in parameter a_j: $\sigma_{a_j} = \epsilon_{jj}$ corresponds to $\Delta \chi^2 = 1$.

EXERCISES

8.1. Use an interpolation method (see Appendix A) to find the equation of the parabola that passes through the three points (x_1, y_1), (x_2, y_2), and (x_3, y_3). Find the value of x at the minimum of the parabola and thus verify Equation (8.12).

8.2. From the results of Exercise 8.1, verify Equation (8.13).

8.3. The following data represent histogram bin counts across a Lorentzian peak:

x_i	1.824	1.828	1.832	1.836	1.840	1.844	1.848	1.852	1.856	1.860
y_i	558	679	696	736	834	812	899	817	767	657

(*a*) Use the grid-search method to fit the equation $y(x) = AP_L(x; \mu, \Gamma)$ to the data and find the maximum-likelihood value of μ, where $P_L(x; \mu, \Gamma)$ is the Lorentzian function of Equation (2.32) and the known parameters are $A = 75$ and $\Gamma = 0.55$. Assume that x is given at the lower edge of each histogram bin and that the errors in y are statistical. Find the uncertainty in μ.

Suggested procedure: (i) Calculate χ^2 at the peak of the distribution and at a value on each side. (ii) Find the minimum of a parabola that passes through the three points. (iii) Repeat the procedure with three points centered on the minimum χ^2 until the value of μ has been determined to ± 0.001.

(*b*) Repeat the procedure for a two-parameter fit, with γ as the second unknown.

8.4. Consider the histogram of measurements of the length of a block displayed in Figure 1.2. The numbers of events in the bins bounded by $x = 18.6$ to 21.4 are

$$1, 3, 4, 7, 13, 14, 11, 12, 16, 11, 4, 1, 1, 2$$

Fit a Gaussian curve [Equation (2.23)] to these data by the least-squares method to find μ, σ, and the amplitude of the curve A. Bins with fewer than seven events should be merged to improve the reliance on Gaussian statistics. Compare the parameters obtained from the fit with those determined by taking the mean and standard deviation of the data.

8.5. The following data correspond to counts recorded in Example 6.2 with the addition of an unknown randomly fluctuating background term a_1. Use the Marquardt method to fit the equation $C = a_1 = a_2/d^2$ to these data to find the parameters a_1 and a_2 and the full error matrix. Assume statistical uncertainties.

i	1	2	3	4	5	6	7	8	9	10
d_i (m)	0.20	0.25	0.30	0.35	0.40	0.45	0.50	0.60	0.75	1.00
C_i	944	688	467	366	316	317	264	251	214	184

8.6. Use the method of least squares to fit the five-parameter equation $y(x) = a_1 + a_2 x + a_3 G(x; a_4, a_5)$ to the following data where $a_4 = \mu$, $a_5 = \sigma$, and $G(x; \mu, \sigma)$ is the Gaussian curve of Equation (2.23).

i	1	2	3	4	5	6	7	8	9	10
x_i	1.0	1.1	1.2	1.3	1.4	1.5	1.6	1.7	1.8	1.9
y_i	31	25	24	30	34	37	31	30	64	54

i	11	12	13	14	15	16	17	18	19	20
x_i	2.0	2.1	2.2	2.3	2.4	2.5	2.6	2.7	2.8	2.9
y_i	95	94	78	79	43	54	58	52	46	41

Use the Marquardt method and find an estimate of the error matrix. The value of x is given at the lower edge of each bin. Assume statistical uncertainties.

CHAPTER
9

FITTING
COMPOSITE
CURVES

9.1 LORENTZIAN PEAK ON QUADRATIC BACKGROUND

Many fitting problems involve determining the parameters of a peak or peaks that are superimposed upon a smoothly varying background. Two such examples are the search for a resonant state in elementary particle interactions and the attempt to pick out an identifying counter signal from a background of random signals and noise.

> **Example 9.1.** As an example of this type of problem, we shall consider the following function, which corresponds to a Lorentzian peak $y_p(E_i)$ on a second-degree polynomial background $y_b(E_i)$. The function is linear in the parameters a_1 through a_4 ($a_4 = A_p$) but not in the parameters μ and Γ.
>
> $$y(E) = a_1 + a_2E + a_3E^2 + A_p \frac{\Gamma/(2\pi)}{(E - \mu)^2 + (\Gamma/2)^2} \tag{9.1}$$
>
> Such distributions are relatively common in nuclear and particle physics, where the

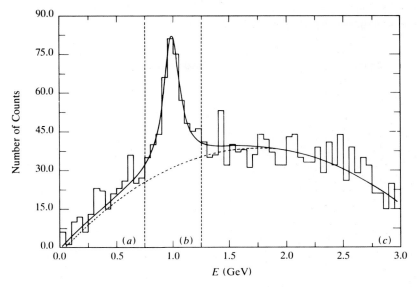

FIGURE 9.1

A histogram of 2000 simulated events corresponding to a Lorentzian peak on a second-degree polynomial background. The curve is the result of fitting Equation (9.1) to the data.

Lorentzian or other peak function may describe a resonant state or a detector signal, and a polynomial function serves as a reasonable approximation to a smooth background. An experimental histogram of 2000 events, drawn from the distribution of Equation (9.1) with $\mu = 1.000$ GeV and $\Gamma = 0.200$ GeV is shown in Figure 9.1.[1]

We used the Marquardt method with numerical derivatives to fit Equation (9.1) to the histogram of Figure 9.1 because this is clearly the most flexible and convenient of the four methods considered in Chapter 8. The amplitudes of the polynomial function (a_1 through a_3), the amplitude of the Lorentzian peak ($a_4 = A_p$), and the mean μ and half-width Γ of the Lorentzian function (a_5 and a_6) were treated as free parameters of the fit. Starting values for a_5 and a_6 were obtained by inspecting the histogram of Figure 9.1; starting values for the other parameters, the coefficients of the various terms, were obtained by trial and error. Because the Marquardt method is exact for a function that is linear in the parameters, convergence for this problem is relatively insensitive to the choice of starting values for a_1 through a_4. Convergence is, however, quite sensitive to the initial choice of a_5 and a_6. The starting value a_5 of the mean should be

---

[1]These "data" were actually generated by the Monte Carlo method from the distribution of Equation (9.1). See Chapter 5.

TABLE 9.1
Results of least-squares fits of Equations (9.1) and (9.3) to data displayed in Figures 9.1 and 9.2

	Six-parameter fit (Figure 9.1)	Seven-parameter fit (Figure 9.2)
dof	54	53
χ^2	59.0	59.1
P_{χ^2}	29.6%	25.8%
a_1	-0.97 ± 0.89	-0.83 ± 0.89
a_2	46.1 ± 2.6	43.3 ± 2.5
a_3	-13.4 ± 1.0	-12.5 ± 0.9
a_4	13.8 ± 2.2	13.7 ± 2.2
$a_5 \ (\mu_1)$	0.987 ± 0.011	1.00 (fixed)
$a_6 \ (\Gamma_1)$	0.172 ± 0.035	0.20 (fixed)
a_7		3.7 ± 1.8
$a_8 \ (\mu_2)$		0.855 ± 0.015
$a_9 \ (\Gamma_2)$		0.086 ± 0.069

within the obvious limits of the peak, and the starting value of the half-width a_6 should be within about a factor of 1.5 of the final value. Starting values outside these ranges may result in the program's coasting to a halt in a shallow local minimum with obviously incorrect values for the parameters and with a higher than expected value of χ^2.

The results of this six-parameter fit are summarized in Table 9.1 and the curve calculated from Equation (9.1) with the parameters found in the fit is plotted on the histogram of Figure 9.1. The dashed curve shows the contribution of background $y_b(E)$ under the peak. The χ^2 probability of the fit (9.3%) is acceptable.

Because one of the objectives of this analysis is to determine μ, the mean of the peak function of Equation (9.1), we must be careful in the choice of the value of the independent variable that we use in the fit. On the histogram of Figure 9.1, the value of E_i at the *left-hand edge* of selected bins is indicated, but for the fit we should use the mean value of E in a bin, which will be approximately equal to the value at the *center* of the bin. Thus, if we use values of E_i from the histogram data in the fitting routines, we shall obtain from the fit a value for μ that is too low by approximately half a bin width, and must make the necessary correction. Alternatively, we can use values of E_i at the center of each bin as the independent variable. For wide bins and a rapidly varying fitting function, it might be advisable to select the value of E_i for each bin by weighting according to the steepness of the function.

Note that this problem was not important in the determination of the mean lifetimes in Example 8.1 because lifetimes are determined effectively from differences, rather than absolute values, of the independent variable. We must, however, always take care when we plot the results that the curve is not displaced half a bin width from the histogram.

9.2   AREA DETERMINATION

When dealing with problems of peaks and backgrounds, we may wish to determine, not only the position and width of a peak in a spectrum, but also the number of events or area of the peak, which may measure the intensity of a transition or the strength of a reaction. When peaks are not well separated, or when the contribution from background is substantial, least-squares fitting can provide a consistent method of extracting such information from the data.

The importance of consistency should not be underestimated. Whether or not the method chosen is the best possible method, as long as it involves a well understood and clearly specified procedure, other experimenters will be able to check and compare the results safe in the knowledge that their comparisons are justified and meaningful. The method of least squares is considered to be an unbiased estimator of the fitting parameters and all parameters are presumed to be estimated as well as possible. This assumption is based on the validity of both the fitting function in describing the data and the least-squares method. If we try to fit the data with an incorrect fitting function, or try to fit data with uncertainties that do not follow the Gaussian distribution, then the fitting procedure may not yield optimum results.

Although we refer to the number of events as the *area* of a peak or plot, the true area is, of course, the number of events multiplied by the data interval or histogram bin width. Thus, to find the area of the peak from the results of the fit in Example 9.1, we calculate

$$A_p = \int_{-\infty}^{+\infty} y_p(E) \, dE = \int_{-\infty}^{+\infty} a_4 \frac{\Gamma/(2\pi)}{(E - \mu)^2 + (\Gamma/2)^2} \, dE \qquad (9.2)$$

Because we used the normalized form of the Lorentzian function, the integral is just the coefficient a_4 obtained in the search, $A_p = a_4$. The area of the peak on the histogram is the product of the number of events N_p in the peak by the width ΔE of the histogram bin

$$A_p = N_p \times \Delta E$$

so the number of events in the peak is given by

$$N_p = A_p/\Delta E \qquad (9.3)$$

The result from Example 9.1 is $N_p = (11.9 \pm 2.4)/0.05 = (240 \pm 50)$ events.

Alternatively, we might plot the background curve on the graph

$$y_b(E) = a_1 + a_2 E + a_3 E^2 \qquad (9.4)$$

and count the number of events in the peak above the background in a region between $\mu - \delta$ and $\mu + \delta$, where the range δ is chosen to encompass the peak, as illustrated by the dotted vertical lines on Figure 9.1.We should be obliged to estimate and correct for events in the tails of the peak distribution; that is, those events outside our arbitrarily selected limits $\mu \pm \delta$. Using this method, ignores some of the improvement in the area estimate resulting from the fitting procedure.

Uncertainties in Areas under Peaks

If we calculate the area of the peak from Equations (9.2) and (9.3), then the uncertainty should be estimated from the uncertainties in the parameters by the error propagation equation. We have used this method to obtain the uncertainty in the number of events of the peak of Figure 9.1 in the calculation that follows Equation (9.3).

The uncertainty σ_A in the area under a peak can also be estimated by considering the uncertainty in the parent distribution. If the data are distributed according to the Poisson distribution, the uncertainty in the area A is given by $\sigma_a^2 \simeq A$. If we obtain the area by counting the number of events above background, then the variance of the difference will be the sum (*not the difference*) of the variance of the total area under the peak and the variance of the subtracted background A_b:

$$\sigma_p^2 = \sigma_t^2 + \sigma_b^2 = A_t + A_b$$

where the subscripts p, b, and t correspond to peak, background, and total (= peak + background). In order to keep $\sigma_t = A_t$ as small as possible, we should count events only in that region where the peak-to-background ratio is large and make corrections for the tails of the distribution.

Area under a Curve with Poisson Statistics

Curiously enough, if the data are distributed around each data point according to the Poisson distribution, as in a counting experiment, the method of least squares consistently *underestimates* the area under a fitted curve by an amount approximately equal to the value of χ^2. To show this, let us consider fitting such data with an arbitrary peak, represented by $bf_p(x; \mu, \sigma)$ plus a polynomial background similar to Example 9.1:

$$y(x) = a + bf(x; \mu, \sigma) \tag{9.5}$$

where we have simplified the background to a single term a for clarity.

Using the method of least squares, we define χ^2 to be the weighted sum of the squares of deviations of the data from the fitted curve

$$\chi^2 = \sum \left[\frac{1}{\sigma_i^2} (y_i - a - bf(E; \mu, \sigma))^2 \right] \tag{9.6}$$

and obtain the solution by minimizing χ^2 simultaneously with respect to each of the parameters. The required derivatives with respect to the two parameters a

and b, in which the function is linear, are

$$\frac{\partial \chi^2}{\partial b} = -2\sum\left[\frac{1}{\sigma_i^2}(y_i - a - bf(E;\mu,\sigma))f(E;\mu,\sigma)\right] = 0$$

$$\frac{\partial \chi^2}{\partial a} = -2\sum\left[\frac{1}{\sigma_i^2}(y_i - a - bf(E;\mu,\sigma))\right] = 0 \tag{9.7}$$

We can write χ^2 in terms of the derivatives of Equation (9.7) as

$$\chi^2 = \sum\left[\frac{y_i}{\sigma_i^2}(y_i - a - bf(E;\mu,\sigma))\right] + \frac{1}{2}\left(a\frac{\partial \chi^2}{\partial a} + b\frac{\partial \chi^2}{\partial b}\right) \tag{9.8}$$

and setting the derivatives to 0 gives

$$\chi^2 = \sum\left[\frac{y_i}{\sigma_i^2}(y_i - a - bf(E;\mu,\Gamma))\right] \tag{9.9}$$

If the data represent the number of counts per unit time in a detector, then they are distributed according to the Poisson distribution and we can approximate $\sigma_i^2 \simeq y_i$. Equation (9.9) becomes

$$\chi^2_{\min} \simeq \sum[y_i - (a + bf(E;\mu,\Gamma))]$$

$$= \text{area(data)} - \text{area(fit)} \tag{9.10}$$

Thus, we observe that the area under the total fit is underestimated by an amount equal to $\chi^2_{\min}$.

For this derivation we require only that the fitting function consist of a sum of terms, each one of which is multiplied by a coefficient

$$y(x) = \sum_{j=1}^{m} a_j f_j(x) \tag{9.11}$$

The function $f_j(x)$ can contain any number of other parameters in nonlinear form, but may not contain any of the coefficients a_j. Even reparameterizing the function of Equation (9.5) [or Equation (9.1)] and minimizing χ^2 with respect to the area explicitly would not affect the discrepancy between the actual and estimated areas.

Note that for data that are distributed with a constant uncertainty $\sigma_i = \sigma$, the second equation of Equations (9.7) is sufficient to ensure that $\sum y(x_i) = \sum y_i$. It is the assumption of a Poisson distribution for the data $\sigma_i^2 = y_i$ that yields the discrepancy between the actual and estimated areas.

If the agreement between the fit and the data should be exact, $\chi^2 = 0$, then the estimated and actual areas would be equal. For a fitting function that is a good representation of the data, the value of χ^2 will approximately equal the number of degrees of freedom, so that if there are many bins and a few parameters to be determined, the average discrepancy will be about 1 per bin.

Thus, the correction may be negligible for distributions with large numbers of events.

We would like to find ways to reduce the discrepancy. The fact that we know the approximate value of the discrepancy in the total histogram is, in itself, not very helpful because we do not know how to allocate the discrepancy between peak and background. We might find the ratio of the integral A_p of the peak [Equation (9.2)] to the integral A of the complete function Equation (9.1) and scale to the total number of events in the plot to estimate the number of events in the peak. This method assumes that the correction is proportional to the area. Another possibility is to make separate fits to the peak and background regions of the plot, so that we can try to assign the estimated correction separately to the two regions of the plot.

One obvious way of reducing the discrepancy between the area of the measured and fitted data is to reduce the value of χ^2 at the minimum so that the correction is small. A method of accomplishing this reduction, which is not universally accepted but that can be justified by practical considerations, is the technique of smoothing the data, averaging in some mathematically acceptable way over adjacent bins. (See Appendix A.) Under any smoothing process there can be no overall gain in information, and a net improvement of the fit to the area must be offset by an increased uncertainty in the estimation of other parameters, such as the width and position of the peak. But smoothing will decrease the value of χ^2 at the minimum and thereby reduce the bias in the estimation of the area.

9.3   COMPOSITE PLOTS

Single Peak and Background

For a fitting function $y(x)$ that is separable into a peak $y_p(x)$ plus a background $y_b(x)$, such as Equation (9.1), it may be convenient to consider at least some facets of the fitting procedure separately. The least-squares procedure for minimizing χ^2 with respect to each of the parameters a_j,

$$\frac{\partial}{\partial a_j} \sum \left\{ \frac{1}{\sigma_i^2} [y_i - y_b(x_i) - y_p(x_i)]^2 \right\} = 0 \tag{9.12}$$

can be considered equally well in terms of fitting the sum of the curves $y(x)$ to the total yield y_i or of fitting one function $y_p(x)$ to the difference spectrum $y_i' = y_i - y_b(x_i)$. The only provision is that the uncertainties in the data points $\sigma_i' = \sigma_i$ must be the same in both calculations.

If the background curve can be assumed to be a slowly varying function under the peak, as in Figure 9.1, and may reasonably be interpolated under the peak from fitting on both sides, it may be preferable to fit the background curve $y_b(x)$ outside the region of the peak and to fit the peak function $y_p(x)$ only in the region of the peak.

One reason for separating the fit in this way may be to attempt to isolate special problems that result from fitting with an incorrect peak or background. The χ^2 function measures not only the deviations of the parameters from an ideal fit, but also the discrepancy between the form chosen for the shape of the fitting function $y(x)$ and the parent distribution of the data. If the shape of the fitting function does not represent that of the parent distribution exactly, the value of χ^2 may have large contributions from local data regions. By fitting separate regions of a plot, it may be possible to discover whether the disagreement is in the background or the peak region. In the histogram of Figure 9.1, our interest is in the properties of the peak function, and not in the background, which we parameterize with a simple power series in E. However, the value of χ^2 for the fit is calculated for the entire plot and includes contributions from discrepancies between the background and the fitted curve, as well as between the peak and curve. We may be able to isolate problems to one or the other region by separating the fit into two parts.

Another reason for making separate fits to regions of a plot is to search for starting values for an overall fit. For example, when fitting a function that consists of peak functions plus background function, it may be useful first to fit the regions outside the peaks to get starting values for the background parameters and then to fit separately the region close to each peak, to find starting values for the peak parameters.

As an example, assume that we wish to find starting values for the fit of Equation (9.1) to the data of Figure 9.1. The following procedure could be used:

1. Separate the curve into three regions (a), (b), and (c) as indicated by the two vertical lines on Figure 9.1.
2. Fit the background polynomial $y_1(x) = a_1 + a_2E + a_3E^2$ simultaneously to regions below and above the peak to obtain provisional values for the parameters a_1 through a_3.
3. Fit the entire function of Equation (9.1) to the central region, with the fixed values of a_1 through a_3 obtained in step 2 to obtain values for the parameters a_4, a_5, and a_6.
4. Fit the entire function of Equation (9.1) simultaneously to regions (a) and (c), with the starting values of the parameters a_4 through a_6 set to the values obtained in steps 2 and 3 to obtain new values of the parameters a_1 through a_3.

If the parameters continue to change significantly on each iteration, the process can be repeated from step 2 as required. Alternatively, it may be sufficient to skip step 3 and to fit for all parameters after step 2.

In fitting the peak and background functions over different parts of the spectrum, it is important to note that the complete function $y(E)$ of Equation (9.1) must be fitted to both regions; that is, in the region outside the peak where the background is being fitted, the calculation of the tail of the peak must be included, and underneath the peak, the background terms must be included.

Multiple Peaks

Separation of closely spaced peaks is an important problem in nuclear and particle physics, as well as other fields.

> **Example 9.2.** Let us assume that the existence of the peak in Example 9.1 has been well established by previous experiments with parameters $\mu_1 = 1.000$ GeV and $\Gamma_1 = 0.200$ GeV, and that there has been a theoretical prediction of the existence of a second, narrow peak with mean $\mu_2 \simeq 0.850$ GeV. The question then arises as to whether our data can be used to support this prediction?
>
> The histogram of Figure 9.2 shows an experimental distribution of data from the distribution

$$y(x) = a_1 + a_2 E + a_3 E^2 + A_1 \frac{\Gamma_1/(2\pi)}{(E - \mu_1)^2 + (\Gamma_1/2)^2}$$

$$+ A_2 \frac{\Gamma_2/(2\pi)}{(E - \mu_2)^2 + (\Gamma_2/2)^2} \qquad (9.13)$$

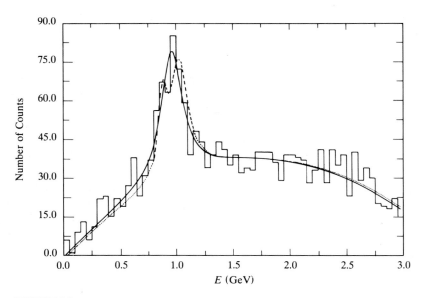

FIGURE 9.2

A histogram of 2000 simulated events corresponding to two Lorentzian peaks on a second-degree polynomial background. The solid curve corresponds to a fit to the data of a single peak plus background terms. The dashed curve is the result of a fit to the parameters of the smaller peak with fixed values for the parameters of the larger peak.

which corresponds to the distribution of Equation (9.1) with the addition of a second peak with $\mu_2 = 0.850$ GeV and $\Gamma_2 = 0.100$ GeV. We shall refer to the original, established peak as peak 1 and the new peak as peak 2. Superimposed on the smooth background and on the tail of peak 1, peak 2 shows up only as an excess of events in the region around 0.800–0.900 GeV.

We first attempt to fit the data with the single-peak Equation (9.1) with six free parameters as in Example 9.1. The result is indicated by the solid curve on Figure 9.2. The fitting routine has created a broad distribution that encompasses both peaks, and although there is a suggestion of structure in the peak, the χ^2 probability of 21.9% for the single-peak fit is satisfactory. Without further information, we would not be justified in claiming the existence of the second peak in our data.

However, because we assume that peak 1 has been established, we try to extract additional information from our data. We begin by fitting Equation (9.1) to our data with the parameters of peak 1 fixed at their known values, $\mu_1 = 1.000$ GeV and $\Gamma_1 = 0.200$ GeV, to look for any difference between the fitted curve $y(x)$ and the histogram. The three parameters of the background, a_1 through a_3, and the area of peak 1, $A_1 = a_4$, are free parameters in this fit. The results is illustrated by the dashed curve on Figure 9.2. As we might expect, the χ^2 probability for the fit (7.9%) is lower than for the corresponding fit to the single peak data of Example 9.1, but it is still acceptable.

The statistical significance of the excess can be studied further by finding the difference between the histogram and the dashed curve of Figure 9.2 in the region of the suspected peak 2. These data are tabulated in Table 9.2. The uncertainty in $y(x)$ was calculated from the uncertainties in the fitted parameters. The uncertainties in the binned data are assumed to be the square root of the number of events in each bin. The final uncertainties were calculated with the error propagation equation and the assumption that all uncertainties are independent.

With these uncertainties, our observation of peak 2 may be characterized at best as a 2.4 standard deviation fluctuation in the region between 0.75 and 0.95 GeV. If we look in Table C.2 for the probability of obtaining a fluctuation greater than 2.4σ, we see that the probability is only about 1.6%—encouragingly low. But, if we treat the four bins as one and replot the data in bins of 0.20

TABLE 9.2
Data and calculation from a selected region of Figure 9.2

E (GeV)	Histogram	$y(x)$	Difference
0.75–0.80	37	34.1 ± 2.4	2.9 ± 6.5
0.80–0.85	56	39.4 ± 2.7	16.6 ± 8.0
0.85–0.90	67	47.9 ± 3.2	19.1 ± 8.8
0.90–0.95	63	61.8 ± 4.3	1.2 ± 9.0
Sum	223	183.2	39.8 ± 16.2

instead of 0.05, then we should have fifteen bins in which such a fluctuation might occur, and the probability for observing a fluctuation of 2.4σ or more climbs to $1 - (1 - 0.016)^{15} = 21\%$. Thus, if we did not have the theoretical suggestion that there should be a narrow peak near 0.850 GeV, we would not be justified in making any claim at all.

Although the evidence for the second peak is very weak in our data, we shall attempt to determine its mean μ_2 and half-width Γ_2 by fitting Equation (9.13) to the histogram with parameters of peak 1 fixed at their nominal values and additional free parameters for the area, mean, and width of peak 2. Results of this seven-parameter fit are listed in Table 9.1. As we expected, the amplitude of the peak (a_7) is about twice its uncertainty, so the significance is no better than about 2 standard deviations. Although this is not sufficient to establish the existence of the peak, we do find a reasonable value for the mean of the peak $(\mu_2 = a_8 = 0.855 \pm 0.015)$, consistent with the predicted value. Thus, we can at least state that our data are consistent with the existence of a narrow peak in the region of 0.850 GeV.

Can we make a stronger statement? For problems such as this, where the statistical significance of a result is in question, the Monte Carlo method (Chapter 5) provides a convenient tool for more detailed examination. We shall use this method in Chapter 11 for further study of these data, to try to determine whether our evidence for peak 2 is indeed significant, or whether it may have been created by our method of analyzing the data.

SUMMARY

Background subtraction:

$$y_p(x) = y(x) - y_b(x) \qquad (y_p \rightarrow \text{peak}; y_b \rightarrow \text{background})$$

Uncertainty in area of peak:

$$\sigma_{A_p}^2 = \sigma_A^2 + \sigma_{A_b}^2 \qquad (\simeq A + A_B \text{ for Poisson statistics})$$

Area under fitted peak curve:

$$A_p = \int_{-\infty}^{+\infty} y_p(x)\, dx$$

Discrepancy in area under a curve with Poisson statistics:

$$\chi_{\min}^2 = \sum \left[\frac{y_i}{\sigma_i^2} (y_i - y(x_i)) \right] \simeq \text{area(data)} - \text{area(fit)}$$

EXERCISES

9.1. Find the area of the peak in Figure 9.1 by counting the area between the vertical dotted lines and subtracting the estimated background. Estimate the correction for the tails. Estimate the uncertainty in your determination of the area (see Exercise 9.3).

9.2. Refer to the data of Exercise 8.6. Fit the histogram by the method outlined in Section 9.3 with separate fits of the linear background polynomial to the region outside the peak and of the Gaussian function to the region of the peak.

9.3. The accompanying table lists the numbers of events in the histogram bins of Example 9.1 bounded by $E = 0.0$ and 3.0 GeV.

(a) Fit Equation (9.1) to the data to obtain the parameters for this distribution listed in Table 9.1.

(b) Repeat the fit with adjacent bins merged and observe the effect on the value of χ^2, the determination of the area of the peak, and the determination of the mean and half-width of the peak. Assume statistical uncertainties.

6	1	10	12	6	13	23	22	15	21	23	26	36	25	27	35	40	44	66	81
75	57	48	45	46	41	35	36	53	32	40	37	38	31	36	44	42	37	32	32
43	44	35	33	33	39	29	41	32	44	26	39	29	35	32	21	21	15	25	15

CHAPTER
10

DIRECT APPLICATION OF THE MAXIMUM-LIKELIHOOD METHOD

The least-squares method is a powerful tool for extracting parameters from experimental data. However, before a least-squares fit can be made to a data set that is made up of individual measurements or events, the events must be sorted into a histogram, which may obscure some detailed structure in the data. Because the least-squares method was derived from the principle of maximum likelihood, it might be better in some instances to use the maximum-likelihood method directly to compare experimental data to theoretical predictions, without the necessity of *binning* data into histograms with the corresponding loss of information.

We have already used the method in Chapter 4 to find estimates for the mean and standard deviation of data obtained in repeated measurements of a single variable, where we have assumed that the measurements were distributed according to Gaussian probability. Now, we extend the method to other distribution functions and to multiparameter fits. Maximum-likelihood methods can be applied directly to many "curve fitting" problems, and such fitting is almost as easy to use as the least-squares method, and considerably more flexible. However, the direct maximum-likelihood method requires computations for each *measured event*, rather than for each *histogram bin* as in least-squares fitting, and therefore the technique may be too slow for very large data samples.

Direct maximum-likelihood calculations have an advantage over the least-squares method for two particular types of problems: (i) low-statistics experiments with insufficient data to satisfy the requirement of Gaussian statistics for individual histogram bins and (ii) experiments in which the fitting function corresponds to a different probability density function for each measured event so that binning the data leads to a reduction in information and a loss of sensitivity in determining the parameters. If the data set is sufficiently large, then the least-squares method can be applied to problems of either type, and that method is generally preferred in view of its smaller computing requirement. At any rate, it is not possible to extract more than minimal information from a very small data set, so we should expect the direct maximum-likelihood method to be most useful for intermediate problems with modest data samples.

10.1   MAXIMUM-LIKELIHOOD METHOD

The basic maximum-likelihood procedure is relatively simple. Assume that we have a collection of N events corresponding to the measurement of an independent variable x_i and a dependent variable y_i, where i runs from 1 to N. We wish to obtain the parameters, $a_1, a_2, \ldots, a_m$, of a fitting function $y(x_i) \equiv y(x_i; a_1, a_2, \ldots, a_m)$ from these data. For each event, we convert $y(x_i)$ to a normalized probability density function

$$P_i \equiv P(x_i; a_1, a_2, \ldots, a_m) \tag{10.1}$$

evaluated at the observed value x_i. The likelihood function $\mathscr{L}(a_1, a_2, \ldots, a_m)$ is the product of the individual probability densities

$$\mathscr{L}(a_1, a_2, \ldots, a_m) = \prod_{i=1}^{N} P_i \tag{10.2}$$

and the maximum-likelihood values of the parameters are obtained by minimizing $\mathscr{L}(a_1, a_2, \ldots, a_m)$ with respect to the parameters.

In many experiments, the probability density function P_i will be made up of two components: a theoretical factor corresponding to the underlying principle being tested and an experimental factor corresponding to the biases introduced by experimental conditions.

> **Example 10.1.** Let us consider an experiment to determine the lifetime of an elementary particle, the short-lived K_s^0 meson (which we shall refer to as the K meson), from measurements of the decay in flight of many such particles. The experimental arrangement is sketched in Figure 10.1. A high energy charged particle p_i enters the apparatus and interacts in the production target at the vertex V_1 to produce secondary particles, including a K meson. The production vertex can be determined by tracing back the trajectories of the charged secondary particles, as measured in the production vertex detector, to their intersections with

the trajectory of the beam particle, as measured in the beam particle detector. The K meson produced in the interaction travels a distance L before decaying at V_2 into two π mesons, π_1 and π_2. For simplicity, we assume that the K^0 mesons are produced at small angles to the x axis. The decay vertex detector array defines the location of the decaying particle from the intersection of the charged decay products. The momentum of the neutral particle is determined from measurements of the momentum vectors of its two decay products, π_1 and π_2.

The dashed rectangle on Figure 10.1 indicates the *fiducial region* for the experiment. Decay vertices are selected only within this region to assure precise measurements of the secondary tracks from the decay of the K meson. In the following examples, we shall assume that the coordinates of the two vertices and the magnitude of the momentum of the decaying K meson have been measured.

The geometry of the detector is critical to the solution of the problem. Only unstable particles that are produced and decay within a defined fiducial region, indicated by the dashed rectangle in Figure 10.1, can be measured precisely, and the rejection of events outside that region introduces a bias into the determination of the mean life.

The probability of observing a single event that survives for a time t_i is

$$P_i = A_i \cdot \rho(t_i; \tau) \tag{10.3}$$

The first factor A_i represents the *detection efficiency*, or probability that the particle will decay within a predefined *fiducial volume* within our apparatus, so that a satisfactory measurement can be made of its flight time. This factor depends upon the coordinates of the production and decay vertices of the decaying particle, its momentum vector, and the geometry of the fiducial

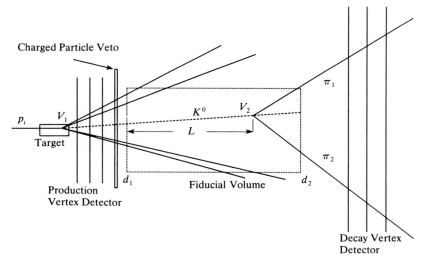

FIGURE 10.1
Experimental arrangement to measure the lifetime of an elementary particle.

volume. The second factor $\not{p}(t_i; \tau)$ is proportional to the probability that a particle of mean lifetime τ will decay between time t_i and $t_i + dt$ and is therefore proportional to $e^{-t_i/\tau}$. Equation (10.3) becomes

$$P_i = A_i e^{-t_i/\tau} \tag{10.4}$$

It might appear that the two factors in Equation (10.3) are independent, so that the detection efficiency factor is independent of the decay probability, but this is not generally true. To see this, consider the measurement of a short-lived particle with apparatus that is sufficiently large so that the number of particles decaying outside the fiducial volume is negligible. In this case, the efficiency factor is unity for all decays. The mean lifetime can be determined by measuring and plotting on a semilogarithmic graph the number of decays as a function of time, and finding the inverse slope from the graph. Now consider using the same apparatus to measure a much longer mean life, such that a significant fraction of the particles decay outside the detection apparatus. For each particle that is observed to decay within the apparatus, we can define a *potential path length* as the distance it would travel if it had not decayed. Unless the decaying particles all have the same potential path length, we must calculate a geometric factor to correct for those particles that decay outside the detector. The correction factor will depend on the parameters and will be a function of the production and decay coordinates and the momentum vectors of each decaying particle. Clearly, one element of good experiment design should be to minimize the dependence of these corrections on the parameters sought in the experiment.

Least-Squares Method

If we wish to apply the least-squares method to this problem, then we must determine an average geometrical efficiency for the experiment. For simple experiments we may be able to determine this from simple geometrical considerations. For more complex experiments, we can simulate the experiment by a Monte Carlo calculation that creates events and traces them through the apparatus, applying geometric cuts that correspond to equipment dimensions and acceptances. The generated sample of events should be large enough so that statistical errors from the Monte Carlo generation do not contribute to the uncertainties in the final parameters. An average, overall detection efficiency for the experiment can be found, and the least-squares method used to fit the generated distribution to the data. For a one-parameter fit, this is a straightforward exercise, even if the efficiency factor is a function of the parameter. If there are several parameters to be determined and if the efficiency is sensitive to these parameters, the procedure can become tedious because the entire Monte Carlo generation may have to be repeated for each trial value of the parameters. Nevertheless, this is a valuable and exceedingly useful method for analyzing experimental data.

Normalization for Maximum Likelihood

The factor A_i in Equation (10.4) corresponds to a normalization for each measurement to assure unit probability for observing in this experiment *any* event that has the mean life, coordinates, and kinematics of the observed decaying particle. To determine the normalizing factor A_i we refer to Figure 10.1 and consider the fiducial volume of our apparatus, indicated by the dashed rectangle. From each particle's production coordinates and momentum vector, we can determine the minimum distance d_1 that the particle must travel to enter the region and the maximum distance d_2 it can travel before leaving the region. (We can, of course, observe some events outside the fiducial volume, but we reject them because they cannot be measured precisely.) These minimum and maximum distances d_1 and d_2 must be converted to times-of-flight t_1 and t_2 in the rest frame of the decaying particles, and the normalizing factors A_i can then be determined from the condition

$$\int_{t_1}^{t_2} P_i \, dt_i = A_i \int_{t_1}^{t_2} e^{-t_i/\tau} \, dt_i = 1 \tag{10.5}$$

With this normalization, the individual event probability P_i of Equation (10.4) becomes the probability density for observing a single event. The normalized joint probability or the likelihood function for observing N such events in our experiment is just the product of the individual probability functions:

$$\mathscr{L}(\tau) = \prod_{i=1}^{N} P_i = \prod_{i=1}^{N} A_i e^{-t_i/\tau} \tag{10.6}$$

Parameter Search

Our object is to find the value of the parameter τ that maximizes this likelihood function. Because the probability of observing any particular event is less than 1, the product of a large number of such probabilities (one for each measured event) may be a very small number, and may, in fact, be too small for the computer to handle. To avoid problems, it is usually preferable to maximize the logarithm of the likelihood function

$$M = \ln \mathscr{L} \tag{10.7}$$

rather than the likelihood function itself, so that the product of Equation (10.6) becomes a sum. The logarithms should be reasonable, negative numbers. For our particular example, the logarithm of the likelihood function of Equation (10.6) is given by

$$M(\tau) = \ln[\mathscr{L}(\tau)] = \sum \left[\ln A_i - \frac{t_i}{\tau} \right] \tag{10.8}$$

with A_i defined by Equation (10.5). Note that A_i is a function of the unknown parameter τ, as well as of the production coordinates, momentum vector, and fiducial volume, and must be calculated separately for each event, *and for every trial value of τ.*

In general, this problem, like the corresponding nonlinear least-squares fitting problem, cannot be solved in closed form. However, either the grid- or gradient-search method of minimizing the χ^2 function discussed in Chapter 8 can be adopted directly. It is only necessary to search for a maximum of M (or a minimum value of $-M$) with the same routines we used in Chapter 8 to find a minimum of χ^2.

We may note a correspondence between the quantity $M(\tau)$, determined in Equation (10.7) from the likelihood function for *individual events*, and the goodness-of-fit parameter χ^2, determined by Equation (8.7) from the likelihood function $P(a)$ for *binned data*:

$$\chi^2 = -2\ln[\mathscr{L}(\tau)] + \text{constant} \tag{10.9}$$

In the limit of a large number of events, the two methods must yield the same value τ' for the maximum-likelihood estimate of the parameter τ. In both cases the likelihood function will be a Gaussian function of the parameter near the optimum value

$$\mathscr{L}(\tau) \propto \exp\left(-\frac{(\tau - \tau')^2}{2\sigma^2}\right) \tag{10.10}$$

so we can expect $M(\tau)$, like $\chi^2(\tau)$, to vary quadratically with the parameter τ in the vicinity of τ'.

> **Example 10.1a.** Let us consider the simplest form of this problem. We assume that the unknown mean lifetime is short and our apparatus is large enough to include many lifetimes so that the loss of particles that decay at very long times is negligible. Let us also assume that our equipment can detect particles at very short as well as very long times. Then the limits on the normalization integral of Equation (10.5) become $t_1 = 0$ and $t_2 = \infty$ and A_i is the same for every event and is given by $A_i = 1/\tau$. The likelihood function becomes
>
> $$\mathscr{L}(\tau) = \prod A_i e^{-t_i/\tau} = \prod \frac{e^{-t_i/\tau}}{\tau} \tag{10.11}$$
>
> with logarithm
>
> $$M(\tau) = \ln[\mathscr{L}(\tau)] = -\frac{1}{\tau}\sum t_i - N\ln\tau \tag{10.12}$$

We can obtain the maximum of Equation (10.12) by taking the derivative of $M(\tau)$ with respect to τ and setting it to 0:

$$\frac{dM(t)}{d\tau} = \frac{d}{d\tau}\left\{-\frac{1}{\tau}\sum t_i - N\ln\tau\right\}$$

$$= \frac{1}{\tau^2}\sum t_i - \frac{N}{\tau} = 0 \tag{10.13}$$

The solution is $\tau = \Sigma t_i/N$; that is, the maximum-likelihood estimate of the mean life is just the mean of the individual lifetime measurements. We should have reached the same result if we had found the maximum of $\mathscr{L}(t)$ from Equation (10.11).

Example 10.1b. Suppose that we repeat the experiment, but with poorer experimental resolution so that we cannot distinguish the decay vertex (x_2, y_2, z_2) from the creation vertex (x_1, y_1, z_1) unless they are separated by a distance d_1. For simplicity, we assume that the decaying particles are all produced with the same velocity, so that the lower cutoff distance d_1 translates into the same lower cutoff in time t_1 for all events. (In an actual experiment, of course, the decaying particles would be produced with various velocities, so that the calculated lower cutoff time t_1 would vary from event to event.)

For this example, the normalization integral of Equation (10.5) becomes

$$A_i \int_{t_1}^{\infty} e^{-t_i/\tau}\, dt_i = 1 \tag{10.14}$$

which gives

$$A_i = \frac{e^{t_i/\tau}}{\tau} \tag{10.15}$$

The likelihood function becomes

$$\mathscr{L}(\tau) = \prod_{i=1}^{N} A_i e^{-t_i/\tau} = \prod_{i=1}^{N} \frac{e^{t_1/\tau}}{\tau} e^{-t_i/\tau} = \prod_{i=1}^{N} \frac{e^{(t_1-t_i)/\tau}}{\tau} \tag{10.16}$$

so that

$$M = \ln \mathscr{L} = \sum \frac{[t_1 - t_i]}{\tau} - \sum \ln \tau \tag{10.17}$$

Setting

$$\frac{dM(\tau)}{d\tau} = 0 \tag{10.18}$$

gives

$$\frac{d}{d\tau} \sum \left\{ \frac{[t_1 - t_i]}{\tau} - \ln \tau \right\} = -\sum \left\{ \frac{t_1 - t_i}{\tau^2} \right\} - \frac{N}{\tau} = 0 \tag{10.19}$$

or

$$\tau = \frac{\Sigma[t_i - t_1]}{N} = \frac{\Sigma t_i}{N} - t_1 \tag{10.20}$$

As we should expect, the lifetime τ would have been underestimated if we had neglected to take account of the cutoff at short times.

Example 10.1c. As a third example we consider a more realistic problem in which we have both short and long cutoffs on the observable path. We also assume that the unstable particles are produced at various locations within the target and with various momentum vectors **p**.

For this example, we must calculate the normalization integral, Equation (10.5), separately for each event with individual values for t_1 and t_2 determined from the minimum and maximum distance cutoffs, d_1 and d_2, respectively. The resulting expression for the likelihood function is

$$\mathscr{L}(\tau) = \prod_{i=1}^{N} A_i e^{-t_i/\tau} = \prod_{i=1}^{N} \left[\frac{e^{-t_i/\tau}}{\tau[e^{-t_1/\tau} - e^{-t_2/\tau}]} \right] \qquad (10.21)$$

with $\qquad M(\tau) = \ln[\mathscr{L}(\tau)]$

Setting to zero the derivative of $M(\tau)$ with respect to τ gives us the equation for the maximum-likelihood value of τ. However, the resulting equation cannot be solved analytically for τ although it could be solved by interpolation (see Appendix A). We choose, rather, to maximize $M(\tau)$ by a one-dimensional grid-search method because search methods are more generally applicable to maximum-likelihood problems and can readily be extended to multiple parameter problems.

10.2   COMPUTER EXAMPLE

Sample Data

We used a Monte Carlo program to generate simulated data for Example 10.1c. Two data sets were produced, a small 50-event set and a large 1000-event set, both with mean lifetime $\tau_0 = 0.8922 \times 10^{-10}$ s, and lower and upper fiducial cutoffs of $d_1 = 10.0$ cm and $d_2 = 40.0$ cm, respectively. To keep the problem simple we assumed that the decaying particles were produced on the x axis with momentum vector **p** along that axis, and that the production coordinate x and magnitude of momentum p followed Gaussian distributions. Figures 10.2 and 10.3 show the distributions of the times-of-flight of the generated particles in the two samples as they would have been determined in the apparatus.

Grid-Search Solution

We use a simple grid-search method to find the maximum-likelihood solution of this single parameter problem. Elements of the computer program are described in the following text and listed in Appendix E. The search begins at a starting value τ, supplied by the user. At each step τ is increment by a present amount $\Delta\tau$ and the logarithm of the likelihood function $M(\tau)$ is calculated from Equation (10.21), until $M(\tau)$ has passed through a maximum and has started to decrease. The program fits a parabola to the three points that bracket the

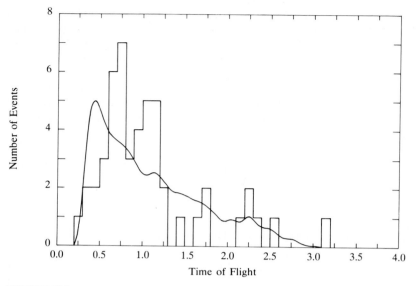

FIGURE 10.2
Histogram of times-of-flight of unstable particles in the apparatus for a 50-event experiment. The curve was generated with the Monte Carlo program based on the parameters obtained in maximum-likelihood fits to the data.

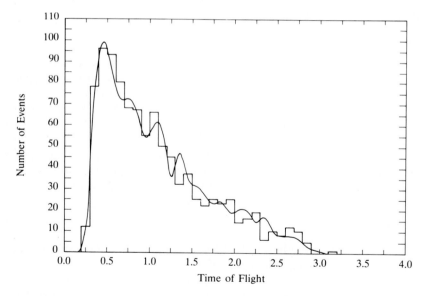

FIGURE 10.3
Histogram of times-of-flight of unstable particles in the apparatus for a 1000-event experiment. The curve was generated with the Monte Carlo program based on the parameters obtained in maximum-likelihood fits to the data.

maximum to find the value τ' at the maximum of $M(\tau)$. For a more complex problem, the program could be written to select the sign of the step size $\Delta\tau$ and to repeat the calculation with smaller values of $\Delta\tau$ to find an optimum estimate of τ', as in the fitting examples of Chapter 8. For multiparameter problems, either the grid- or gradient-search method of Chapter 8 could be adopted.

> **Program 10.1.** MaxLike A grid-search method to maximize the logarithm of the likelihood function of Equation (10.21). The routines have been written specifically for Example 10.1c.
>
> StartUp Assigns the input data files and sets the range of the search and graph parameters. (Not listed.)
>
> FetchData Reads production and decay coordinates xProds and xDecays, and pLab, the momentum three-vector of the decaying particle. Because we have assumed that the unstable particle is produced and travels along the x axis, only the x components of these vectors are nonzero. (Not listed.)
>
> Search Sets and increments the parameter tau and calls LogLike to calculate the logarithm of the likelihood function M. Terminates the search when M stops decreasing and starts to rise, indicating that the search has passed through a maximum. Fits a parabola to the last three points to find a better estimate of tau at the maximum.
>
> LogLike Calculates the logarithm of the likelihood function LogLikelihood Called from LogLike to calculate the logarithms of the probability densities for each measure variable.
>
> Error Calculates the uncertainty sigTau in tauAtMin, the maximum likelihood value of the parameter tau, by finding the change in tau needed to decrease M by $\Delta M = \frac{1}{2}$.
>
> PlotLikeCurve Calculates and plots as crosses the shape of the likelihood function in the region of the maximum. Calculates and plots a Gaussian curve with mean and standard deviation equal to tauMin and dTau. (Not listed.)

Results of the Fit

The fitting program gave $\tau' = 0.98 \pm 0.22$ for the best-fit value from the 50-event sample and $\tau' = 0.945 \pm 0.046$ for the 1000-event sample. A plot of the relative values of the likelihood function versus trial values of the parameter τ for the 1000-event sample is shown in Figure 10.4, with the data points indicated by crosses.

We have observed that, for a sufficiently large event sample, the likelihood function becomes Gaussian in the parameters in the vicinity of a χ^2 minimum (or a maximum of the likelihood function):

$$\mathscr{L}(\tau) \propto \exp\left(-\frac{(\tau - \tau')^2}{2\sigma^2}\right) \qquad (10.22)$$

where τ' is the value of the parameter τ that maximizes the likelihood function.

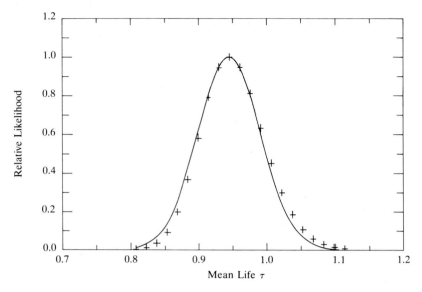

FIGURE 10.4
Relative values of the likelihood function versus trial values of the parameter τ for the 1000-event sample. The data points are indicated by crosses. The smooth Gaussian curve was calculated from Equation (10.10) with values of the mean and standard deviation of the fitted parameter obtained in the fit, $\tau' = 0.945$ and $\sigma = 0.046$.

This is illustrated by the Gaussian curve on Figure 10.4 that was calculated from Equation (10.22) with parameters found in the fit to the 1000-event sample. Both the data points and the Gaussian curves have been scaled to unit height at $\tau = \tau'$. The corresponding plot for the smaller 50-event sample (see Figure 11.1) departs considerably from the Gaussian form.

Uncertainties

We estimated the uncertainties in the parameters from our sample data by considering the parameter change necessary to decrease M by $\Delta M = \frac{1}{2}$ from its value at the maximum (corresponding to an increase of χ^2 by 1, or a change of $e^{-1/2}$ in the likelihood function $\mathscr{L}$). For multiparameter fits it is often useful to plot contours of χ^2 (or of M) as a function of pairs of the parameters to make this estimate. (See Chapter 11.) Because the likelihood function for the 1000-event sample closely followed the Gaussian form, the estimate should be satisfactory. The smaller 50-event sample (Figure 11.1) was skewed from the Gaussian, indicating that our estimate of the standard deviation might be somewhat low.

There are several other ways to estimate the uncertainty in a parameter after performing a maximum-likelihood fit. If the distribution of the likelihood function is sufficiently close to a Gaussian, we can find σ_τ from Equation (8.11):

$$\sigma_\tau^2 = \left(\frac{\partial^2 M(\tau)}{\partial \tau^2} \right)^{-1} \tag{10.23}$$

If it is not possible to calculate Equation (10.23) exactly (although it is possible for our example), we can find the second derivative by taking finite differences as discussed in Appendix A.

   If the likelihood function does not follow the Gaussian distribution, we can try a numerical integration of the likelihood function to find limiting values that include ~ 68.3% of the total area, corresponding to the 1 standard deviation limits. Alternatively, we may use a method suggested by Orear who points out that, for small event samples, where the likelihood function may not be very Gaussian-like, it may be preferable to calculate an average value of the second derivative through the equation

$$\overline{\frac{\partial^2 M}{\partial a^2}} = \frac{\int \left[\partial^2 M / \partial a^2 \right] \mathscr{L}(a) \, da}{\int \mathscr{L}(a) \, da} \tag{10.24}$$

where a is the unknown parameter and the integrals are over the allowable range of the parameter. This procedure has the advantage over the method of Equation (10.23), of giving more weight to the tails of the distribution in cases where they drop off more slowly than those of a Gaussian curve.

   Another method of determining the uncertainties in the parameters is to use a Monte Carlo calculation to produce simulated data sets, comparable to our measured data, and to use the method outlined in Chapter 11 for determining confidence levels for our results. This method has the advantage that it depends only on the assumptions made in the Monte Carlo generation, and not on any statistical expectations about the shape of the likelihood function. In many experiments, especially with those low statistics, it provides the most reliable estimate of parameter uncertainties.

Goodness of Fit

One disadvantage of the direct maximum-likelihood method is that there it does not provide a convenient test of the quality of the fit. The value at the peak of the likelihood function itself is not useful, because it represents only the maximized probability for obtaining our particular experimental result and we have no way of predicting the expected probability.

   An estimate of the goodness-of-fit can be obtained by making a histogram of the data and comparing it to a prediction based on our best estimate of the parameters. A Monte Carlo simulation of the experiment may be required to calculate the predicted distribution, with a χ^2 test to compare the data to the prediction.

   It is not always clear just which data variable should be histogrammed for this purpose. We would like to find that variable on which the parameters depend most strongly. For our sample problem, the lifetime τ in the rest frames of the particles is an obvious choice, because that is the variable we would choose if we were to solve the problem by the least-squares method. However, it might be wise to try plots of several variables to be sure that the fit is satisfactory. We should be aware that, because we did not actually minimize χ^2 with respect to the parameters in this method, a satisfactory value of χ^2 may be at best an indication that nothing is drastically wrong with the solution.

To illustrate the χ^2 comparison, we generated two 10,000 event "parent" distributions, one based on the parameters obtained from the fit to the 1000-event data sample and the other based on the parameters from the 50-event sample. The results were normalized to the corresponding data samples and plotted as smoothed curves on Figures 10.2 and 10.3. The χ^2 probabilities of $\sim 19\%$ for the 1000-event sample and $\sim 9\%$ for the 50-event sample are reasonable, and are probably acceptable as figures of merit. However, we should not rely heavily on the actual values of χ^2 obtained by this method, but rather treat them as guides to the quality of the fit. To explore the sensitivity of χ^2 to our fitted parameters, we could rerun our Monte Carlo program, varying the parameters to observe the effect on the calculated value of χ^2.

SUMMARY

Normalized probability density function:

$$P_i \equiv P(x_i; a_1, a_2, \ldots, a_m).$$

Likelihood function:

$$\mathcal{L}(a_1, a_2, \ldots, a_m) = \prod_{i=1}^{N} P_i$$

Single event probability density: $P_i = A_i \cdot p(x_i; a)$ where A_i is the detection efficiency and $p(x_i; a)$ is proportional to the interaction probability.
Logarithm of likelihood function: $M = \ln \mathcal{L} = \Sigma \ln P_i$.
Maximization of $\mathcal{L}$ or of M: $\partial \mathcal{L}/\partial a_j = 0$ or $\partial M/\partial a_j = 0$ for all a_j.
Gaussian form of likelihood function for large data sample:

$$\mathcal{L}(a_j) \propto \exp\left(-\frac{(a_j - a_j')^2}{2\sigma^2}\right)$$

Uncertainties in parameters:

$$\sigma_j^2 = \left(\frac{\partial^2 M(a_j)}{\partial a_j^2}\right)^{-1}$$

Method for low statistics:

$$\overline{\frac{\partial^2 M}{\partial a^2}} = \frac{\int[\partial^2 M/\partial a^2]\mathcal{L}(a)\,da}{\int\mathcal{L}(a)\,da}$$

EXERCISES

10.1. In a scattering experiment, the angles of the scattered particles are measured and the cosines of the angles in the center-of-mass rest frame of the incident and target particles are calculated and recorded. Fifty such measurements, drawn from the distribution $y(x) = a_1 + a_2 \cos^2 \theta$, are listed in the table. Use the direct maximum-likelihood method to determine the values of the parameters a_1 and a_2. Note that it is necessary to convert the distribution function $y(x_i)$ to a normalized probability function and that the normalization constant will be different for each pair of trial values of a_1 and a_2.

−0.999	−0.983	−0.956	−0.946	−0.933	−0.925	−0.916	−0.910
−0.881	−0.739	−0.734	−0.717	−0.715	−0.675	−0.665	−0.649
−0.621	−0.537	−0.522	−0.508	−0.499	−0.471	−0.460	−0.419
−0.403	−0.311	−0.305	−0.281	−0.170	−0.162	−0.063	0.214
0.438	0.444	0.508	0.586	0.638	0.677	0.721	0.730
0.768	0.785	0.790	0.793	0.877	0.896	0.931	0.938
0.948	0.993						

Because of the small amount of data, the uncertainties in the parameters a_1 and a_2 are so large that the values of the parameters are not very meaningful. Therefore, to complete the problem, you should use the Monte Carlo program written for Exercise 5.8 to generate 500 events and use your calculation to find the parameters from those data.

10.2. Students in an undergraduate physics laboratory determined the mass of the Λ hyperon by measuring graphically the energies and the momentum vectors of the proton and π meson into which the Λ hyperons decayed. Because of the large uncertainties in the measurements, the calculated square of the masses of the decaying particles forms a truncated Gaussian distribution that is limited on the low-mass side by $(M_p + M_\pi)^2 = 1.1617$ $(GeV/c^2)^2$, but is not limited on the high-mass side. The following 50 numbers represent squares of the calculated masses in units of $(GeV/c^2)^2$.

1.2981	1.2618	1.2145	1.2539	1.4230	1.3963	1.3701	1.2303	1.3655	1.2042
1.3190	1.2086	1.2118	1.2078	1.2726	1.2438	1.1838	1.1666	1.1908	1.1922
1.2525	1.3615	1.1855	1.2697	1.2044	1.3397	1.4317	1.2713	1.2203	1.2817
1.2046	1.2856	1.1980	1.2595	1.1721	1.2608	1.1689	1.4838	1.1743	1.2954
1.2586	1.2655	1.2316	1.2372	1.2969	1.2015	1.2000	1.1677	1.2080	1.1893

Use the direct maximum-likelihood method to fit a truncated Gaussian to these data to determine the maximum-likelihood value of the squared mass of the particle. A search in two-parameter space will be required since neither the mean nor the width of the distribution is known.

   Note that it is necessary to calculate numerically the normalization of the truncated Gaussian for each pair of trial values of the mean and standard deviation of the Gaussian function. It is advisable to set up a table of the integral of the standard Gaussian and to use interpolation to find the desired normalizations. A simple automatic or manual grid search will suffice for maximizing the likelihood function.

CHAPTER
11

TESTING
THE FIT

11.1 χ^2 TEST FOR GOODNESS OF FIT

The method of least squares is based on the hypothesis that the optimum description of a set of data is one that minimizes the weighted sum of the squares of the deviation of the data y_i from the fitting function $y(x_i)$. The sum is characterized by the variance of the fit s^2, which is an *estimate* of the variance of the data σ^2. For a function $y(x_i)$, which is linear in m parameters and is fitted to N data points, we have

$$s^2 = \frac{1}{N-m} \frac{\Sigma\{(1/\sigma_i^2)[y_i - y(x_i)]^2\}}{(1/N)\Sigma(1/\sigma_i^2)} = \frac{1}{N-m}\Sigma w_i[y_i - y(x_i)]^2 \quad (11.1)$$

where the factor $\nu = N - m$ is the number of degrees of freedom for fitting N data points (implied in the unlabelled sums) with m parameters and the weighting factors for each measurement are given by

$$w_i = \frac{1/\sigma_i^2}{(1/N)\Sigma(1/\sigma_i^2)} \quad (11.2)$$

for each data point are the inverse of the variances σ_i^2 that describe the uncertainties in each point, normalized to the average of all the weighting factors.

The variance of the fit s^2 is also characterized by the statistic χ^2 defined in Equation (7.5) for polynomials:

$$\chi^2 \equiv \sum \left\{ \frac{1}{\sigma_i^2} [y_i - y(x_i)]^2 \right\} \tag{11.3}$$

with

$$y(x_i) = \sum_{k=1}^{m} a_k f_k(x_i)$$

The relationship between s^2 and χ^2 can be seen most easily by comparing s^2 with the reduced chi-square χ_ν^2,

$$\chi_\nu^2 = \frac{\chi^2}{\nu} = \frac{s^2}{\langle \sigma_i^2 \rangle} \tag{11.4}$$

where $\langle \sigma_i^2 \rangle$ is the weighted average of the individual variances

$$\langle \sigma_i^2 \rangle = \frac{(1/N)\Sigma((1/\sigma_i^2)\sigma_i^2)}{(1/N)\Sigma(1/\sigma_i^2)} = \left[\frac{1}{N} \sum \frac{1}{\sigma_i^2} \right]^{-1} \tag{11.5}$$

and is equivalent to σ^2 if the uncertainties are all equal, $\sigma_i = \sigma$.

The parent variance of the data σ^2 is a characteristic of the dispersion of the data about the parent distribution and is not descriptive of the fit. The estimated variance of the fit about s^2, however, is characteristic of both the spread of the data and the accuracy of the fit. The definition of χ^2, as the ratio of the estimated variance s^2 to the parent variance σ^2 times the number of degrees of freedom ν, makes it a convenient measure of the goodness of fit.

If the fitting function is a good approximation to the parent function, then the estimated variance s^2 should agree well with the parent variance σ^2, and the value of the reduced chi-square should be approximately unity, $\chi_\nu^2 = 1$. If the fitting function is not appropriate for describing the data, the deviations will be larger and the estimated variance will be too large, yielding a value of χ_ν^2 greater than 1. A value of χ_ν^2 less than 1 does not necessarily indicate a better fit, however; it is simply a consequence of the fact that there exists an uncertainty in the determination of s^2, and the observed values of χ_ν^2 will fluctuate from experiment to experiment. A value of χ_ν^2 that is very small may indicate an error in the assignment of the uncertainties in the measured variables.

Distribution of χ^2

The probability distribution function for χ^2 with ν degrees of freedom is given by

$$P_x(x^2; \nu) = \frac{(x^2)^{1/2(\nu-2)} e^{-x^2/2}}{2^{\nu/2}\Gamma(\nu/2)} \tag{11.6}$$

The chi-square distribution of Equation (11.6) is derived in many texts on statistics[1] but we shall simply quote the results here.

The gamma function $\Gamma(n)$ is equivalent to the factorial function $n!$ extended to nonintegral arguments. It is defined for integral and half-integral arguments by the values at arguments of 1 and $\frac{1}{2}$ and a recursion relation:

$$\Gamma(1) = 1 \qquad \Gamma(\tfrac{1}{2}) = \sqrt{\pi} \qquad \Gamma(n + 1) = n\Gamma(n)$$

For integral values of n

$$\Gamma(n + 1) = n! \qquad n = 0, 1, \ldots$$

For half integral values of n

$$\Gamma(n + 1) = n(n - 1)(n - 2) \cdots (\tfrac{3}{2})(\tfrac{1}{2}\sqrt{\pi})$$

$$n = \tfrac{1}{2}, \tfrac{3}{2}, \tfrac{5}{2}, \ldots$$

$$(11.7)$$

Calculating factorial functions can lead to computer overflow problems. For computational purposes it is convenient to replace the factorial form of the gamma function by a form of Stirling's approximation[2]:

$$\Gamma[n] = \sqrt{2}\, e^{-n} n^{(n-1/2)}(1 + 0.0833/n) \tag{11.8}$$

This approximation, which is accurate to $\sim 0.1\%$ for all $n \geq \frac{1}{2}$, avoids both the problems of overflow in calculating factorials and the necessity of testing and choosing the appropriate form for integral or half-integral argument. The trade-off is computer speed. Calculating exponentials is rather slow compared to calculating factorials (especially in Pascal with its unfortunate lack of a power function), but high speed usually is not required for nonrepetitive calculations.

If the function of the parent population is denoted by $y_0(x)$, the value of χ_0^2 determined from the parameters of the parent function

$$\chi_0^2 = \sum \left\{ \frac{1}{\sigma_i^2} [y_i - y_0(x_i)]^2 \right\} \tag{11.9}$$

is distributed according to Equation (11.6) with $\nu = N$ degrees of freedom. If the function $y(x)$ used in the determination of χ^2 contains m parameters, the value of χ^2 calculated from Equation (11.3) is distributed according to Equation (11.6) with $\nu = N - m$ degrees of freedom.

More important for our purposes than the probability distribution $P_x(x^2; \nu)$ of Equation (11.6) is the integral probability $P(\chi^2; \nu)$ between $x^2 = \chi^2$ and

---

[1] See Pugh and Winslow, Section 12-5.

[2] "Review of Particle Properties," *Physics Letters*, vol. 170B, p. 53 (1986).

$x^2 = \infty$:

$$P_\chi(\chi^2; \nu) = \int_{\chi^2}^{\infty} P_x(x^2; \nu) \, dx^2 \tag{11.10}$$

Equation (11.10) describes the probability that a random set of n data points drawn from the parent distribution would yield a value of χ^2 equal to or greater than the tabulated value.

> **Program 11.1.** Chi2Prob Computation of the probability of Equation (11.10) by numerical integration. The program and subroutine are listed in Appendix E in the program unit FitUtil. Input to the program are the variables chi2 and nFree, corresponding to the value of χ^2 at the lower integration limit and the number of degrees of freedom, respectively.
>
> ChiProbDens Computes the function $P_x(x^2; \nu)$ of Equation (11.6) using Gamma to approximate the gamma function. This routine is provided as an example, but is not actually used in calculating the integrated probability.
> ChiProb Computes $P_\chi(\chi^2; \nu)$ of Equation (11.10) by numerically integrating the χ^2-probability density function by Simpson's rule with the routine Simpson. To reduce the number of computations and to take advantage of the logarithms in reducing variable overflow problems, the routine ChiProbDens is written as two separate routines, Chi, which contains the variation with the variable χ^2, and ChiProb, which is a function only of ν, and thus need only be called once by the routine. If variable overflow is still a problem (overflow occurs for $\nu > 56$), double-precision variables could be employed.
> The calculation returns the integral to an accuracy of about $\pm 0.1\%$. The trade-off on accuracy versus speed of computation is controlled by the value of the constant dx, the integration step.
> For the special case of 1 degree of freedom, $\nu = 1$, the χ^2-probability density function of Equation (11.6) takes the form
>
> $$P_x(x^2; \nu) = e^{-x^2/2}/(2\pi x^2)^{1/2}$$
>
> which is difficult to integrate numerically near $x = 0$. However, the integral is finite, and the function can be expanded in a Taylor series about $x = 0$ and integrated analytically. We use that technique for $\nu = 1$ and $\chi^2 < 2$.
> Similarly, for $\nu = 2$, where the function takes the form
>
> $$P_x(x^2; \nu) = e^{-x^2/2}/2$$
>
> the analytic form of the integral is used.

For a fitting function that is a good approximation to the parent function, the experimental value of χ_ν^2 should be average and the probability from Equation (11.10) should be approximately 0.5. For poorer fits, the values of χ_ν^2 will be larger and the associated probability will be smaller. There is an ambiguity in interpreting the probability because χ_ν^2 is a function of the quality of the data as well as the choice of parent function, so that even correct fitting functions occasionally yield large values of χ_ν^2. However, the probability of

Equation (11.10) is generally either reasonably close to 0.5, indicating a reasonable fit, or unreasonably small, indicating a bad fit. In fact, for most purposes, the reduced chi-square χ_ν^2 is an adequate measure of the probability directly. The probability will be reasonably close to 0.5 so long as χ_ν^2 is reasonably close to 1; that is, less than about 1.5.

> **Example 11.1.** Consider the solution of the problem of fitting two exponential curves plus a linear background to the data from the radioactive silver decay of Example 8.1. The fit (see Table 8.5) gave $\chi^2 = 66.1$ for 54 degrees of freedom, or $\chi_\nu^2 = 1.22$, with $P_\chi(\chi^2; \nu) = 12.4\%$. We can interpret this result in the following way. Assume that the parameters we found are, indeed, the parameters of the parent distribution. Then, suppose that we were to repeat our experiment many times, drawing many different data samples from that parent distribution. Our result indicates that in 12.4% of those experiments we should expect to obtain fits that are no better than those listed in Table 8.5.

11.2 LINEAR-CORRELATION COEFFICIENT

Let us assume that we have made measurements of pairs of quantities x_i and y_i. We know from the previous chapters how to fit a function to these data by the least-squares method, but we should stop and ask whether the fitting procedure is justified and whether, indeed, there *exists* a physical relationship between the variables x and y. What we are asking here is whether or not the variations in the observed values of one quantity y are *correlated* with the variations in the measured values of the other quantity x.

For example, if, as in Example 6.1, we were to measure the potential difference across segments of a current-carrying wire as a function of the segment length, we should find a definite and reproducible correlation between the two quantities. But if we were to measure the potential of the wire as a function of time, even though there might be fluctuations in the observations, we should not find any significant reproducible long-term relationship between the pairs of measurements.

On the basis of our discussion in Chapter 6, we can develop a quantitative measure of the degree of correlation or the probability that a linear relationship exists between two observed quantities. We can construct a linear-correlation coefficient r that will indicate quantitatively whether or not we are justified in determining even the simplest linear correspondence between the two quantities.

Reciprocity in Fitting x versus y

Our data consist of pairs of measurements (x_i, y_i). If we consider the quantity y to be the dependent variable, then we want to know if the data correspond to a straight line of the form

$$y = a + bx \tag{11.11}$$

We have already developed the analytical solution for the coefficient b, which represents the slope of the fitted line given in Equation (6.12):

$$b = \frac{N\Sigma x_i y_i - \Sigma x_i \Sigma y_i}{N\Sigma x_i^2 - (\Sigma x_i)^2} \qquad (11.12)$$

where the weighting factors in σ_i have been omitted for clarity. If there is no correlation between the quantities x and y, then there will be no tendency for the values of y to increase or decrease with increasing x, and, therefore, the least-squares fit must yield a horizontal straight line with a slope $b = 0$. But the value of b by itself cannot be a good measure of the degree of correlation because a relationship might exist that included a very small slope.

Because we are discussing the interrelationship between the variables x and y, we can equally well consider x as a function of y and ask if the data correspond to a straight line form

$$x = a' + b'y \qquad (11.13)$$

The values of the coefficients a' and b' will be different from the values of the coefficients a and b in Equation (11.11), but they are related if the variables x and y are correlated.

The analytical solution for the inverse slope b' is similar to that for b in Equation (11.12):

$$b' = \frac{N\Sigma x_i y_i - \Sigma x_i \Sigma y_i}{N\Sigma y_i^2 - (\Sigma y_i)^2} \qquad (11.14)$$

If there is no correlation between the quantities x and y, then the least-squares fit must yield a horizontal straight line with a slope $b' = 0$.

If there is a complete correlation between x and y, then there exists a relationship between the coefficients a and b of Equation (11.11) and between a' and b' of Equation (11.13). To see what this relationship is, we rewrite Equation (11.13):

$$y = -\frac{a'}{b'} + \frac{1}{b'}x = a + bx \qquad (11.15)$$

and equate coefficients

$$a = -\frac{a'}{b'} \qquad b = \frac{1}{b'} \qquad (11.16)$$

We see from Equation (11.16) that $bb' = 1$ for complete correlation. If there is no correlation, both b and b' are 0 and Equations (11.16) do not apply. We therefore define, as a measure of the degree of linear correlation, the experimental linear-correlation coefficient $r \equiv \sqrt{bb'}$:

$$r \equiv \frac{N\Sigma x_i y_i - \Sigma x_i \Sigma y_i}{\left[N\Sigma x_i^2 - (\Sigma x_i)^2\right]^{1/2}\left[N\Sigma y_i^2 - (\Sigma y_i)^2\right]^{1/2}} \qquad (11.17)$$

The value of r ranges from 0, when there is no correlation, to ± 1, when there is complete correlation. The sign of r is the same as that of b (and b'), but only the absolute magnitude is important.

The correlation coefficient r cannot be used directly to indicate the degree of correlation. A probability distribution for r can be derived from the two-dimensional Gaussian distribution, but its evaluation requires a knowledge of the correlation coefficient ρ of the parent population. A more common test of r is to compare its value with the probability distribution for the parent population that is completely uncorrelated; that is, for which $\rho = 0$. Such a comparison will indicate whether or not it is probable that the data points could represent a sample derived from an uncorrelated parent population. If this probability is small, then it is more probable that the data points represent a sample from a parent population where the variables are correlated.

For a parent population with $\rho = 0$, the probability that any random sample of uncorrelated expeimental data points would yield an experimental linear-correlation coefficient equal to r is given by[3]

$$P_r(r;\nu) = \frac{1}{\sqrt{\pi}} \frac{\Gamma[(\nu + 1)/2]}{\Gamma(\nu/2)} (1 - r^2)^{(\nu-2)/2} \tag{11.18}$$

where $\nu = N - 2$ is the number of degrees of freedom for an experimental sample of N data points. The gamma function for integral and half-integral values was defined in Equation (11.7).

Integral Probability

A more useful distribution than that of Equation (11.18) is the probability $P_c(r; N)$ that a random sample of N uncorrelated experimental data points would yield an experimental linear-correlation coefficient as large as or larger than the observed value of $|r|$. This probability is the integral of $P_r(r;\nu)$ for $\nu = N - 2$:

$$P_c(r; N) = 2 \int_{|r|}^{1} P_r(r;\nu)\, dr \qquad \nu = N - 2 \tag{11.19}$$

With this definition, $P_c(r; N)$ indicates the probability that the observed data could have come from an uncorrelated ($\rho = 0$) parent population. A small value of $P_c(r, N)$ implies that the observed variables are probably correlated.

Because Equation (11.19) cannot be integrated analytically, the function must be integrated either by making a series expansion of the argument and integrating term-by-term or by performing a numerical integration. With fast computers, the latter method is more convenient and generally applicable to such problems.

---

[3] For a derivation see Pugh and Winslow, Section 12-8.

Program 11.2. LcorProb Computation of the probability of Equation (11.19) by numerical integration. Input to the program are the variables rCorr and nObserv corresponding to the value of the experimental linear-correlation coefficient and the number of observations, respectively. (The number of degrees of freedom is the number of observations minus 2.) The program uses the following routines, which are listed in Appendix E and are part of the program unit Lcorlate.

LinCorrel Compute the function $P_r(r; N)$ of Equation (11.18) using the approximation of Equation (11.8) for the gamma function (calculated by the function Gamma in the program unit GenUtil). Because LinCorProb is intended to be used as an argument to the integration routine Simpson, it can have only one argument. The parameter ν is passed in the global variable pSimps by the calling routine.

LinCorProb Computes $P_c(r; N)$ of Equation (11.19) by numerically integrating LinCorrel by Simpson's rule. The calculation returns the integral to an accuracy of about ± 0.01. The trade-off on accuracy versus speed of computation is controlled by the value of the constant dx, the integration step.

Example 11.2. For the data of Example 6.1, the linear-correlation coefficient r can be calculated from Equation (11.17) with the data of Table 6.1:

$$r = \frac{9 \times 779.3 - 450.0 \times 12.44}{\sqrt{(9 \times 28,500 - 450.0^2) \times (9 \times 21.32 - 12.44^2)}}$$

$$= 0.9994$$

The probability for determining, from an uncorrelated population with $9 - 2 = 7$ degrees of freedom, a value of r equal to or larger than the observed value, can be calculated from Equation (11.19) (see Table C.3). The result $P_c(r; N) < 0.001\%$ indicates that it is extremely improbable that the variables x and V are linearly uncorrelated. Thus, the probability is high that the variables are correlated and the linear fit is justified.

Similarly, in the experiment of Example 6.2, the linear-correlation coefficient can be calculated from Equation (11.17) by including the weighting factors $\sigma_i^2 = y_i$ as in Table 6.2, so that, for example, N is replaced by Σw_i and Σx_i is replaced by $\Sigma w_i x_i$, and so forth:

$$r = \frac{0.03570 \times 81.02 - 0.1868 \times 10.}{\sqrt{(0.03570 \times 1.912 - 0.1868^2) \times (0.03570 \times 3693. - 10.^2)}}$$

$$= 0.9938$$

Again, the probability $P_c(r; N)$ for $r = +0.9938$ with $\nu = 10 - 2 = 8$ degrees of freedom is very small ($< 0.001\%$), indicating that the change in counting rate C is linearly correlated to a high degree of probability with $x = 1/r^2$, the inverse square of the distance between the source and counter.

11.3 MULTIVARIABLE CORRELATIONS

If the dependent variable y_i is a function of more than one variable,

$$y_i = a + b_1 x_{i1} + b_2 x_{i2} + b_3 x_{i3} + \cdots \tag{11.20}$$

we might investigate the correlation between y_i and each of the independent variables x_{ij} or we might also enquire into the possibility of correlation between different variables x_{ij}. Here, we use the first subscript i to represent the observation, as in the previous discussions, and the second subscript j to represent the particular variable under investigation. The variables x_{ij} could be different variables, or they could be functions of x_i, $f(x_i)$, as in Chapter 7. We shall rewrite Equation (11.17) for the linear-correlation coefficient r in terms of another quantity s_{jk}^2.

We define the *sample covariance* s_{jk}^2:

$$s_{jk}^2 \equiv \frac{1}{N-1} \sum \left[(x_{ij} - \bar{x}_j)(x_{ik} - \bar{x}_k) \right] \tag{11.21}$$

where the means $\bar{x}_j$ and $\bar{x}_k$ are given by

$$\bar{x}_j \equiv \frac{1}{N} \sum x_{ij} \quad \text{and} \quad \bar{x}_k = \frac{1}{N} \sum x_{ik} \tag{11.22}$$

and the sums are taken over the range of the subscript i from 1 to N. The weights have been omitted for clarity. With this definition, the sample variance for one variable s_j^2,

$$s_j^2 \equiv s_{jj}^2 = \frac{1}{N-1} \sum (x_{ij} - \bar{x}_j)^2 \tag{11.23}$$

is analogous to the sample variance s^2 defined in Equation (1.9):

$$s^2 = \frac{1}{N-1} \sum (x_{ij} - \bar{x}_j)^2 \tag{11.24}$$

It is important to note that the sample variances s_j^2 defined by Equation (11.23) are measures of the ranges of variation of the variables and not of the uncertainties in the variables.

Equation (11.21) can be rewritten for comparison with Equation (11.17) by substituting the definitions of Equation (11.22):

$$\begin{aligned}
s_{jk}^2 &\equiv \frac{1}{N-1} \sum \left[(x_{ij} - \bar{x}_j)(x_{ik} - \bar{x}_k) \right] \\
&= \frac{1}{N-1} \sum (x_{ij} x_{ik} - \bar{x}_j \bar{x}_k) \\
&= \frac{1}{N-1} \sum \left(x_{ij} x_{ik} - \frac{1}{N} \sum x_{ij} \sum x_{ik} \right)
\end{aligned} \tag{11.25}$$

If we substitute x_{ij} for x_i and x_{ik} for y_i in Equation (11.17), we can define the *sample linear-correlation coefficient* between any two variables x_j and x_k as

$$r_{jk} = \frac{s_{jk}^2}{s_j s_k} \tag{11.26}$$

with the covariances and variances s_{jk}^2, s_j^2, and s_k^2 given by Equations (11.23) and (11.25). Thus the linear-correlation coefficient between the jth variable x_j and the dependent variable y is given by

$$r_{jy} = \frac{s_{jy}^2}{s_j s_y} \tag{11.27}$$

Similarly, the linear-correlation coefficient of the parent population of which the data are a sample is defined as

$$\rho_{jk} = \frac{\sigma_{jk}^2}{\sigma_j \sigma_k} \tag{11.28}$$

where σ_j^2, σ_k^2, and σ_{jk}^2 are the true variances and covariances of the parent population. These linear-correlation coefficients are also known as product-moment correlation coefficients.

With these definitions we can consider either the correlation between the dependent variable and any other variable r_{jy} or the correlation between any two variables r_{jk}.

Polynomials

In Chapter 7 we investigated functional relationships between y and x of the form

$$y = a_0 + a_1 x + a_2 x^2 + a_3 x^3 + \cdots \tag{11.29}$$

In a sense, this is a variation on the linear relationship of Equation (11.20) where the powers of the single independent variable x are considered to be various variables $x_j = x^j$. The correlation between the independent variable y and the mth term in the power series of Equation (11.29), therefore, can be expressed in terms of Equations (11.23) through (11.27):

$$r_{my} = \frac{s_{my}^2}{s_m s_y}$$
$$s_m^2 = \frac{1}{N-1}\left[\sum x_i^{2m} - \frac{1}{N}\left(\sum x_i^m\right)^2\right]$$
$$s_y^2 = \frac{1}{N-1}\left[\sum y_i^2 - \frac{1}{N}\left(\sum y_i\right)^2\right] \tag{11.30}$$
$$s_{my}^2 = \frac{1}{N-1}\left[\sum x_i^m y_i - \frac{1}{N}\sum x_i^m \sum y_i\right]$$

Weighted Fit

If the uncertainties in the data points are not all equal ($\sigma_i \neq \sigma$), we must include the individual standard deviations σ_i as weighting factors in the definition of variances, covariances, and correlation coefficients. From Chapter 6 the prescription for introducing weighting is to multiply each term in the sum by $1/\sigma_i^2$.

The formula for the correlation remains the same as Equations (11.26) and (11.27), but the formulas of Equations (11.21) and (11.23) for calculating the variances and covariances must be modified:

$$s_{jk}^2 \equiv \frac{1/(N-1)\Sigma\left[(1/\sigma_i^2)(x_{ij}-\bar{x}_j)(x_{ik}-\bar{x}_k)\right]}{(1/N)\Sigma(1/\sigma_i^2)}$$

$$s_j^2 \equiv s_{jj}^2 = \frac{1/(N-1)\Sigma\left[(1/\sigma_i^2)(x_{ij}-\bar{x}_j)^2\right]}{(1/N)\Sigma(1/\sigma_i^2)}$$

(11.31)

where the means $\bar{x}_j$ and $\bar{x}_k$ are also weighted means

$$\bar{x}_j = \frac{\Sigma x_{ij}w_i}{N} = \frac{\Sigma(x_{ij}/\sigma_i^2)}{\Sigma(1/\sigma_i^2)}$$

The weighting factors

$$w_i = \frac{1/\sigma_i^2}{(1/N)\Sigma(1/\sigma_i^2)}$$

(11.32)

for each data point are the inverse of the variances σ_i^2 that describe the uncertainties in each point, normalized to the average of all the weighting factors.

Multiple-Correlation Coefficient

We can extrapolate the concept of the linear-correlation coefficient, which characterizes the correlation between two variables at a time, to include multiple correlations between groups of variables taken simultaneously. The linear-correlation coefficient r of Equation (11.17) between y and x can be expressed in terms of the variances and covariances of Equation (11.31) and the slope b of a straight-line fit given in Equation (11.12):

$$r^2 = \frac{s_{xy}^4}{s_x^2 s_y^2} = b\frac{s_{xy}^2}{s_y^2}$$

(11.33)

In analogy with this definition of the linear-correlation coefficient, we define the *multiple-correlation coefficient* R to be the sum over similar terms for the variables of Equation (11.20):

$$R^2 \equiv \sum_{j=1}^{n}\left(b_j\frac{s_{jy}^2}{s_y^2}\right) = \sum_{j=1}^{n}\left(b_j\frac{s_j}{s_y}r_{jy}\right)$$

(11.34)

The linear-correlation coefficient r is useful for testing whether one particular variable should be included in the theoretical function that is fitted to the data. The multiple-correlation coefficient R characterizes the fit of the data to the entire function. A comparison of the multiple-correlation coefficient for different functions is therefore useful in optimizing the theoretical functional form.

We shall discuss in the following sections how to use these correlation coefficients to determine the validity of including each term in the polynomial of Equation (11.29) or the series of arbitrary functions of Equation (11.20).

11.4   *F* TEST

As noted in the Section 11.1, the χ^2 test is somewhat ambiguous unless the form of the parent function is known, because the statistic χ^2 measures not only the discrepancy between the estimated function and the parent function, but also the deviations between the data and the parent function simultaneously. We would prefer a test that separates these two types of information so that we can concentrate on the former type. One such test is the *F* test, which combines two different methods of determining a χ^2 statistic and compares the results to see if their relation is reasonable.

F distribution

If two statistics χ_1^2 and χ_2^2, which follow the χ^2 distribution, have been determined, the ratio of the reduced chi-square $\chi_{\nu 1}^2$ and $\chi_{\nu 2}^2$ is distributed according to the *F* distribution[4]

$$P_f(f; \nu_1, \nu_2) = \frac{\chi_1^2/\nu_1}{\chi_2^2/\nu_2} \tag{11.35}$$

with probability density function

$$P_f(f; \nu_1, \nu_2) = \frac{\Gamma[(\nu_1 + \nu_2)/2]}{\Gamma(\nu_1/2)\Gamma(\nu_2/2)} \left(\frac{\nu_1}{\nu_2}\right)^{\nu_1/2} \frac{f^{1/2(\nu_1 - 2)}}{(1 + f\nu_1/\nu_2)^{1/2(\nu_1 + \nu_2)}} \tag{11.36}$$

where ν_1 and ν_2 are the numbers of degrees of freedom corresponding to χ_1^2 and χ_2^2. By the definition of χ^2 [see Equation (11.3)], a ratio of ratios of variances

$$\frac{\chi_{\nu_1}^2}{\chi_{\nu_2}^2} = \frac{s_1^2/\sigma_1^2}{s_2^2/\sigma_2^2} \tag{11.37}$$

is also distributed as *F*, where s_1 and s_2 are experimental estimates of standard deviations σ_1 and σ_2 pertaining to some characteristic of the same or different distributions.

---

[4] See Pugh and Winslow, Section 12-7, for a derivation.

As with our tests of χ^2 and the linear-correlation coefficient r, we shall be more interested in the integral probability

$$P_F(F; \nu_1, \nu_2) = \int_F^\infty P_f(f; \nu_1, \nu_2)\, df \tag{11.38}$$

which describes the probability of observing such a large value of F from a random set of data when compared to the correct fitting function. The integral function $P_F(F; \nu_1, \nu_2)$ is tabulated and graphed in Table C.5 for a wide range of F, ν_1, and ν_2.

A word of caution is in order concerning the use of these tables. Because the statistic F in Equation (11.35) is defined as the ratio of two determinations of χ^2 without specifying which must be in the numerator, we can define two statistics F_{12} and F_{21},

$$F_{12} = \frac{\chi_{\nu 1}^2}{\chi_{\nu 2}^2} \qquad F_{21} = \frac{\chi_{\nu 2}^2}{\chi_{\nu 1}^2} = \frac{1}{F_{12}} \tag{11.39}$$

which must both be distributed according to the F distribution.

If in some experiment our calculations yield a particular value of F_{12}, we can use Table C.5 to determine whether such a large value is less than 5% probable (Table C.6 and Figure C.6) or less than 1% probable (Table C.7 and Figure C.7). If the test value is less than the tabulated values, we must also make sure that it is not too small. To do this, we compare the value

$$F_{21} = 1/F_{12} \tag{11.40}$$

to the same tables and graphs, noting that the values of ν_1 and ν_2 are reversed. The values of ν_1 and ν_2 specified in Table C.5 correspond to the degrees of freedom for the numerator and denominator of Equation (11.39), respectively.

Example 11.3. Suppose that $F_{12} = 0.2$ with $\nu_1 = 2$ and $\nu_2 = 10$. For Table C.6, the observed value of F_{12} may be as high as 4.10 and still be exceeded by about 5% of random observations. Similarly, we compare $F_{21} = 1/F_{12} = 5.0$ with the 5% point for $\nu_1 = 10$ and $\nu_2 = 2$, which has a value of 19.4. Because the values of F_{12} and F_{21} are well within the 5% limits, we can have confidence in the fit.

What we are estimating in this example is the probability $P_F(F_{12}; \nu_1, \nu_2)$ that F_{12} is not too large and the probability $P_F(1/F_{12}; \nu_2, \nu_1)$ that F_{12} is not too small. It is tempting to simplify this procedure by assuming that

$$P_F(1/F_{12}; \nu_2, \nu_1) = P_F(F_{12}; \nu_1, \nu_2) \tag{11.41}$$

so that our test consists of determining F such that

$$P_F(F; \nu_1, \nu_2) = 0.05$$

with the requirement that

$$F > F_{12} > 1/F$$

This approximation is valid for reasonably large values of ν_1 and ν_2 but not for small values of either, as in the preceding example, where we have $4.10 > F_{12} > 1/19.4$.

Multiple-Correlation Coefficient

There are two types of F tests that are normally performed on least-squares fitting procedures. One is designed to test the entire fit and can be related to the multiple-correlation coefficient R. The other, to be discussed later, tests the inclusion of an additional term in the fitting function.

If we consider the sum of squares of deviations S_y^2 associated with the spread of the data points around their mean (omitting factors of $1/\sigma_i^2$ for clarity),

$$S_y^2 = \sum (y_i - \bar{y})^2 \tag{11.42}$$

this is a statistic that follows the χ^2 distribution with $N - 1$ degrees of freedom (only one parameter $\bar{y}$ must be determined from the N data points). It is a characteristic of quantities that follow the χ^2 distribution that they may be expressed as the sum of other quantities that also follow the χ^2 distribution such that the number of degrees of freedom of the original statistic is the sum of the numbers of degrees of freedom of the terms in the sum.

By suitable manipulation and rearrangement, it can be shown that S_y^2 can be expressed as the sum of the two terms,

$$S_y^2 = \sum (y_i - \bar{y})^2 = \sum_{j=1}^{m} \left[(y_i - \bar{y}) \sum_{j=1}^{m} a_j (f_j - \bar{f}_j) \right] + \sum_{j=1}^{m} \left(y_i - \sum a_j f_j \right)^2$$

$$= \sum_{j=1}^{m} \left[a_j \sum \left[(y_i - \bar{y})(f_j - \bar{f}_j) \right] \right] + \sum [y_i - y(x_i)]^2 \tag{11.43}$$

where the fitting function is of the form

$$y(x_i) = \sum_{j=1}^{m} a_j f_j (x_i) \tag{11.44}$$

and we have

$$\bar{f}_j = \frac{1}{N} \sum f_j (x_i) \tag{11.45}$$

The left-hand side of Equation (11.43) is distributed as χ^2 with $N - 1$ degrees of freedom. The right-hand term is our definition of χ^2 from the Equation (11.3) and has $N - m$ degrees of freedom. Consequently, the middle term must be distributed according to the χ^2 distribution with $m - 1$ degrees of freedom.

By comparison with our definition of the multiple-correlation coefficient R in Equation (11.34), we can express this middle term as a fraction R^2 of the statistic S_y^2:

$$\sum_{j=1}^{m} a_j \sum \left[(y_i - \bar{y})(f_j - \bar{f}_j) \right] = R^2 \sum (y_i - \bar{y})^2 \tag{11.46}$$

Equation (11.43) becomes

$$\sum (y_i - \bar{y})^2 = R^2 \sum (y_i - \bar{y})^2 + (1 - R^2) \sum (y_i - \bar{y})^2 \tag{11.47}$$

or
$$S_y^2 = R^2 S_y^2 + (1 - R^2) S_y^2 \tag{11.48}$$

where, as before, both terms on the right-hand side are distributed as χ^2, the first with $m - 1$ degrees of freedom and the second with $N - m$ degrees of freedom.

Thus, the physical meaning of the multiple-correlation coefficient becomes evident. It divides the total sum of squares of deviations S_y^2 into two parts. The first reaction of S_y^2 is a measure of the spread of the dependent and independent variable data space. The second fraction, $(1 - R^2) S_y^2$, is the sum of squares of the deviations about the regression and represents the agreement between the fit and the data.

From the definition of Equation (11.35), we can define a ratio F_R of the two terms in the right-hand side of Equation (11.47) that follow the F distribution with $\nu_1 = m - 1$ and with $\nu_2 = N - m$ degrees of freedom,

$$F_R = \frac{R^2/(m-1)}{(1 - R^2)/(N - m)} = \frac{R^2}{(1 - R^2)} \times \frac{(N - m)}{(m - 1)} \tag{11.49}$$

From this definition of F_R in terms of the multiple-correlation coefficient R, it is clear that a large value of F_R corresponds to a good fit, where the multiple correlation is good and $R \simeq 1$. The F test for this statistic is actually a test that the coefficients are 0 ($a_j = 0$). So long as F_R exceeds the test value for F, we can be fairly confident that our coefficients are nonzero. If, on the other hand, $F_R < F$, we may conclude that at least one of the terms in the fitting function is not valid, is decreasing the multiple correlation by its inclusion, and should have a coefficient of 0.

The routine MultCor, listed in Appendix E, calculates the multiple-regression coefficient R (rMul) from Equation (11.34) and calculates F_R from Equation (11.49).

Test of Additional Term

Because of the additive nature of functions that obey the χ^2 statistics, we can form a new χ^2 statistic by taking the difference of two other statistics that are distributed as χ^2. In particular, if we fit a set of data with a fitting function with m terms, the resulting value of chi-square associated with the deviations about the regression $\chi^2(m)$ has $N - m$ degrees of freedom. If we add another term to the fitting function, the corresponding value of chi-square $\chi^2(m + 1)$ has $N - m - 1$ degrees of freedom. The difference between these two must follow the χ^2 distribution for 1 degree of freedom.

If we form the ratio of the difference $\chi^2(m) - \chi^2(m + 1)$ to the new value $\chi^2(m + 1)$, we can form a statistic F_χ that follows the F distribution with $\nu_1 = 1$ and $\nu_2 = N - m - 1$:

$$F_\chi = \frac{\chi^2(m) - \chi^2(m + 1)}{\chi^2(m)/(N - m - 1)} = \frac{\Delta\chi^2}{\chi_\nu^2} \qquad (11.50)$$

This ratio is a measure of how much the additional term has improved the value of the reduced chi-square and should be small when the function with the $m + 1$ terms does not significantly improve the fit over the function with m terms. Thus, we can be confident in the relative merit of the new terms if the value of F_χ is large. As for F_R, this is really a test of whether the coefficient for the new term is 0 ($a_{m+1} = 0$). If F_χ exceeds the test value for F, we can be fairly confident that the coefficient should not be 0 and the term, therefore, should be included. Table C.5 and Figure C.5 are useful for testing F_χ. They give the value of F corresponding to various values of the probability $P_F(F; 1, \nu_2)$ and various values of ν_2 for the case where $\nu_1 = 1$. Thus, rather than evaluating F for critical values of the probability (for example, 5% or 1%), we can evaluate the probability corresponding to the observed value of F_χ.

A calculation of F_χ could be built into a linear regression program and the resulting value compared to a supplied test value F, to indicate whether or not the last term in the series is justified, and therefore, to determine how many terms in the series should be included in the fit. However, it is probably safer, except possibly in a large, well debugged production run involving fitting polynomials to many similar data sets, to examine the individual values of χ^2 along with F_χ and to adjust the number of terms in the calculation manually. One should, however, be aware that the important figure of merit for added terms is the difference of the two values of χ^2 divided by the new value χ_ν^2 of the *reduced* chi-square.

11.5 CONFIDENCE INTERVALS

The object of data fitting is to obtain values for the parameters of the fitted function, and the uncertainties in the parameters. The quality of the fit is indicated by χ^2 and its associated probability, and the uncertainties give the probabilities that our values of the fitted parameters are good estimates of the parent parameters. Whether we estimate our parameters by the least-squares method or by direct application of the maximum-likelihood method, as discussed in Chapter 10, we must always estimate the uncertainty in our parameters to indicate numerically our confidence in our results.

Generally, we assume Gaussian statistics and quote the standard deviation σ in a result, where σ appears in the Gaussian probability density function

$$P_G(x;\mu,\sigma) = \frac{1}{\sigma\sqrt{2\pi}} \exp\left[-\frac{1}{2}\left(\frac{x-\mu}{\sigma}\right)^2\right] \tag{11.51}$$

and determines the width of the distribution. As noted in Chapter 2, approximately 68.3% of the events of the Gaussian distribution fall within $\pm\sigma$ of the mean μ and approximately 95.4% fall within $\pm 2\sigma$.

Confidence Level for One-Parameter Fit

One way of looking at the 1 standard deviation limit is to consider that, in a series of repeated experiments, there is approximately a 68% chance of obtaining values within $\pm\sigma$ of the mean μ. Of course, we usually do not know μ, and perhaps not σ either, but have determined experimentally only $\bar{x}$ and s, our estimate of the parameters. However, as long as our experimental estimates $\bar{x}$ and s are reasonably close to the true values μ and σ, we can state that there is approximately a 68% probability that the true value of the measured parameter lies between $\bar{x} - s$ and $\bar{x} + s$, or that at the 68.3% *confidence level*, the true value of the parameter lies between these two limits.

We may wish to quote all results in terms of other confidence levels. For example, we refer to the $\pm 2\sigma$ limit as the 95.4% confidence interval, or we may quote a 99% or 99.9% confidence level for a high precision experiment. The conventional 1σ and 2σ limits are based on the Gaussian distribution, which may or may not apply to the data in question, and even an experimental distribution that nominally follows Gaussian statistics is apt to deviate in the tails.

For any distribution, represented by the normalized probability density function, $P_x(x;\mu)$, we determine the probability that a measurement of the parameter will fall between $\bar{x} - a$ and $\bar{x} + b$ by the integral

$$P_x = \int_{\bar{x}-a}^{\bar{x}+b} P_x(\bar{x};x)\, dx \tag{11.52}$$

and could quote a confidence level of P_x that the "true" value of the measured parameter is between these two values. Note that we have not specified a region that is symmetrical about the mean. The uncertainties in our measurements may not be symmetrical, although the asymmetry may be hidden if we assume Gaussian statistics in our calculations. For example, the routines for finding uncertainties in parameters found by least-squares fitting (Chapters 7 and 8) generally assume a Gaussian distribution of the parameters and hence produce a single number of the uncertainties.

11.4. As an example of an asymmetrical probability distribution, consider the 50-event data sample of Example 10.1c. In Figure 11.1 we plot as crosses the

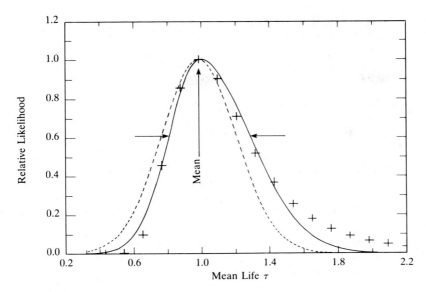

FIGURE 11.1
Relative values of the likelihood function versus trial values of the mean lifetime parameter τ for the 50-event sample from Chapter 10. The data points are indicated by crosses. The dashed curve was calculated from Equation (10.10) with parameters obtained in the fit. The solid curve is a composite of two Gaussian curves as explained in the text.

scaled values of the likelihood function for these data as a function of trial values of the parameter τ. The data points exhibit a marked asymmetry about the mean τ'. The dashed curve was calculated from Equation (10.10) with parameters obtained from the fit.

To make a better determination of σ from this curve, we consider the regions on each side of the mean separately and estimated two separate standard deviations, σ_L and σ_R, with the aid of Equation (1.11). To reduce the effect of the right-hand side tail on the value of σ_r, we imposed a cutoff at $\tau = 1.6$ and used only those data points below the cutoff in this calculation.

A composite curve formed of two Gaussians with the same mean τ but different values of σ is shown as the solid curve in Figure 11.1. It would be reasonable to consider the two values of σ obtained in this way as appropriate estimates of the uncertainty in τ, so that we could report $\tau' = 0.98^{+0.30}_{-0.24}$, as indicated by the arrows on Figure 11.1 rather than $\tau' = 0.98 \pm 0.22$ as we did in Chapter 10. This is equivalent to finding the two positions at which the logarithm of the likelihood function has decreased by $\Delta M = \frac{1}{2}$ as discussed in Section 10.2. Clearly this result is somewhat subjective if either side of the curve does not follow the Gaussian form. For this example, the value of σ_R depends on how much of the tail is included in the calculation.

Confidence Levels for Multiparameter Fits

The definition of the confidence level in a one-parameter experiment is generally straightforward. We can plot our data and observe if the distribution is Gaussian and estimate directly from the distribution of the probability that the true result lies between two specified values. When two or more variables have been determined and those variables exhibit some correlation, the definition of the confidence level becomes a little more difficult. Consider, for example, the determination of the mean lives τ_1 and τ_2 of two unstable silver isotopes of Example 8.1. The problem was treated in Chapter 8 as a five-parameter problem, with parameters a_4 and a_5 corresponding to the two mean lives, τ_1 and τ_2, respectively, and parameters a_1, a_2, and a_3 corresponding to the amplitudes of a uniform background and the two decaying states. The parameters of most interest in the experiment are a_4 and a_5, and we want to define a joint confidence interval for these two variables.

Figure 11.2 shows two set of contours for the variation of χ^2 as a function of a_4 and a_5 from the least-squares fit by the Marquardt method discussed in Chapter 8. The small contours, drawn with solid lines, were calculated by holding the parameters a_1, a_2, and a_3 fixed at their optimum values (see Table 8.5) and varying a_4 and a_5 to obtain increases in χ^2 of 1, 2, and 3 from the

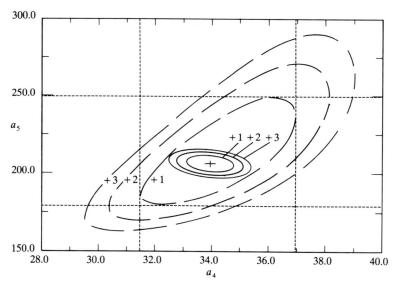

FIGURE 11.2

Two sets of contours for the variation of χ^2 with parameters a_4 and a_5 in the region of the χ^2 minimum. Data are from the least-squares fit by the Marquardt method discussed in Chapter 8. The small contours, drawn with solid lines, were calculated by holding parameters a_1 through a_3 fixed at their optimum while varying a_4 and a_5 to obtain increases in χ^2 of 1, 2, and 3 from the minimum values. The large contours, shown as dashed lines, were calculated by allowing a_1, a_2, and a_3 to vary to minimize χ^2 for each pair of values of a_4 and a_5.

minimum value. The large contours, shown as dashed lines, were calculated by allowing a_1, a_2, and a_3 to vary to minimize χ^2 for each pair of values of a_4 and a_5. The contour plots cover very different ranges because of the correlations of the displayed parameters, a_4 and a_5, with the remaining parameters a_1 through a_3. The tilt of the closed figures on each plot indicates the degree of correlation of parameters a_4 and a_5 with each other. In an ideal experiment, the contours are ellipses in the region of the χ^2 minimum and if a_4 and a_5 are not correlated, then, with suitable scaling of the axes, the ellipses are circles.

Which plot should we use? Additionally, how do we determine a confidence interval; that is, a region of the a_4-a_5 space in which we estimate there is, for example, a $\sim 68\%$ probability of finding the true values of the two parameters?

First, we should note that, because the fitting function, Equation (8.2), is not linear in the parameters, the methods of testing described in the previous sections strictly do not apply. However, we are much more likely to run into nonlinear fitting problems than the easier linear problems, so we shall continue with this example. At any rate, the function is linear in parameters a_1 through a_3, and we could make a linear expansion of it, over a limited region, in the parameters a_4 and a_5. In fact, this was the basis of a method of fitting nonlinear functions in Chapter 8.

Then, we should use the larger of the two contour diagrams to define our confidence intervals. That implies that if we wish to find the standard deviation of a_4 from the contour plot, we should consider the full range of the outer limit of the $\Delta\chi^2 = 1$ contour, and not the intersection of that contour with the a_4 axis. This is equivalent to allowing a_5 to assume its best values for each chosen value of a_4, as we have already assumed for the parameters a_1 through a_3. The two dashed vertical lines indicate the two limits on a_4 that include the 1 standard deviation, or 68.3% of the probability, and the two horizontal lines indicate the 1 standard deviation limits for a_5.

How do we know that the vertical lines enclose 68.3% of the probability? By allowing the four parameters a_1, a_2, a_3, and a_5 to find their optimum values for each chosen value of a_4 and varying a_4, we have separated our χ^2 fitting problem into two parts: a fit of N data points to $m - 1$ parameters with $N - m - 1$ degrees of freedom and a variation of $\Delta\chi^2$ with a_4 about the minimum χ^2, with 1 degree of freedom. As we observed in the previous section, the two variations separately must follow their appropriate χ^2 distributions, so our variation of $\Delta\chi^2$ obeys the χ^2 probability distribution for 1 degree of freedom. If we look at the integrated probability distribution P_χ for 1 degree of freedom [Table C.4, or calculated from Equation (11.10)], we see that $\chi^2 \geq 1$ corresponds to 31.7% of the probability, or $\Delta\chi^2 < 1$ corresponds to 68.3%. Similarly, if we wish to find the limits for 2 standard deviations, we should find the limits of a_4 on the $\Delta\chi^2 = 4$ contour, with all other parameters optimized.

To find the 1 standard deviation region encompassed by the *joint variation* of two parameters, a_4 and a_5, with all other parameters optimized, we must draw the contour corresponding to that value of $\Delta\chi^2$ for 2 degrees of freedom that includes 68.3% of the probability. Referring again to Table C.4 or Equation (11.10), we find that we should draw the contour for $\Delta\chi^2 = 2.30$, and for the 2 standard deviation contour, we should choose $\Delta\chi^2 = 6.14$. Joint confidence intervals with more than two parameters are often of interest, but are difficult to display and are represented best by two-dimensional projections of contours for pairs of variables.

Confidence Level for a Predicted Value

Suppose the predicted value of a physical quantity is $\mu = 1000.0$, and we have made a measurement and obtained the value $\bar{x} = 999.4 \pm 2.0$. At what confidence level is the predicted value consistent with our measurement? The question could be rephrased as, "What is the probability of obtaining from the predicted parent distribution a distribution that is as bad as the one we got, or worse?" Because the shape of the parent distribution was not predicted, but only the value of the mean, we must use our value of the standard deviation, $\mu \simeq 2.0$, as an estimate of that of the parent distribution. If the distribution is known to follow Gaussian statistics, then the required confidence is twice the integral of the standard Gaussian probability function from $x = \delta$ to ∞, where $\delta = (10002 - 9994)/2.0$.

Now, suppose that the predicted value was necessarily positive—an intensity, for example. Then, we might again assume a Gaussian distribution, but only for positive values of the variable x, and therefore our confidence integral becomes the integral of the standard Gaussian from δ to ∞. However, because the total probability must be normalized to 1, we again multiply the integral by 2 so that the probability or confidence level is the same for both problems.

The method of determining the confidence level thus depends on the type of problem as well as the probability function that is applicable to the problem. For distributions that are symmetrical about their means, such as the Gaussian distribution, we generally consider the probability of obtaining a result that is the specified number of standard deviations from the mean, without regard to sign, unless a particular sign is excluded by the physical problem. For distributions such as the chi-square and Poisson distributions, which are only defined for positive values of their arguments, it is conventional to find a "one-sided" probability as in the case of the χ^2 distribution where we quote the probability of obtaining a value as large as or larger than the value we obtained for a given number of degrees of freedom.

11.6   MONTE CARLO TESTS

A Monte Carlo calculation can help us understand the statistical significance of our results and possibly obtain a better estimate of some of the parameters of the experiment. As a by-product, the Monte Carlo program may also help us identify biases in our analysis procedure.

Suppose, for example, that we have measured a quantity x that is predicted to have a value μ. From our experiment we obtain the value $\bar{x}$ for our estimate of μ. We want to find the probability of obtaining in our experiment a value x that differs from the predicted value μ by

$$\Delta x \geq |\mu - \bar{x}| \tag{11.53}$$

We can set up a Monte Carlo program to simulate our experiment and to generate events with the parameters predicted by the theoretical principle that we are testing and with the same cuts as those imposed by our experimental apparatus. Such a program can be quite complex, but it may already exist at the time of analysis, if, for example, a Monte Carlo program was written to help plan the experiment. Or it might be possible to use some geometric and kinematic quantities from the actual experiment and only generate those parts of each event that are affected by the parameters in question.

After the Monte Carlo program has been written and debugged, we can simulate repeated experiments with the same parent parameters and the same number of final measurements as in our real experiment. The data from each of these simulated experiments is processed by our regular analysis program to obtain the final result, $\bar{x}$ in our example.

It is then an easy matter to examine our results to see what fraction of the simulated experiments satisfies the inequality of Equation (11.53), and thus to find an estimate of the required probability.

> **Example 11.5.** Let us use the Monte Carlo method to try to learn more about the significance of the small peak (peak 2) in our data of Example 9.2. We assumed that main peak (peak 1) in Figure 9.2 had been established to have a mean $\mu_1 = 1.000$ GeV and a width $\Gamma_1 = 0.200$ GeV, and that peak 2 has been predicted to have a mean $\mu_2 = 0.850$ GeV and a width of $\Gamma_2 \simeq 0.100$ GeV. The summary of the results of the least-squares fit to the distribution in Table 9.1 lists the mean of the peak from the fit to be $\mu_2 = a_8 = (0.855 \pm 0.015)$ GeV and the width to be $\Gamma_2 = a_9 = (0.086 \pm 0.069)$ GeV, consistent with the predicted values. The fitted amplitude of the peak was $A_2 = a_7 = 3.7 \pm 1.8$, corresponding to a 2 standard deviation hump on the side of the main, established peak.

Because our objective is to discover if our observation of the smaller peak (peak 2) could have been spurious, we begin with the hypothesis that the peak does *not* exist. We generate with a Monte Carlo program two-hundred 2000-event "single-peak experiments" from the distribution of Equation (9.1), which includes only the established peak 1. The mean energy (μ_1) and half-width (Γ_1) of peak 1 and the amplitudes a_1 through a_3 of the quadratic background, and A_1 of the peak were set to the values obtained in the six-parameter fit, listed in Table 9.1.

We attempted to fit to each data set the double-peak Equation (9.13) and to determine the parameters of peak 2 in exactly the same manner as we did for the original data in Example 9.4. The coefficients a_1 through a_3 of the

TABLE 11.1
Results of fitting two sets of 200 Monte Carlo experiments, each with 2000 events, to Equation (9.13)†

	Single-peak data		Double-peak data	
	Mean value	Standard deviation	Mean value	Standard deviation
μ_2	0.86	1.83	0.85	0.02
Γ_2	0.044	0.061	0.095	0.076
A_2	0.86	1.83	3.96	2.02
A_2/σ	0.20	0.66	1.4	1.0

† The single-peak data were generated from Equation (9.1) and the double-peak data from Equation (9.13).

polynomial background and the amplitude $a_7 = A_2$, mean $a_8 = \mu_2$, and half-width $a_9 = \Gamma_2$ of the possible second peak were treated as free parameters of the fit, as they were in the analysis of Example 9.4. We began each search with the starting values of the peak and half-width of peak 2 at the predicted values, $a_8 = 0.850$ and $a_9 = 0.100$, just as we did in analyzing the original expeirmental data.

For comparison, we generated and fitted two-hundred 200-event "double-peak experiments" from the distribution of Equation (9.13) using the predicted values of the mean and half-width of peak 2 and an amplitude $A_2 = 3.7$ corresponding to the value observed in our experiment.

The mean values and standard deviations of some of the parameters from the fit to the two distributions are listed in Table 11.1. The distributions of values of the mean energy parameter a_8 (μ_2) and the half-width parameter a_9 (Γ_2) found by the fitting program from the two data sets are essentially indistinguishable. Both are broad distributions, centered on the predicted mean mass μ_2 or half-width Γ_2. The two distributions of the amplitude factors a_7 (A_2) are different with the distribution from the single-peak experiment centered at $a_7 \simeq 0$ and the distribution from the double-peak experiment centered at $a_7 \simeq 4$. However, the distributions are so broad that in any one observation from the single-peak experiment there is a high probability of observing a spurious peak of reasonable magnitude.

The distributions of the ratios of the fitted amplitude a_7 to its uncertainty σ_7 suggests a way of distinguishing a real from a spurious peak. The ratio from the experiment was $3.7/1.8 = 2.1$, and only about $1/200$ of the simulated single-peak experiments produce a ratio as large as 2. (There is a somewhat higher probability of obtaining a low ratio from the double-peak experiment.) This result is consistent with our estimate in Chapter 9 of a 1.6% probability that our observed bump in the 0.75- to 0.95-GeV region is a fluctuation in the background.

The fact that our fitting program finds a peak at just the predicted mean energy should make us suspicious of a bias in our fitting procedure, and, indeed, if we repeat the fits to our Monte Carlo data with different starting values of $a_8(\mu_2)$, say $a_8 = 0.70$, then we get the same sort of results but with mean values of a_8 from the fits close to the new starting values.

It would, at this point, be wise to refit the real data of Example 9.2 with different starting values for the parameters, particularly for a_8. The result of such a test indicates that, as long as our selection of starting values is within the region of the predicted peak, say between 0.70 and 0.95 GeV, then the fitting routine does find the predicted peak at 0.85 GeV, but if the starting value is outside this region, the program finds a peak at one of the statistical fluctuations that are evident in Figure 9.2.

We are thus lead to essentially the same conclusion that we drew in Chapter 9, that if there is strong theoretical or other reason for believing that there is a peak near 0.850 GeV, then the effect we observe in our data is consistent with that prediction and we are justified in quoting the parameters of the peak. However, without extra information to select the energy region we should be obliged to conclude that our result is consistent with a statistical fluctuation in the background events.

We offer a final word of caution on using the Monte Carlo technique to study the statistical significance of experimental results. For Examples 9.2 and 11.5 we used a very simple problem to illustrate this technique. Yet, there are many opportunities for errors, which can lead to erroneous conclusions about the significance of our Chapter 9 data. In a larger study, it would be very easy to make a simple mistake that might lie undetected in the program and have a subtle effect on the results. It is important to test the program under a variety of conditions, and to examine results at intermediate stages before drawing conclusions from the result. In particular, if the results of the program lead to conclusions that violate intuition about the experiment, we should check and recheck the calculation. The Monte Carlo method is very powerful, and can enable us to solve very difficult statistical problems in a straightforward manner, but like all powerful tools, it must be used with care.

SUMMARY

Variance of the fit:

$$s^2 = \frac{1}{N - M} \frac{\Sigma\{(1/\sigma_i^2)[y_i - y(x_i)]^2\}}{(1/N)\Sigma(1/\sigma_i^2)} = \frac{1}{N - m} \Sigma w_i[y_i - y(x_i)]^2$$

Weighting factors:

$$w_i = \frac{1/\sigma_i^2}{(1/N)\Sigma(1/\sigma_i^2)}$$

Relationship between s^2 and χ^2:

$$\chi_\nu^2 = \frac{\chi^2}{\nu} = \frac{s^2}{\langle \sigma_i^2 \rangle}$$

where

$$\langle \sigma_i^2 \rangle = \left[\frac{1}{N} \sum \frac{1}{\sigma_i^2} \right]^{-1}$$

Probability $P_\chi(\chi^2; \nu)$ that any random set of N data points will yield a value of chi-square as large as or larger than χ^2:

$$P_\chi(\chi^2; \nu) = \int_{\chi^2}^{\infty} \frac{z^{1/2(\nu-2)} e^{-z/2}}{2^{\nu/2} \Gamma(\nu/2)} \, dz$$

Linear-correlation coefficient:

$$r \equiv \frac{N \sum x_i y_i - \sum x_i \sum y_i}{\left[N \sum x_i^2 - (\sum x_i)^2 \right]^{1/2} \left[N \sum y_i^2 - (\sum y_i)^2 \right]^{1/2}}$$

Probability $P_c(r, N)$ that any random sample of uncorrelated experimental data points would yield an experimental linear-correlation coefficient as large as or larger than $|r|$:

$$P_c(r; \nu + 2) = \int_{|r|}^{1} \frac{1}{\sqrt{\pi}} \frac{\Gamma[(\nu+1)/2]}{\Gamma(\nu/2)} (1 - r^2)^{(\nu-2)/2}$$

Sample covariance:

$$s_{jk}^2 \equiv \frac{1/(N-1) \sum \left[(1/\sigma_i^2)(x_{ij} - \bar{x}_j)(x_{ik} - \bar{x}_k) \right]}{(1/N) \sum (1/\sigma_i^2)} \quad \text{with} \quad \bar{x}_j = \frac{\sum (x_{ij}/\sigma_i^2)}{\sum (1/\sigma_i^2)}$$

Sample variance: $\sigma_j^2 = \sigma_{jj}^2$.
Sample linear-correlation coefficient:

$$r_{jk} = \frac{s_{jk}^2}{s_j s_k}$$

Multiple-correlation coefficient:

$$R^2 \equiv \sum_{j=1}^{n} \left(b_j \frac{s_{jy}^2}{s_y^2} \right) = \sum_{j=1}^{n} \left(b_j \frac{s^j}{s_y} r_{jy} \right)$$

F test:

$$F = \frac{\chi_{\nu1}^2}{\chi_{\nu2}^2}$$

$$P_F(F; \nu_1, \nu_2) = \int_{F}^{\infty} P_f(f; \nu_1, \nu_2) \, df$$

F test for multiple-correlation coefficient R (for $\nu = N - m$):

$$F_R = \frac{R^2/(m-1)}{(1-R^2)/(N-m)} = \frac{R^2}{(1-R^2)} \times \frac{(N-m)}{(m-1)}$$

F test for χ^2 validity of adding $(m+1)$th term:

$$F_\chi = \frac{\chi^2(m) - \chi^2(m+1)}{\chi^2(m)/(N-m-1)} = \frac{\Delta\chi^2}{\chi_\nu^2}$$

Confidence limits: $1\sigma \to 68.3\%$; $2\sigma \to 95.4\%$; $3\sigma \to 99.7\%$.

EXERCISES

11.1. Discuss the meaning of χ^2 and justify the relationship between it and the sample variance $s^2 = \chi_\nu^2$.

11.2. Compare the exact calculation of the gamma function $\Gamma(n)$ of Equation (11.7) with the approximate calculation of Equation (11.8) for $n = \frac{1}{2}, 1, \frac{5}{2}, 4, \frac{9}{2}, 10$.

11.3. From Equation (11.6), show that the χ^2-probability density for 1 degree of freedom can be written as

$$P(x^2) = \frac{e^{-x^2/2}}{\sqrt{2\pi x^2}}$$

Calculate to 1% the probability of obtaining a value of χ^2 that is less than 2.00 by expanding the function in a Taylor series and integrating term by term.

11.4. For a typical number of degrees of freedom ($\nu \simeq 10$), find, by numerically integrating Equation (11.6), the range of probability $P_\chi(x^2, \nu)$ for finding χ^2 as small as 0.5 or as large as 1.5. Use the approximation for the gamma function of Equation (11.8).

11.5. By numerically integrating Equation (11.6), find the probability of finding a value of $\chi_\nu^2 = 1.5$ with $\nu = 100$ degrees of freedom. (Note that double-precision variables must be used.) Would you consider this to be a reasonably good fit?

11.6. Express the linear-correlation probability density of Equation (11.18) in terms of the approximation for the gamma function of Equation (11.8).

11.7. Work out the details of the calculation of the linear-correlation coefficients r for Examples 6.1 and 6.2.

11.8. If a set of data yields a zero slope $b = 0$ when fitted with Equation (11.11), what can you say about the linear-correlation coefficient r? Justify this value in terms of the correlation between x_i and y_i.

11.9. Find the linear-correlation coefficient r_1 between the independent variable T_i and the dependent variable V_i for the data of Example 7.1.

11.10. Find the correlation coefficient r_2 between T_i^2 and V_i for the data of Example 7.1. Does the correlation justify the use of a quadratic term?

11.11. Express the multiple correlation R in terms of x_{ij}, y_i, and their averages.

11.12. Evaluate the multiple-correlation coefficient R for the data of Example 7.1.

11.13. Is a large value of F good or bad? Explain.

11.14. If we wish to set as an arbitrary criterion a probability of 0.01 for the F_χ test, what would be the reasonable average value for F test?

11.15. What different aspects of a fit do the F_R and F_χ tests represent?

11.16. Apply the F_χ test for the quadratic term to the data of Example 7.1 and state your conclusions. (Refer to Table 7.4.)

11.17. Show the intermediate steps in the derivation of Equation (11.43).

11.18. Estimate from Figure 11.2 the 90% confidence limit for each of the two mean lifetimes (a_4 and a_5) of Example 8.1 when all variables are allowed to find their optimum values.

APPENDIX
A
NUMERICAL
METHODS

There are several reasons why we might want to fit a function to a data sample, and several different techniques that we might use. If we wish to estimate parameters that describe the parent population from which the data are drawn, then the maximum-likelihood or least-squares method is best. If we wish to interpolate between entries in data tables to find values at intermediate points or to find numerically derivatives or integrals of tabulated data, then an interpolation technique will be more useful. Additionally, if we wish to obtain intermediate values between calculated coordinate pairs in order to plot a smooth curve on a graph, then we may wish to use a spline fitting method. In this appendix we shall summarize some standard methods for treating the latter two types of problems, as well as some methods of finding the roots of nonlinear functions, a different sort of interpolation problem.

A.1 POLYNOMIAL INTERPOLATION

With modern fast computers, the need for interpolating within tables to find intermediate values of tabulated functions has reduced markedly. Nevertheless, there are situations in which it may be convenient to represent a complicated function by a simple approximation over a limited range. For example, in a large Monte Carlo calculation, where computing time is a significant consideration, we may approximate a complex function by a simpler polynomial that can be calculated quickly. Alternatively, we may save time by creating a probability integral once at the beginning of the program and interpolating to find values of x corresponding to the randomly chosen values of y.

For many purposes a linear or quadratic interpolation is satisfactory; that is, we fit a straight line to two coordinate pairs, or a parabola to three, and use the equation of the fitted polynomial to find values of y at nearby values of x. Higher orders may be necessary for functions that have strong variations, but in general, it is better and more convenient to represent a function over a limited region by a series of low-order approximations.

Lagrange's Interpolation Method

Here is a method that is easy to remember and can be used to expand a function to any order. We know it works because of the theorem that states that if you can find any nth-degree polynomial that passes exactly through $n + 1$ points, then you have found the one and only nth-degree polynomial that passes through those points. Think about it. It is obvious for $n = 1$ (2 points).

Let us start with an easy problem. Suppose we have two coordinate pairs (x_0, y_0) and (x_1, y_1), and we want to find the straight line that passes through both of them. We write a function of the form

$$P(x) = y_0 A_0(x) + y_1 A_1(x) \tag{A.1}$$

and search for a function $A_0(x)$ that is 1 when $x = x_0$ and 0 when $x = x_1$, and a function $A_1(x)$ that is 1 when $x = x_1$ and 0 when $x = x_0$. We can guess the form. If we write $A_0(x)$ as a fraction and set its numerator to $(x - x_1)$, then $A_0(x)$ will be 0 for $x = x_1$ and will be $(x_0 - x_1)$ for $x = x_0$. But we want $A_0(x) = 1$ for $x = x_0$, so the denominator of A_0 must be $(x_0 - x_1)$. We can make similar arguments for $A_2(x)$ and thus write as our interpolation equation

$$P(x) = y_0 \frac{(x - x_2)}{(x - x_2)} + y_1 \frac{(x - x_1)}{(x_1 - x_0)} \tag{A.2}$$

Suppose we want a parabola that passes through three points. Then we simply write

$$P(x) = y_0 A_0(x) + y_1 A_1(x) + y_2 A_2(x) \tag{A.3}$$

and, following the previous arguments, write

$$P(x) = y_0 \frac{(x - x_1)(x - x_2)}{(x_0 - x_1)(x_0 - x_2)} + y_1 \frac{(x - x_0)}{(x_1 - x_0)} \frac{(x - x_2)}{(x_1 - x_2)}$$
$$+ y_2 \frac{(x - x_0)}{(x_2 - x_0)} \frac{(x - x_1)}{(x_2 - x_1)} \tag{A.4}$$

The expansion to higher orders should be obvious. The kth term in an mth order expansion is given by the following product in which the $i = k$ term must be omitted:

$$\prod_{j=0}^{m-1} \frac{(x - x_j)}{(x_k - x_j)} y_k \quad (\text{excluding } j = k) \tag{A.5}$$

Note that the intervals in x need not be equally spaced. The interpolation for a well behaved function $y = f(x)$ is completely general.

Newton's Divided Differences

Although the Lagrange interpolation method is especially easy to derive and provides a convenient way of interpolating between points in a function or table, it is not very convenient for repetitive calculations. It is not very convenient as an expansion either, because increasing the order of the expansion requires adding another factor to each term as well as adding another term. What we require is a more familiar form—a discrete analog of the Taylor expansion. For this we turn to Newton's method of divided differences.

There are several forms of the divided differences expansion, roughly characterized by the method we choose to define the differences, forward, backward, or about a central point. We shall restrict ourselves here to forward differences; that is, we calculate the variation of y with respect to x by taking increments in the positive x direction.

Again, consider a set of data points, (x_0, y_0), (x_1, y_1), (x_2, y_2), ... Let us assume that we wish to make a linear interpolation from x_0 to some point x with a first-degree polynomial. We define the zeroth divided difference as the function itself $f(x)$ evaluated at $x = x_0$:

$$f[x_0] \equiv f(x_0) = y_0 \tag{A.6}$$

The first divided difference is defined to be

$$f[x_0, x_1] \equiv \frac{f[x_1] - f[x_0]}{(x_1 - x_0)} \tag{A.7}$$

which is the slope of a linear function. Then, for a linear function,

$$f[x, x_0] = f[x_0, x_1] \tag{A.8}$$

or
$$\frac{f[x_0] - f[x]}{(x_0 - x)} = \frac{f[x_1] - f[x_0]}{(x_1 - x_0)} \tag{A.9}$$

which, on rearrangement of the terms, gives the first-order expansion

$$P_1(x) = f[x_0] + (x - x_0)\frac{f[x_1] - f[x_0]}{(x_1 - x_0)}$$

$$= f[x_0] + (x - x_0)f[x_0, x_1] \tag{A.10}$$

where we have written $P_1(x)$ instead of $f(x)$ to indicate that the expansion is a polynomial approximation to the function $f(x)$.

To find the second-order expansion, we consider the second divided differences

$$f[x_0, x_1, x_2] \equiv \frac{f[x_2, x_1] - f[x_1, x_0]}{(x_2 - x_1)(x_1 - x_0)} \tag{A.11}$$

which corresponds to the slope of the slope, or the second derivative. This must be constant for a second-order function, so we have

$$f[x, x_0, x_1] = f[x_0, x_1, x_2] \tag{A.12}$$

which leads to the second-order expansion

$$P_2(x) = f[x_0] + (x - x_0)f[x_0, x_1] + (x - x_0)(x - x_1)f[x_0, x_1, x_2] \tag{A.13}$$

The general form for the nth-order expansion should again be obvious.

Remainders

The extrapolation formula for an nth-order expansion is only exact when the function itself is an nth-degree polynomial. Otherwise, the *remainder* at x after n terms $R_n(x)$, defined as the difference between the original function $f(x)$ and the expansion $P_n(x)$, is given by

$$R_n(x) = f(x) - P_n(x)$$
$$= (x - x_0)(x - x_1) \cdots (x - x_n)f[x, x_0, x_1, \ldots, x_n] \tag{A.14}$$

Calculation of the remainder requires the value of the function $f(x)$ at x, which is generally not available. (If it were, we might not be doing this expansion.) However, it may be possible to make an estimate of $f_n(x)$, or to use a nearby value, and thus find an estimate of $R_n(x)$. An expression for the remainder can also be obtained in terms of the $(n + 1)$th derivative of the function.[1]

Uniform Spacing

The divided difference expressions have a particular convenient form when the intervals in x are uniform; that is, if $x_2 - x_1 = x_3 - x_2 = x_i - x_{i-1} = h$. The divided difference of the previous discussion can be written

$$f[x_0, x_1] = \frac{f[x_1] - f[x_0]}{(x_1 - x_0)} = \frac{\Delta f(x_0)}{h}$$

or $\quad \Delta f(x_0) \equiv f(x_1) - f(x_0) \quad$ and $\quad h = x_1 - x_2 \tag{A.15}$

---

[1] See Hildebrand for a derivation.

and higher-order differences become

$$\Delta^2 f(x_0) \equiv \Delta(\Delta f(x_0)) = \Delta f(x_1) - \Delta f(x_0), \text{ etc.} \qquad (A.16)$$

If we define the relative distance along the interval by

$$\alpha = (x - x_0)/h \qquad (A.17)$$

we can write for the nth-order expansion,

$$P_n(x) = f(x_0) + \alpha \, \Delta f(x_0) + \alpha(\alpha - 1) \, \Delta^2 f(x_0)/2! + \cdots$$
$$+ \alpha(\alpha - 1) \cdots (\alpha - n - 1) \, \Delta^n f(x_0)/n! \qquad (A.18)$$

Equation (A.18) is a finite difference analog of the familiar Taylor expansion with the important difference that the factors multiplying the coefficients $\Delta^k f(x_0)/n!$ are not successive powers of the relative distance from the starting point, but rather the product of relative distances from successive points used in the expansion, because $(\alpha - 1) = (x - x_0 - h)/h = (x - x_1)/h$, and so forth.

Extrapolation

Equations (A.15) through (A.18) are perfectly general for fitting exactly n sequential equally spaced data points with a polynomial of degree $n - 1$. In principle, the position of the first data point (x_0, y_0) can be anywhere, but for optimum interpolation, the values of x_0 and x_n should straddle the interpolation point x and be approximately equidistant from it.

The same formula can be used for extrapolating to values beyond the region of data, but the uncertainties in the validity of the approximation increases as x gets farther from the average of x_1 and x_n. The approximation is limited by both the degree of the interpolating polynomial and by uncertainties in the coefficients of the polynomial resulting from fluctuations in the data.

> **Example A.1.** Table A.1 shows a uniform divided difference table for the cosine function for a range of the argument θ between 0 and 90°. Table A.2 shows values of $\cos \theta$ for $\theta = 10$ and 75° calculated from the divided difference table in orders 1 through 5. The interpolation starts at 0° so that only the top row of Table A.1 is

TABLE A.1
Uniform differences for cos θ

θ (degrees)	y	Δ_1	Δ_2	Δ_3	Δ_4	Δ_5
0	1.0000	−0.0489	−0.0931	0.0139	0.0078	−0.0021
18	0.9511	−0.1420	−0.0792	0.0217	0.0056	
36	0.8090	−0.2212	−0.0575	0.0273		
54	0.5878	−0.2788	−0.0302			
72	0.3090	−0.3090				
90	−0.0000					

TABLE A.2
Extrapolation from 0 to 10° and from 0 to 75° in various orders

θ (degrees)	cos θ	Order				
		1	2	3	4	5
10	0.9848	0.9728	0.9843	0.9851	0.9848	0.9848
75	0.2588	0.7961	0.1819	0.2481	0.2589	0.2588

used and thus, $\theta > 18°$, the calculation is an extrapolation. The true value of $\cos \theta$ is also listed. As we should expect, the large extrapolation to 75° is very poor in low order. Usually, an approximation can be improved by increasing the number of terms in the expansion. However, the better method would be to drop to a different line of the table; that is, to ensure that the calculation is an interpolation rather than an extrapolation.

A.2   INTEGRATION AND DIFFERENTIATION

Once we have obtained interpolation expressions, it is relatively straightforward to obtain expressions for integrals and derivatives in terms of the expansion of order n.

Integration

Integrating Equation (A.18) leads to expressions for calculating the numerical integral in various orders, depending on the number of terms in the polynomial approximation. There are various forms for each order, depending on how we choose the limits of integration. We quote three of the most useful forms with the remainder estimates.

First-order, endpoint trapezoidal

$$\int_{x_0}^{x_1} f(x)\, dx = \frac{h}{2}[f(x_0) + f(x_1)] - \frac{h^3}{12}f^{(2)}(\xi)$$

(first-order closed-end trapezoidal)   (A.19)

$$\int_{x_0}^{x_2} f(x)\, dx = 2hf(x_1) + \frac{h^3}{3}f^{(2)}(\xi) \quad \text{(first-order open end)} \qquad (A.20)$$

$$\int_{x_0}^{x_2} f(x)\, dx = \frac{h}{3}[f(x_0) + 4f(x_1) + f(x_2)] - \frac{h^5}{90}f^{(4)}(\xi)$$

(second-order closed-end Simpson's rule)   (A.21)

The factors $f^{(n)}(\xi)$ in the remainder estimates represent the nth derivative of the function evaluated at some (unknown) value of x in the range of the integral. Note the large reduction on the error estimate in going from either of the first-order approximations to the second-order approximation.

For an integral over an extended range of x, it is usually advisable to employ a series of first- or second-order integrals over sections of the function, rather than to attempt to fit a large region with a higher-order function. In fact, it can be shown that the gain in accuracy in going from a second- to a third-order numerical integral is relatively small, and, for the same number of calculations of the ordinate y_j, the second-order Simpson rule may be more accurate than the third-order form. This relation applies in general to even and odd orders, so that, to make a significant improvement in the numerical integration of a function, one should advance to the next higher *even* order.

Thus, to find the integral by Simpson's rule of $f(x)$ over an extended range between $x = x_0$ and $x = x_n$, we divide the region into n equal intervals in x, with $nh = (x_n - x_0)$, to obtain

$$\int_{x_0}^{x_n} f(x)\, dx = \frac{h}{3} [f(x_0) + 4f(x_1) + 2f(x_2) + 4f(x_3) + \cdots$$

$$+ 4f(x_{n-1}) = f(x_n)] - \frac{nh^5}{180} f^{(4)}(\xi) \quad \text{(A.22)}$$

where ξ is the value of x somewhere in the range of integration.

> **Program example A.1.** Simpson calculates an extended integral by the second-order approximation of Equation (A.22). The program is listed in Appendix E and included in the program unit GenUtil. See Programs 11.1 and 11.2 for examples of the use of this routine. The user supplies four arguments: (i) Funct: the name of the function to be integrated. The function must have one real argument. If other arguments are required, they must be made accessible to the function as global variables. (ii) nInt: the number of *double* intervals. The interval is calculated as dx = (hiLim - LoLim) / (2*nInt). (iii) LoLim and (iv) hiLim: the integration limits.

Differentiation

We can differentiate Equation (A.18) to find approximations for the derivatives of the function $f(x)$. We obtain

$$\frac{dP_n(x)}{dx} = \frac{1}{h} \frac{dP_n(x)}{d\alpha} = [\Delta f(x_0) + (2\alpha - 1) \Delta^2 f(x_0)/2!$$

$$+ (3\alpha^2 - 6\alpha + 2) \Delta^3 f(x_0)/3! + \cdots]/h \quad \text{(A.23)}$$

and

$$\frac{d^2 P_n(x)}{dx^2} = \frac{1}{h^2} \frac{d}{d\alpha} \left[\frac{dP_n(x)}{d\alpha} \right] = [\Delta^2 f(x_0) + (\alpha - 1) \Delta^3 f(x_0) + \cdots]/h^2$$

$$\text{(A.24)}$$

We should note that the use of forward differences introduces an asymmetry in the calculation. For a general solution, we could replace the forward differences by central differences, which are taken symmetrically about a central starting point. For a particular problem, we can usually arrange the expansion to provide reasonable symmetry of the differences about the point of interest. Thus, we can replace Equations (A.23) and (A.24) by

$$\frac{dP_n(x)}{dx} = \Delta f(x_0)/h = \frac{f(x + h/2) - f(x - h/2)}{h} \qquad (A.25)$$

and

$$\frac{d^2 P_n(x)}{dx^2} = \Delta^2 f(x_0)/h^2 = \frac{f(x + h) - 2f(x) + f(x - h)}{h^2} \qquad (A.26)$$

A.3   CUBIC SPLINES

If we attempt to represent by an nth-degree polynomial a function that is tabulated at $n + 1$ points, we are apt to obtain disappointing results if n is large. The polynomial will necessarily coincide with the data points, but may exhibit large oscillations between points. In addition, if there are many data points, the calculations can become rather cumbersome. It is often better to make several low-order polynomial fits to separate regions of the function, and this procedure is usually satisfactory for simple interpolation in tables. However, if we want a smooth function, which passes through the data points, the results may not be satisfactory.

Suppose we have calculated a function at $n + 1$ points, and want to represent the function as a smooth curve on a graph. The nth-order polynomial is out—too wiggly. Breaking the curve up into small sections produces disjointed segments on the plot. It is unlikely that they will combine to form a smooth curve. What do we do now? Reach for our pencil and trusty drafting spline? No, we call up our spline fitting subroutine and let it join up the separate fits for us.

Spline fitting procedures have other uses besides plotting pretty curves on graphs, but the plotting function is of interest to us and is easily illustrated. Suppose we choose to make a series of cubic fits to successive groups of data points. What conditions do we need to produce a smooth curve that passes through the data points? We want the first and second derivatives, as well as the function itself, to be continuous at the data points. Suppose we consider a separate cubic polynomial for each interval on the graph, or a total of n polynomials for the $n + 1$ points. Then we write the polynomial equation, take derivatives, and, at each data point, equate the first and second derivatives of the left-side polynomial to those of the right-side polynomial.

Following the method discussed in Thompson, we begin by writing the Taylor series for the cubic polynomial for interval i, expanded about the point x_i

$$y(x) = y(x_i) + (x - x_i)\frac{dy(x_i)}{dx} + (x - x_i)^2\frac{d^2y(x_i)}{dx^3}\bigg/2!$$

$$+ (x - x_i)^3\frac{d^3y(x_i)}{dx^3}\bigg/3! \tag{A.27}$$

where the function and derivatives are evaluated at x_i. This can be written in a more concise form as

$$y(x) = y_i + (x - x_i)y_i' + (x - x_i)^2 y_i''/2$$

$$+ (x - x_i)^3(y_{i+1}'' - y_i'')/6h \tag{A.28}$$

where y_i' and y_i'' stand for the first and second derivatives evaluated at $x = x_i$ and the third derivative has been replaced by its divided difference form, which is exact for a cubic function. At $x = x_i$, we have $y = y_i$, as required. We can also set $x = x_{i+1} = x_i + h$ and solve the equation

$$y(x_{i+1}) = y_i + (x_{i+1} - x_i)y_i' + (x_{i+1} - x_i)^2 y_i''/2$$

$$+ (x_{i+1} - x_i)^3(y_{i+1}'' - y_i'')/6h \tag{A.29}$$

to obtain

$$y_{i+1} - y_i = hy_i' + h^2[2y_i'' + y_{i+1}'']/6 \tag{A.30}$$

We repeat the calculation, using the equation for $y(x)$ in interval $i - 1$ [i.e., we replace i by $i - 1$ in Equation (A.29)],

$$y(x) = y_{i-1} + (x - x_{i-1})y_{i-1}' + (x - x_{i-1})^2 y_{i-1}''/2$$

$$+ (x - x_{i-1})^3(y_i'' - y_{i-1}'')/6h \tag{A.31}$$

and again require that $y(x) = y(x_i)$ at the ith data point and obtain

$$y_i - y_{i-1} = hy_{i-1}' + h^2[2y_{i-1}'' + y_i'']/6 \tag{A.32}$$

To establish the continuity conditions at the data points, we need the first derivative in the interval i,

$$y'(x) = y_i' + (x - x_i)y_i'' + (x - x_i)^2(y_{i+1}'' - y_i'')/2h \tag{A.33}$$

which we equate to the first derivative in the interval $i - 1$ at the position $x = x_i$, to obtain

$$y_i' - y_{i-1}' = h[y_i'' + y_{i-1}'']/2 \tag{A.34}$$

Similarly, equating the derivatives at the boundary $x = x_{i+1}$ gives

$$y_{i+1}' - y_i' = h[y_{i+1}'' y_i'']/2 \tag{A.35}$$

(Repeating the procedure with the second derivative leads to an identity, because our use of the divided difference form for the third derivative assures continuity of the second derivative across the boundaries.) Eliminating the first derivatives from Equations (A.30), (A.32), (A.34), and (A.35) gives us the spline equation

$$y''_{i-1} + 4y''_i + y''_{i+1} = D_i \qquad \text{(A.36)}$$

with

$$D_i = y[y_{i+1} - 2y_i + y_{i-1}]/h^2 \qquad \text{(A.37)}$$

Note that the D_i's are proportional to the second differences of the tabulated data and are all known. We can write Equation (A.36) as a set of linear equations relating the unknow variables y'', beginning with $i = 2$ and ending with $i = n - 1$:

$$y''_1 + 4y''_2 + y''_3 \qquad\qquad = D_2 \qquad \text{(A.38a)}$$
$$y''_2 + 4y''_3 + y''_4 \qquad\qquad = D_3 \qquad \text{(A.38b)}$$
$$\vdots$$
$$y''_{n-3} + 4y''_{n-2} + y''_{n-1} \qquad = D_{n-2} \qquad \text{(A.38c)}$$
$$y''_{n-2} + 4y''_{n-1} + y''_n = D_{n-1} \qquad \text{(A.38d)}$$

These equations can be solved for the second derivatives y''_i, as long as we know the values of y''_1 and y''_n. One possibility is to set them to 0 to obtain *natural splines*. Alternatively, we may use the true second derivatives, if they are known, or a numerical approximation.

For example, suppose we have only four points to consider. Then, if we know y''_1 and y''_4, we can solve the simultaneous Equations (A.38a) and (A.38b) for y_2 and y_3. Similarly, if we have a full set of n equations, we can rewrite Equation (A.38a) to express $y''_2 = (D_2 - y''_1 - y''_3)/4$, and substitute this expression into Equation (A.38b) to eliminate y''_2. Then, we repeat the procedure to eliminate y''_3 from the next equation. We continue this procedure until we reach the last equation, which will contain only terms in y''_1, y''_{n-1}, and y''_n. Because y''_1 and y''_n are known, we can solve this equation for y''_{n-1}, and then work back down the chain determining successively y''_{n-2}, y''_{n-3} and so forth, until we reach Equation (A.38a) from which we determine the last unknown y''_2. Once we have found the values of the y''_i's, we can find the y'_i's from Equation (A.30) or (A.32), and use Equation (A.28) to interpolate in each interval.

The solution of Equation (A.38) is discussed in several textbooks. Essentially, one sets up a recursion relations to build a table of the second derivatives y''. The method is illustrated by the computer routines SplineMake listed in Appendix E.

An interesting alternative method of solving the set of simultaneous equations, Equations (A.38), is to set them up in a spreadsheet program. Then, when the boundary values y''_1 and y''_n are supplied, the program will readjust the variables until they stabilize at the solutions to Equations (A.38). Although this method is not very practical for graphical applications where we want to build the solution into our plotting program, it does provide a quick way of finding the second derivatives and an interesting illustration of the solution.

As with all techniques, a certain amount of care must be exercised in using spline routines. The choice of a second derivative at the boundary may have an important effect on the interpolation at the ends of the function, and a wrong choice, for example, can produce undesirable shapes at the edges of a plot. Then too, although the spline routine assures a smooth variation between the data points, with continuity of the function and first and second derivative across the points, it cannot guarantee that there will be no peculiar oscillation between the points.

> **Program example A.2.** `Spline interpolation` Spline interpolation routines are listed in Appendix E.

> `SplineMake` Constructs a table of second derivatives for a spline interpolation by the method discussed in the previous paragraphs.
>
> `SplineInt` Performs the interpolation. For simplicity, we have chosen to store only the second derivatives and to calculate the first and third derivatives as needed in functions `D1ydX1` and `D3ydx3`. If speed is important, the derivatives could be computed and stored in arrays.

> The spline interpolation routine is called by the routine `Scurve` in the program unit `QuikPlots` to calculate and plot a smoothed curve by the spline method. The routine has been used to produce many of the graphs in this book.

A.4 ROOTS OF NONLINEAR EQUATIONS

Finding roots of nonlinear equations is essentially the reverse of an interpolation problem. When we interpolate a function, our object is to find a value of the dependent variable y at a specific value of the independent variable x. When we are searching for the root of a function, we are trying to find the value of x at a particular value, usually 0, of y. However, interchanging the variables completely changes the nature of the problem. Interpolation involves straightforward application of well defined equations that are independent of the form of the original function: Finding roots of nonlinear equations may require different equations for different problems and almost always requires some sort of a search and iteration procedure.

The diffraction of light by a single split provides an interesting example of a nonlinear equation. It is well known that the position of the interference maxima and minima from double slits and diffraction gratings can be determined analytically from consideration of the phase difference between the rays that pass through each slit, but only the minima of the diffraction pattern of a single slit can be found in this way. To find the position of a maximum, with the exception of the central one, we must differentiate the expression for the intensity with respect to the phase α:

$$I = I_0 \left(\frac{\sin \alpha}{\alpha} \right)^2 \quad \text{with} \quad \alpha = \frac{\pi a}{\lambda} \sin \theta \qquad (\text{A.39})$$

In Equation (A.39), I_0 is the intensity of the light at the central maximum ($\theta = 0$), I is the intensity at angle θ, λ is the wavelength of the light, and a is the slit width. The position of the maximum is given by solving

$$\frac{dI}{d\alpha} = 2 I_0 \left(\frac{\sin \alpha}{\alpha^3} \right)(\alpha \cos \alpha - \sin \alpha) = 0 \qquad (\text{A.40})$$

to obtain the value α_r at the root of the equation

$$f(\alpha) = \alpha_r - \tan \alpha_r = 0 \qquad (\text{A.41})$$

The first root is at $\alpha_r = 0$. The other roots cannot be calculated analytically and must be found by an iterative method. An approximate solution can be obtained by rewriting Equation (A.41) as

$$\alpha_r = \tan \alpha_r \qquad (\text{A.42})$$

and plotting separately the left and right sides to find the intersection of the straight line and the tangent curves. There are several mathematical ways to solve the problem, but making a plot of the function is always a good starting procedure.

Trial-and-Error: The Half-Interval Method

With a personal computer, trial-and-error may be a suitable method for solving the occasional root finding problem. An orderly approach is advisable and the half-interval method is convenient. The procedure is to write a little program that requests a trail value of the root and calculates the function and displays its value. The initial trial value might be obtained from a graph, or perhaps by mapping the function for various values of the independent variable x, until a reasonable estimate of the root has been obtained. Then, a second trail x is submitted, which produces a value of y on the other side of the root. The half-interval method begins at this point. The procedure is to select a third trial value that is midway between the two that bracket the root. For the fourth trial value, we use the mean of the most recent value, and whichever of the two previous trials was on the other side of the root. The process continues until the root is found to the desired accuracy.

This rather primitive method of root finding could be improved with a little programming to let the program decide which root to choose, to calculate the mean, and perform the next trial. The program could proceed in a loop until the root had been found to a predefined degree of accuracy, or the calculation could be stopped manually. However, if we are willing to program that little bit of logic, slightly more effort will produce a much faster root finding program.

Secant Methods

The gain in speed comes from using the slope of the function in the calculation. We begin with two trial estimates of the root, x_k and x_{k+1}, preferably, but not necessarily, on either side of the root. Then we write an expression for a linear interpolation between the two points. Equation (A.10) gives

$$f(x) = y_k + (x - x_k)\frac{(y_{k+1} - y_k)}{(x_{k+1} - x_k)} \tag{A.43}$$

where we have written $y_k = f(x_k)$ and so forth. Setting $f(x) = 0$ and solving for x gives us an approximation to the value of x at the root:

$$x = x_k - y_k\frac{x_{k+1} - x_k}{y_{k+1} - y_k} = \frac{x_k y_{k+1} - x_{k+1} y_k}{y_{k+1} - y_k} \tag{A.44}$$

For the next trial, we replace x_{k+1} or x_{k+2} by the value x found in Equation (A.44) and repeat the calculation. The process can be repeated until the root is approximated as closely as desired. This is the first-order secant method.

There are various ways of choosing which of the previous values of x (x_k or x_{k+1}) to keep for the next iteration. The simplest is to keep the most recent value and discard the older value. Another way is to choose whichever is closer to the root [i.e., gives a smaller value of $f(x)$]. A third is to start the process with two values that straddle the root (i.e., give opposite signs for y_1 and y_2) and to continue to choose values that straddle the root after each iteration. This is the Regulo-Falsi method.

Clearly any method will find the root most quickly if the starting values are close to the root, but, in principle, the secant methods will almost always find a root of the function, eventually. With some functions, such as those that are antisymmetric about the root, there is the possibility that the search by the Regulo-Falsi method, for example, will jump back and forth across the root and never approach it. Additionally, for functions with several roots, we may not always find the one we want. Problems may also arise if two roots are very close together.

Newton-Raphson Method

Instead of calculating the slope by finite differences, as in the secant method, we could use the tangent, or derivative of the function, if it can be calculated. Then, we can replace Equation (A.43) by

$$f(x) = y_k + (x - x_k)\frac{df(x_k)}{dx} \tag{A.45}$$

where x_k and y_k are the values of x and $f(x)$ after the kth iteration. We find the next estimate x_{k+1} for the root, as before, by setting $f(x)$ in Equation

TABLE A.3
Determination of the first two nonzero roots of $\alpha = \tan \alpha$

	First root		Second root	
trial	x	y	x	y
	(a) Newton's Method†			
0	4.40000	1.30368	7.70000	1.25713
1	4.53598	− 1.07376	7.73028	− 0.31270
2	4.50186	− 0.17769	7.72545	− 0.01188
3	4.49375	− 0.00679	7.72525	− 0.00002
4	4.49341	− 0.00001	7.72525	− 0.00000
5	4.49341	− 0.00000		
	(b) Secant Method‡			
0	4.40000	1.30368	7.80000	− 10.70682
0	4.50000	− 0.13733	7.70000	1.25713
1	4.49047	0.05854	7.71051	0.78849
2	4.49332	0.00184	7.72819	− 0.17931
3	4.49341	− 000003	7.72491	0.02025
4	4.49341	0.00000	7.72524	0.00047
5			7.72525	− 0.00000

†The calculation continues without assistance after the initial trial value has been selected
‡Two x, y pairs are required for each stage of the calculation. After the first trial, the most recently calculated x, y pair was used with whichever of the two previous pairs was closer to the root.

(A.45) to zero to obtain

$$x_{k+1} = x_k - y_k \div \frac{df(x_k)}{dx} \qquad (A.46)$$

Example A.2. Table A.3 shows steps in an iterative calculation of the second and third roots of Equation (A.41) by the secant and Newton-Raphson methods. Starting values were chosen by examining a plot of $\tan x$ versus x.

Simultaneous Nonlinear Equations

In the examples of alternate fitting methods in Section 6.6, we obtained two pairs of coupled, nonlinear equations, Equations (6.24) and (6.27), which we wished to solve for the parameters a and b. We used the secant method to solve these equations.

Consider the two equations

$$f_a(u, v) = 0 \quad \text{and} \quad f_b(u, v) = 0 \qquad (A.47)$$

which we wish to solve for u and v. We define the first *partial divided differences*,

$$f_{au} = f_a[u_0, v_0; u_1, v_0] \equiv \frac{f_a(u_1, v_0) - f_a(u_0, v_0)}{u_1 - u_0}$$

$$f_{av} = f_a[u_0, v_0; u_0, v_1] \equiv \frac{f_a(u_0, v_1) - f_a(u_0, v_0)}{v_1 - v_0}$$

$$f_{bu} = f_b[u_0, v_0; u_1, v_0] \equiv \frac{f_b(u_1, v_0) - f_b(u_0, v_0)}{u_1 - u_0}$$ (A.48)

$$f_{bv} = f_b[u_0, v_0; u_0, v_1] \equiv \frac{f_b(u_0, v_1) - f_b(u_0, v_0)}{v_1 - v_0}$$

and, following Equation (A.43), write for a first-order expansion

and
$$f_a(u, v) = f_a(u_0, v_0) + (u - u_0)f_{au} + (v - v_0)f_{av}$$ (A.49)
$$f_b(u, v) = f_b(u_0, v_0) + (u - u_0)f_{bu} + (v - v_0)f_{bv}$$

If we assume that f_a and f_b are linear in u and v, we can find a first approximation to the roots by setting $f_a(u, v)$ and $f_b(u, v)$ to zero in Equation (A.49) and solving the two coupled linear equations for u and v:

$$uf_{au} + vf_{av} - u_0 f_{au} - v_0 f_{av} + f_a(u_0, v_0) = 0$$
$$uf_{bu} + vf_{bv} - u_0 f_{bu} - v_0 f_{bv} + f_b(u_0, v_0) = 0$$ (A.50)

Solution by the determinant method gives

and
$$u_2 = u = (Af_{bv} - Bf_{av})/D$$ (A.51)
$$v_2 = v = (Bf_{au} - Af_{bu})/D$$

with

$$D = f_{au}f_{bv} - f_{av}f_{bu}$$
$$A = -u_0 f_{au} - v_0 f_{av} + f_a(u_0, v_0)$$ (A.52)
$$B = -u_0 f_{bu} - v_0 f_{bv} + f_b(u_0, v_0)$$

We then repeat the procedure with coordinate pairs (u_1, v_1) and (u_2, v_2), to obtain the next approximation, until the roots have been found to the desired degree of accuracy.

A.5 DATA SMOOTHING

The concept of smoothing is not one that meets with universal approval. The discussion that follows should be considered with the admonition of a caveat emptor: For rigorously valid least-squares fitting, smoothing is neither desirable nor permissible; however, there are cases where smoothing can be beneficial, and, therefore, the techniques are introduced.

Consider, for example, the discussion of Section 9.2 of the determination of the area under a peak from a least-squares fit to a histogram of the data. Least-squares fitting techniques applied to data that are distributed according to Poisson distributions, rather than Gaussian distributions, underestimate the area of a peak by an amount equal to the value of χ^2. We have seen that we can improve the result by decreasing the value of χ^2 at its minimum. Similarly, if the shape of the fitting function does not exactly simulate that of the parent distribution, a better fit to the data by decreasing χ^2 can yield an improved estimate of the area under a peak.

Another example that might benefit from application of a smoothing algorithm is the parameterization of data for use in a Monte Carlo or other program. In preparing experimental proposals, it is often necessary to estimate yields and distributions based on currently available data. Such data are often sparse and generally must be expressed in parametric form for ease and speed of use in the Monte Carlo simulation program. Smoothing can be useful to average out fluctuations and allow the data to be expressed with a few parameters by a least-squares fit or an interpolation procedure.

In other words, if rigorously valid results are not required, but rather an averaged estimate of the distribution, smoothing may help obtain more reliable estimates. The improvement in the estimate of one parameter must, of course, be accompanied by a decrease in information of some other parameter or parameters. For example, an improved estimate of the area under a peak would be accompanied by an increased uncertainty in the estimates of the width and position of the peak.

Whatever smoothing or other manipulation is done must conserve the information pertaining to the desired parameters. The averaging techniques that we shall discuss, for example, conserve the area under a peak but not the width of the peak. Similarly, this method would be useful for improving the estimate of the constant term of a polynomial but not the coefficients of the other terms.

Data smoothing is similar to the data "smearing" introduced in Chapter 5 to simulate measuring uncertainties in "measurements" generated by a Monte Carlo program. In the Monte Carlo program we used Gaussian smearing; that is, we allowed each event a Gaussian probability distribution about its mean.

In this section, we are dealing with binned data, and thus, for Gaussian smoothing, could consider a Gaussian integration that spreads each event over adjacent bins. Because our object here is to smooth the data, we are at liberty to choose the width of the smearing function to produce the desired degree of uniformity in the data, limited by the requirement that we do not damage the very variable we are trying to study.

The binomial distribution is a useful smoothing function. Suppose we want to smooth low statistics experimental data that follow a Gaussian peak in a way that preserves the area under the peak. Let us assume that the background slope is gentle enough so that smoothing will not affect its determination drastically.

We can approximate the Gaussian peak with a binomial distribution with $p = \frac{1}{2}$ (see Section 2.1):

$$y(x) = \frac{1}{\sigma\sqrt{2\pi}} e^{-1/2\left(\frac{x - \mu}{\sigma}\right)^2} \simeq \left(\frac{1}{2}\right)^n \frac{n!}{X!(n - X)!} \qquad (A.53)$$

We can relate the widths σ and the means of the two distributions

$$\sigma_B^2 = np(1 - p) = n/4 = \sigma^2 \qquad \overline{X} = np = n/2 \qquad \bar{x} = \mu \qquad (A.54)$$

to find the relationships among the parameters

$$n = 4\sigma^2 \qquad X = x - \mu + n/2 = x - \mu + 2\sigma^2 \qquad (A.55)$$

We can then express the binomial distribution of Equation (A.53) as

$$y(x) = \left(\frac{1}{2}\right)^n \frac{n!}{(n/2 + x - \mu)!(n/2 - x + \mu)!} \qquad (A.56)$$

Let us smooth the data by averaging over adjacent channels with a binomial distribution spanning three channels:

$$y'(x) = 1/4 y(x - 1) + 1/2 y(x) + 1/4 y(x + 1) \qquad (A.57)$$

If we fold this averaging into the distribution of Equation (A.53), the result is also binomial:

$$y'(x) = \left(\frac{1}{2}\right)^{n+2} \frac{(n + 2)!}{(n/2 + 1 + x - \mu)!(n/2 + 1 - x + \mu)!} \qquad (A.58)$$

The new distribution has the same mean $\bar{x} = \mu$ but a larger width $\sigma'^2 = n'/4 = (n + 2)/4$ with the variance increased by $\frac{1}{2}$:

$$\sigma'^2 = \sigma^2 + \tfrac{1}{2} \qquad (A.59)$$

Similarly, we could smooth over five channels by using a formula similar to Equation (A.57) but with five terms with coefficients given by the binomial expansion

$$y''(x) = 1/16 y(x - 2) + 1/4 y(x - 1) + 3/8 y(x)$$
$$+ 1/4 y(x + 1) + 1/16 y(x + 2) \qquad (A.60)$$

A five-channel smoothing is identical to two successive smoothings over three channels and yields a variance that is increased accordingly, $\sigma''^2 = \sigma^2 + 1$. Any such smoothings over $2n + 1$ adjacent channels is equivalent to n smoothings over three channels.

If we apply the smoothing of Equation (A.57) to a Gaussian distribution, the resulting distribution will also be nearly Gaussian because the shapes of the binomial and Gaussian distributions are nearly alike. In fact, if we are applying the smoothing because the original shape is not Gaussian enough, the averaging may make the shape more nearly Gaussian. If we apply binomial smoothing to a distribution that is not Gaussian, we should be aware that we are distorting the shape of the peak and making it more Gaussian-like.

If the width of the original Gaussian is not too small ($\sigma < 1$), the increase of Equation (A.59) should not be drastic because the addition is in quadrature. For a width $\sigma = 2$, for example, the new width $\sigma' = 2.1$ is only 5% larger. If the original width is very small ($\sigma < 1$), the approximation of Equation (A.53) is not valid because the Gaussian and binomial distributions are only similar in the limits of large n. A Gaussian fit to the data without smoothing would not be valid either, however, because the parameters of the fit are only meaningful if $\sigma \geq 1$. Because the averaging itself is a binomial distribution, the result is still expected to be a better approximation to a Gaussian distribution than the original data. For a smoothing over three channels, a Gaussian fit requires $\sigma \geq \sqrt{1/2}$ for the original data.

B.1 DETERMINANTS

In applying the method of least squares to both linear and nonlinear functions, we required the solution of a set of n simultaneous equations in n unknowns a_i similar to the following:

$$y_1 = a_1 X_{11} + a_2 X_{12} + a_3 X_{13}$$
$$y_2 = a_1 X_{21} + a_2 X_{22} + a_3 X_{23} \qquad \text{(B.1)}$$
$$y_3 = a_1 X_{31} + a_2 X_{32} + a_3 X_{33}$$

where the constants y_i and X_{ij} are known quantities calculated from the data.

The symmetry of the right-hand side suggests that we write elements of the equations in a two-dimensional array

$$\boldsymbol{\alpha} = \begin{bmatrix} X_{11} & X_{12} & X_{13} \\ X_{21} & X_{22} & X_{23} \\ X_{31} & X_{32} & X_{23} \end{bmatrix} \qquad \text{(B.2)}$$

and separate the other terms and coefficients into one-dimensional arrays.

$$\mathbf{a} = \begin{bmatrix} a_1 \\ a_2 \\ a_3 \end{bmatrix} \quad \text{and} \quad \boldsymbol{\beta} = \begin{bmatrix} y_1 \\ y_2 \\ y_3 \end{bmatrix} \qquad \text{(B.3)}$$

Such arrays are called matrices, and we can write Equations (B.1) in matrix form as

$$\boldsymbol{\beta} = \boldsymbol{\alpha} \cdot \mathbf{a} \qquad \text{(B.4)}$$

Alternatively, because in our problems the matrix $\boldsymbol{\alpha}$ is always symmetric, that is,

239

the element α_{ij} is equal to the element α_{ji}, we can write the matrices **a** and **β** as row matrices

$$a = [a_1 \quad a_2 \quad a_3] \quad \text{and} \quad b = [y_1 \quad y_2 \quad y_3] \tag{B.5}$$

and express Equation (B.1) as

$$\beta = \mathbf{a} \cdot \boldsymbol{\alpha} \tag{B.6}$$

We shall be concerned primarily with linear one-dimensional matrices and with symmetric square two-dimensional matrices that have the same number of rows and columns and are mirror-symmetric about the diagonal. Consider a square matrix **A**:

$$\mathbf{A} = \begin{bmatrix} A_{11} & A_{12} & \cdots & A_{1k} & \cdots & A_{1n} \\ A_{21} & A_{22} & \cdots & A_{2k} & \cdots & A_{2n} \\ \vdots & \vdots & \vdots & \vdots & \vdots & \vdots \\ A_{j1} & A_{j2} & \cdots & A_{jk} & \cdots & A_{jn} \\ \vdots & \vdots & \vdots & \vdots & \vdots & \vdots \\ A_{n1} & A_{n2} & \cdots & A_{nk} & \cdots & A_{nn} \end{bmatrix} \tag{B.7}$$

The *degree* of the matrix **A** is the number n of rows and columns; the jkth *element* (or *component*) of the matrix is A_{jk}; the *diagonal terms* are A_{jj}. If the matrix is diagonally *symmetric*, $A_{jk} = A_{kj}$ and there are n^2 elements but only $n(n + 1)/2$ different elements.

Matrix Algebra

If **A** and **B** are two square symmetric matrices of degree n, then their sum **S** is a square symmetric matrix of degree n with elements that are the sums of the corresponding elements of the two matrices

$$\mathbf{A} + \mathbf{B} = \mathbf{S} \qquad S_{ij} = A_{jk} + B_{jk} \tag{B.8}$$

The product **P** of the matrices **A** and **B** is a square matrix of degree n, with elements determined in the following way:

$$\mathbf{AB} = \mathbf{P} \qquad P_{jk} = \sum_{m=1}^{n} (A_{jm} B_{mk}) \tag{B.9}$$

The elements of the jth row of **A** are multiplied by the elements of the kth column of **B** and the products are summed to obtain the jkth element of **P**. In general, the matrix **P** will not be symmetric.

If **a** is a linear one-dimensional matrix, the product of **A** and **a** is only well defined if the product is taken in a particular order. If **a** is a column matrix, it must be multiplied on the left by the square matrix to yield another column matrix **c**:

$$\begin{bmatrix} A_{11} & \cdots & & \cdots & A_{1n} \\ \vdots & \cdots & & & \vdots \\ A_{j1} & \cdots & A_{jk} & \cdots & A_{jn} \\ \vdots & \cdots & & & \vdots \\ A_{n1} & \cdots & & \cdots & A_{nn} \end{bmatrix} \begin{bmatrix} a_1 \\ \vdots \\ a_k \\ \vdots \\ a_n \end{bmatrix} = \begin{bmatrix} c_1 \\ \vdots \\ c_k \\ \vdots \\ c_n \end{bmatrix} \qquad c_j = \sum_{k=1}^{n} (A_{jk} a_k) \quad \text{(B.10)}$$

If **a** is a row matrix, it must multiply the square matrix on the left to yield another row matrix **r**:

$$[a_1 \quad \cdots \quad a_j \quad \cdots \quad a_n] \begin{bmatrix} A_{11} & \cdots & & \cdots & A_{1n} \\ \vdots & \cdots & & & \vdots \\ A_{j1} & \cdots & A_{jk} & \cdots & A_{jn} \\ \vdots & \cdots & & & \vdots \\ A_{n1} & \cdots & & \cdots & A_{nn} \end{bmatrix}$$

$$= [r_1 \quad \cdots \quad r_k \quad \cdots \quad r_n] \qquad r_j = \sum_{j=1}^{n} (a_j A_{jk}) \qquad \text{(B.11)}$$

The product of two linear matrices depends on the order of multiplication. The product of a row matrix **a** times a column matrix **b** is a scalar. If the order is reversed, the result is a square matrix that is *diagonal*; that is, for which only the diagonal terms are nonzero:

$$[a_1 \quad \cdots \quad a_n] \begin{bmatrix} b_1 \\ \vdots \\ b_n \end{bmatrix} = \sum_{j=1}^{n} (a_j b_j)$$

$$\begin{bmatrix} b_1 \\ \vdots \\ b_n \end{bmatrix} [a_1 \quad \cdots \quad a_n] = \begin{bmatrix} a_1 b_1 & \cdots & 0 & \cdots & 0 \\ \vdots & \cdots & \vdots & \cdots & \vdots \\ 0 & \cdots & a_j b_j & \cdots & 0 \\ \vdots & \cdots & \vdots & \cdots & \vdots \\ 0 & \cdots & 0 & \cdots & a_n b_n \end{bmatrix} \quad \text{(B.12)}$$

Determinants

The *determinant* of a matrix square is defined in terms of its algebra. The *order* of the determinant of a square matrix is equal to the degree n of the matrix. In this section, we shall mainly use determinants of order 3 as examples, although, unless otherwise specified, the comments apply to matrices of all orders. Manipulation of the rows may be substituted for columns throughout.

1. The determinant of the *unity matrix* is 1 where the unity matrix is defined as the diagonal matrix with all diagonal elements equal to 1:

$$|\mathbf{1}| = \begin{vmatrix} 1 & 0 & 0 \\ 0 & 1 & 0 \\ 0 & 0 & 1 \end{vmatrix} = 1 \tag{B.13}$$

2. If a column matrix of degree n is added to one column of a square matrix of degree n, the determinant of the result is the sum of the determinant of the original square matrix plus that of another square matrix obtained by substituting the column matrix for the modified column:

$$\begin{vmatrix} A_{11} + a_1 & A_{12} & A_{13} \\ A_{21} + a_2 & A_{22} & A_{23} \\ A_{31} + a_3 & A_{32} & A_{33} \end{vmatrix} = \begin{vmatrix} A_{11} & A_{12} & A_{13} \\ A_{21} & A_{22} & A_{23} \\ A_{31} & A_{32} & A_{33} \end{vmatrix} + \begin{vmatrix} a_1 & A_{12} & A_{13} \\ a_2 & A_{22} & A_{23} \\ a_3 & A_{32} & A_{33} \end{vmatrix} \tag{B.14}$$

3. If one column of a square matrix is multiplied by a scalar, the determinant of the result is the product of the scalar and are determinant of the original matrix:

$$\begin{vmatrix} cA_{11} & A_{12} & A_{13} \\ cA_{21} & A_{22} & A_{23} \\ cA_{31} & A_{32} & A_{33} \end{vmatrix} = c \begin{vmatrix} A_{11} & A_{12} & A_{13} \\ A_{21} & A_{22} & A_{23} \\ A_{31} & A_{32} & A_{33} \end{vmatrix} \tag{B.15}$$

4. If two columns of a square matrix are interchanged, the determinant retains the same magnitude but changes sign:

$$\begin{vmatrix} A_{12} & A_{11} & A_{13} \\ A_{22} & A_{21} & A_{23} \\ A_{32} & A_{31} & A_{33} \end{vmatrix} = - \begin{vmatrix} A_{11} & A_{12} & A_{13} \\ A_{21} & A_{22} & A_{23} \\ A_{31} & A_{32} & A_{33} \end{vmatrix} \tag{B.16}$$

5. The *minor A^{jk}* of an element A_{jk} of a square matrix of degree n is defined as the determinant of the square matrix of degree $n - 1$ formed by removing the jth row and the kth column:

$$\mathbf{A} = \begin{vmatrix} A_{11} & A_{12} & A_{13} \\ A_{21} & A_{22} & A_{21} \\ A_{31} & A_{32} & A_{33} \end{vmatrix} \qquad A^{21} = \begin{vmatrix} A_{12} & A_{13} \\ A_{32} & A_{33} \end{vmatrix} \tag{B.17}$$

6. The *cofactor* $\text{cof}(A_{jk})$ of an element A_{jk} of a square matrix of degree n is defined as the product of the minor and a phase factor:

$$\text{cof}(A_{jk}) \equiv (-1)^{j+k} A^{jk} \tag{B.18}$$

7. With the preceding definitions 5 and 6, the determinant of a square matrix of degree n can be expressed in terms of cofactors of minors:

$$|\mathbf{A}| = \sum_{k=1}^{n} \left[A_{jk} \, \text{cof}(A_{jk}) \right] = \sum_{k=1}^{n} \left[(-1)^{j+k} A_{jk} A^{jk} \right] \tag{B.19}$$

Equation (B.19) is an iterative definition, because the cofactor is itself a determinant. The determinant of a matrix of degree 1, however, is equal to the single element of that matrix. The determinant of a square matrix of degree 2 is encountered often enough to make its explicit formula useful:

$$\begin{vmatrix} a & b \\ c & d \end{vmatrix} = ad - bc \qquad (B.20)$$

and we can evaluate the determinant of a third-order matrix with the help of Equation (B.19):

$$\begin{vmatrix} A_{11} & A_{12} & A_{13} \\ A_{21} & A_{22} & A_{23} \\ A_{31} & A_{32} & A_{33} \end{vmatrix} = A_{11}\begin{vmatrix} A_{22} & A_{23} \\ A_{32} & A_{33} \end{vmatrix} - A_{12}\begin{vmatrix} A_{21} & A_{23} \\ A_{31} & A_{33} \end{vmatrix} + A_{13}\begin{vmatrix} A_{21} & A_{22} \\ A_{31} & A_{32} \end{vmatrix}$$

$$(B.21)$$

Computation

Matrix computation is generally simpler if we can manipulate matrices into diagonal form in which only the diagonal elements A_{jj} are nonzero. The determinant of a diagonal matrix is equal to the product of all the diagonal elements and the *trace* is their sum:

$$|\mathbf{A}_{\text{diag}}| = \prod_{j=1}^{n} A_{jj} \qquad (B.22)$$

If we combine rules 2, 3, and 4 of the algebra for determinants, we can show that the determinant of a matrix is unchanged if the elements of any column, multiplied by an arbitrary scalar, are added to the elements of any other column. The determinant of the sum is equal to the sum of the two determinants, but one of these determinants has two identical columns except for a scalar factor that may be extracted, and is therefore equal to 0:

$$\begin{vmatrix} A_{11} + cA_{12} & A_{12} & A_{13} \\ A_{21} + cA_{22} & A_{22} & A_{23} \\ A_{31} + cA_{32} & A_{32} & A_{33} \end{vmatrix} = \begin{vmatrix} A_{11} & A_{12} & A_{13} \\ A_{21} & A_{22} & A_{23} \\ A_{31} & A_{32} & A_{33} \end{vmatrix} + c\begin{vmatrix} A_{12} & A_{12} & A_{13} \\ A_{22} & A_{22} & A_{23} \\ A_{32} & A_{32} & A_{33} \end{vmatrix}$$

$$= |\mathbf{A}| \qquad (B.23)$$

Thus, it is possible to eliminate all elements except one of one row by successively subtracting one column, appropriately scaled, from each of the others. For example, if we perform the subtraction

$$A'_{jk} = A_{jk} - A_{1j}\frac{A_{j1}}{A_{11}} \qquad (B.24)$$

on each row except the first, we eliminate all elements of the first column except

A_{11} to obtain

$$\mathbf{A'} = \begin{vmatrix} A_{11} & A_{12} & A_{13} \\ 0 & A'_{22} & A'_{23} \\ 0 & A'_{32} & A'_{33} \end{vmatrix} \tag{B.25}$$

Similarly, if we subsequently start with element A'_{22} and subtract an appropriately scaled second row from the rest of the rows,

$$A''_{jk} = A'_{jk} - A'_{2k} A'_{jk}/A'_{22} \tag{B.26}$$

all the elements of the second column vanish except A'_{22}:

$$\mathbf{A''} = \begin{vmatrix} A_{11} & 0 & A''_{13} \\ 0 & A'_{22} & A''_{23} \\ 0 & 0 & A''_{33} \end{vmatrix} \tag{B.27}$$

Note that A'_{22} is not the original value A_{22}, but is modified as a result of the first subtraction.

By successively subtracting rows (or columns) scaled to their diagonal elements, we can produce a matrix that is diagonal. In practice, it is sufficient to eliminate only half of the nondiagonal elements so that all elements on one side of a diagonal are 0:

$$\mathbf{A} = \begin{vmatrix} A_{11} & 0 & 0 \\ A_{21} & A_{22} & 0 \\ A_{31} & A_{32} & A_{33} \end{vmatrix} = \begin{vmatrix} A_{11} & A_{12} & A_{13} \\ 0 & A_{22} & A_{23} \\ 0 & 0 & A_{33} \end{vmatrix} = \begin{vmatrix} A_{11} & 0 & 0 \\ 0 & A_{22} & 0 \\ 0 & 0 & A_{33} \end{vmatrix}$$

$$= A_{11} A_{22} A_{33} \tag{B.28}$$

Program example B.1. The procedure `MatInv` in program unit `Matrix` (Chapter 6), listed in Appendix E, evaluates a determinant as a by-product of matrix inversion.

B.2 SOLUTION OF SIMULTANEOUS EQUATIONS BY DETERMINANTS

Consider the following set of three equations in three coefficients a_1, a_2, and a_3. We shall consider the y_k and X_{jk} to be known quantities; that is, constants:

$$y_1 = a_1 X_{11} + a_2 X_{12} + a_3 X_{13}$$
$$y_2 = a_1 X_{21} + a_2 X_{22} + a_3 X_{23} \tag{B.29}$$
$$y_3 = a_1 X_{31} + a_2 X_{32} + a_3 X_{33}$$

Let us consider the set of equations as if they were one matrix equation as in Equation (B.10):

$$
\begin{bmatrix} y_1 \\ y_2 \\ y_3 \end{bmatrix} = \begin{bmatrix} X_{11} & X_{12} & X_{13} \\ X_{21} & X_{22} & X_{23} \\ X_{31} & X_{32} & X_{33} \end{bmatrix} \begin{bmatrix} a_1 \\ a_2 \\ a_3 \end{bmatrix} \tag{B.30}
$$

with $\mathbf{a}$ and $\mathbf{y}$ represented by linear matrices and $\mathbf{X}$ represented by a square matrix. If we multiply the first equation of Equations (B.29) by the cofactor of X_{11} in the matrix of Equation (B.30), multiply the second equation by the cofactor of X_{21}, and multiply the third by the cofactor of X_{31}, then the sum of the three equations is an equation involving determinants according to Equation (B.18):

$$
\begin{vmatrix} y_1 & X_{12} & X_{13} \\ y_2 & X_{22} & X_{23} \\ y_3 & X_{32} & X_{33} \end{vmatrix} = a_1 \begin{vmatrix} X_{11} & X_{12} & X_{13} \\ X_{21} & X_{22} & X_{23} \\ X_{31} & X_{32} & X_{33} \end{vmatrix} + a_2 \begin{vmatrix} X_{12} & X_{12} & X_{13} \\ X_{22} & X_{22} & X_{23} \\ X_{32} & X_{32} & X_{33} \end{vmatrix}
$$

$$
+ a_3 \begin{vmatrix} X_{13} & X_{12} & X_{13} \\ X_{23} & X_{22} & X_{23} \\ X_{33} & X_{32} & X_{33} \end{vmatrix} \tag{B.31}
$$

The determinants in the two rightmost terms of Equation (B.31) both vanish because they have two columns that are identical. Thus, the solution for the coefficient a_1 is the ratio of the two determinants:

$$
a_1 = \frac{\begin{vmatrix} y_1 & X_{12} & X_{13} \\ y_2 & X_{22} & X_{23} \\ y_3 & X_{32} & X_{33} \end{vmatrix}}{\begin{vmatrix} X_{11} & X_{12} & X_{13} \\ X_{21} & X_{22} & X_{23} \\ X_{31} & X_{32} & X_{33} \end{vmatrix}} \tag{B.32}
$$

The denominator is the determinant of the square matrix $\mathbf{X}$ of Equation (B.30) and the numerator is the determinant of a matrix that is formed by substituting the column matrix y for the first column of the $\mathbf{X}$ matrix.

Similarly, *Cramér's rule* gives the solution for the jth coefficient a_j of a set of n simultaneous equations as the ratio of two determinants:

$$
y_k = \sum_{j=1}^{n} (a_j X_{kj}) \qquad k = 1, n
$$

$$
\tag{B.33}
$$

$$
a_j = \frac{|\mathbf{X}'(j)|}{|\mathbf{X}|}
$$

The denominator is the determinant of the **X** matrix. The numerator $|\mathbf{X}'(j)|$ is the determinant of the matrix formed by substituting the **y** matrix for the jth column.

A matrix is singular if its determinant is 0. If the **X** matrix is singular, there is no solution for Equation (B.33). For example, if two of the n simultaneous equations are identical, except for a scale factor, there are really only $n - 1$ independent simultaneous equations, and therefore no solution for the n unknowns. In this case, the **X** matrix has two identical rows and therefore a 0 determinant.

Solution by Matrix Equations

Let us consider Equation (B.33) as if it were a matrix equation as in Equation (B.30). If the **X** matrix is square, we can consider the **y** and **a** linear matrices as either column matrices as in Equation (B.10) or row matrices as in Equation (B.11):

$$[y_k] = [a_j][X_{kj}] \tag{B.34}$$

If we could multiply this matrix by another matrix **X**' such that the right-hand side becomes just the linear matrix **a**, then we will have our solution for the coefficients a_j directly. The multiplication of matrices is associative; that is,

$$\mathbf{A}(\mathbf{B}\mathbf{C}) = (\mathbf{A}\mathbf{B})\mathbf{C} \tag{B.35}$$

Therefore, we require a matrix **X**' such that if it is multiplied by the matrix **X**, the result is the unity matrix:

$$[X_{kj}][X'_{kj}] = \mathbf{1} \tag{B.36}$$

The matrix **X**' that satisfies Equation (B.36) is called the inverse matrix $\mathbf{X}^{-1}$ of **X**. Equation (B.34) multiplied from the right by $\mathbf{X}^{-1}$ gives the coefficients a_j explicitly, because any matrix is unchanged when multiplied by the unity matrix:

$$[y_k][X_{jk}]^{-1} = [a_j]\mathbf{1} = [a_j] \tag{B.37}$$

We can express Equation (B.37) in more conventional form to give the solution for each of the coefficients a_j:

$$a_j = \sum_{k=1}^{n} \left(y_k X_{kj}^{-1} \right) \tag{B.38}$$

Thus, the solution for the n unknowns with n simultaneous equations is reduced to evaluating the elements of the inverse matrix $\mathbf{X}^{-1}$.

B.3 MATRIX INVERSION

The adjoint $A^\dagger$ of a matrix A is defined as the matrix obtained by substituting for each element A_{jk} the cofactor of the transposed element A_{kj}:

$$A^\dagger_{jk} = \text{cof}(A_{kj}) \tag{B.39}$$

For a square symmetric matrix, the transposition makes no difference.

The inverse matrix A^{-1} defined in Equation (B.36) may be evaluated by dividing the adjoint matrix $A^\dagger$ by the determinant of A:

$$A^{-1}_{jk} = \frac{A^\dagger_{jk}}{|A|} \tag{B.40}$$

To show that this equality holds, we multiply both sides of Equation (B.40) by $|A|A$.

$$|A|AA^{-1} = |A|1 = AA^\dagger \tag{B.41}$$

Diagonal terms of the matrices in Equation (B.41) are equivalent to the formula of Equation (B.19) for evaluating the determinant:

$$|A| = \sum_{k=1}^{n} \left(A_{jk} A^\dagger_{jk} \right) = \sum_{k=1}^{n} \left[A_{jk} \text{cof}(A_{jk}) \right] \tag{B.42}$$

Off-diagonal elements can be shown to vanish like those of the determinants of Equation (B.31). If the matrix A is singular (that is, if $|A| = 0$), the inverse matrix A^{-1} does not exist and there is no solution to the matrix equation of Equation (B.34).

Gauss-Jordan Elimination

The formula of Equation (B.40) is generally too cumbersome for use in computing the inverse of a matrix. Instead, the Gauss-Jordan method of elimination is used to invert a matrix by building up the inverse matrix from a unity matrix while reducing the original matrix to unity.

Consider the inverse matrix A^{-1} as the ratio of the unity matrix divided by the original matrix, $A^{-1} = 1/A$. If we manipulate the numerator and denominator of this ratio in the same manner (multiplying rows or columns by the same constant factor and adding the same rows scaled to the same constants), the ratio remains unchanged. If we perform the proper manipulation, we can change the denominator into the unity matrix; the numerator must then become equal to the inverse matrix A^{-1}.

Let us write the 3×3 matrix A and the 3×3 unity matrix side by side and manipulate both to reduce the matrix A to the unity matrix. We start by using the formula of Equation (B.24) to eliminate the two off-diagonal elements of the first column:

$$
\begin{bmatrix} A_{11} & A_{12} & A_{13} \\ A_{21} & A_{22} & A_{23} \\ A_{31} & A_{32} & A_{33} \end{bmatrix} \qquad \begin{bmatrix} 1 & 0 & 0 \\ 0 & 1 & 0 \\ 0 & 0 & 1 \end{bmatrix}
$$

$$
\begin{bmatrix} A_{11} & A_{12} & A_{13} \\ 0 & A_{22} - A_{12}\dfrac{A_{21}}{A_{11}} & A_{23} - A_{13}\dfrac{A_{21}}{A_{11}} \\ 0 & A_{32} - A_{12}\dfrac{A_{31}}{A_{11}} & A_{33} - A_{13}\dfrac{A_{31}}{A_{11}} \end{bmatrix} \qquad \begin{bmatrix} 1 & 0 & 0 \\ -\dfrac{A_{21}}{A_{11}} & 1 & 0 \\ -\dfrac{A_{21}}{A_{11}} & 0 & 1 \end{bmatrix} \tag{B.43}
$$

Now, we divide the first row by A_{11} to get a diagonal element of

$$
\begin{bmatrix} 1 & \dfrac{A_{12}}{A_{11}} & \dfrac{A_{13}}{A_{11}} \\ 0 & A_{22} - A_{12}\dfrac{A_{21}}{A_{11}} & A_{23} - A_{13}\dfrac{A_{21}}{A_{11}} \\ 0 & A_{32} - A_{12}\dfrac{A_{31}}{A_{11}} & A_{33} - A_{13}\dfrac{A_{31}}{A_{11}} \end{bmatrix} \begin{bmatrix} \dfrac{1}{A_{11}} & 0 & 0 \\ -\dfrac{A_{21}}{A_{11}} & 1 & 0 \\ -\dfrac{A_{21}}{A_{11}} & 0 & 1 \end{bmatrix} \tag{B.44}
$$

The left matrix now has the proper first column. Let us relabel the matrices **B** (on the left) and **B'** (on the right) and perform the corresponding manipulations to obtain zeros in place of B_{12} and B_{32}, and then divide the second row by B_{22}:

$$
\begin{bmatrix} 1 & 0 & B_{13} - B_{23}\dfrac{B_{12}}{B_{22}} \\ 0 & 1 & \dfrac{B_{23}}{B_{22}} \\ 0 & 0 & B_{33} - B_{23}\dfrac{B_{32}}{B_{22}} \end{bmatrix} \begin{bmatrix} B'_{11} - B'_{21}\dfrac{B_{12}}{B_{22}} & -\dfrac{B_{12}}{B_{22}} & 0 \\ \dfrac{B'_{21}}{B_{22}} & \dfrac{1}{B_{22}} & 0 \\ B'_{31} - B'_{21}\dfrac{B_{32}}{B_{22}} & -\dfrac{B_{32}}{B_{22}} & 1 \end{bmatrix} \tag{B.45}
$$

After similar manipulation of the third column, the matrix on the left becomes the unity matrix and that on the right, therefore, must be the inverse matrix.

For computational purposes, even this method is somewhat inefficient in that two matrices must be manipulated throughout. Note, however, that at each stage of the reduction, there are only n (or three) useful columns of information in the two matrices. As each column is eliminated from the left matrix, the corresponding column is accumulated on the right.

Therefore, we can combine the manipulation into the range of a single matrix. We start with the matrix **A** and use the formula of Equation (B.24) as for Equation (B.43), but instead of applying this formula to the first column, we divide the first column by $-A_{11}$ to get the first column on the right of Equation (B.43); the diagonal element must be divided twice to become $1/A_{11}$. Divide the rest of the first row by A_{11} to get the composite of the two matrices of Equation (B.44):

$$
\begin{bmatrix}
\dfrac{1}{A_{11}} & \dfrac{A_{12}}{A_{11}} & \dfrac{A_{13}}{A_{11}} \\[3mm]
-\dfrac{A_{21}}{A_{11}} & A_{22} - A_{12}\dfrac{A_{21}}{A_{13}} & A_{23} - A_{13}\dfrac{A_{21}}{A_{11}} \\[3mm]
-\dfrac{A_{21}}{A_{11}} & A_{32} - A_{12}\dfrac{A_{31}}{A_{11}} & A_{33} - A_{13}\dfrac{A_{31}}{A_{11}}
\end{bmatrix}
\tag{B.46}
$$

A corresponding manipulation of the second column yields a matrix with the first two columns identical to those of the right side of Equation (B.45) whereas the last column is identical to that of the left side of Equation (B.45). Thus the inverse matrix is accumulated in the space vacated by the original matrix.

Program example B.2. Matinv inverts a square matrix and calculate its determinant, substituting the inverted matrix into the same array as the original matrix. Input variables are array, the matrix to be inverted, and nOrder, the order of its determinant. The routine is part of the program unit Matrix (Chapter 6) and is listed in Appendix E.[1]

The initial program loop iterates through the n columns of the matrix, reorganizing the matrix to get the largest element in the diagonal in order to reduce rounding errors and improve computational precision. The inversion procedure discussed above is then carried out and the determinant det of the matrix is calculated from the diagonalized matrix. After inversion, the inverted matrix is stored back in array and the variable det, the value of the determinant of the original matrix, is returned.

---

[1]The subroutine matInv follows the procedure of the subroutine MINV of the IBM System/360 Scientific Subroutine Package.

APPENDIX
C

GRAPHS
AND
TABLES

C.1 GAUSSIAN PROBABILITY DISTRIBUTION

The probability function $P_G(x; \mu, \sigma)$ for the Gaussian or normal error distribution is given by

$$P_G(x; \mu, \sigma) = \frac{1}{\sigma\sqrt{2\pi}} \exp\left[-\frac{1}{2}\left(\frac{x - \mu}{\sigma}\right)^2\right]$$

If measurements of a quantity x are distributed in this manner around a mean μ with a standard deviation σ, the probability $dQ_G(x; \mu, \sigma)$ for observing a value of x, within an infinitesimally small interval dx, in a random sample measurement is given by

$$dQ_G(x; \mu, \sigma) = P_G(x; \mu, \sigma)\, dx$$

Values of the probability function $P_G(x; \mu, \sigma)$ are tabulated in Table C.1 as a function of the dimensionless deviation

$$z = |x - \mu|/\sigma$$

for z ranging from 0.0 to 3.0 in increments of 0.01 and up to 5.9 in increments

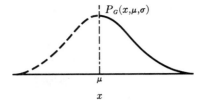

TABLE C.1
Gaussian probability distribution. The Gaussian or normal error distribution
$P_G(x; \mu, \sigma)$ **vs.** $z = |x - \mu| / \sigma$

z	0.00	0.01	0.02	0.03	0.04	0.05	0.06	0.07	0.08	0.09
0.0	0.39894	0.39892	0.39886	0.39876	0.39862	0.39844	0.39822	0.39797	0.39767	0.39733
0.1	0.39695	0.39654	0.39608	0.39559	0.39505	0.39448	0.39387	0.39322	0.39253	0.39181
0.2	0.39104	0.39024	0.38940	0.38853	0.38762	0.38667	0.38568	0.38466	0.38361	0.38251
0.3	0.38139	0.38023	0.37903	0.37780	0.37654	0.37524	0.37391	0.37255	0.37115	0.36973
0.4	0.36827	0.36678	0.36526	0.36371	0.36213	0.36053	0.35889	0.35723	0.35553	0.35381
0.5	0.35207	0.35029	0.34849	0.34667	0.34482	0.34294	0.34105	0.33912	0.33718	0.33521
0.6	0.33322	0.33121	0.32918	0.32713	0.32506	0.32297	0.32086	0.31874	0.31659	0.31443
0.7	0.31225	0.31006	0.30785	0.30563	0.30339	0.30114	0.29887	0.29659	0.29431	0.29200
0.8	0.28969	0.28737	0.28504	0.28269	0.28034	0.27799	0.27562	0.27324	0.27086	0.26848
0.9	0.26609	0.26369	0.26129	0.25888	0.25647	0.25406	0.25164	0.24923	0.24681	0.24439
1.0	0.24197	0.23995	0.23713	0.23471	0.23230	0.22988	0.22747	0.22506	0.22266	0.22025
1.1	0.21785	0.21546	0.21307	0.21069	0.20831	0.20594	0.20357	0.20122	0.19887	0.19652
1.2	0.19419	0.19186	0.18955	0.18724	0.18494	0.18265	0.18038	0.17811	0.17585	0.17361
1.3	0.17137	0.16915	0.16694	0.16475	0.16256	0.16039	0.15823	0.15609	0.15395	0.15184
1.4	0.14973	0.14764	0.14557	0.14351	0.14147	0.13944	0.13742	0.13543	0.13344	0.13148
1.5	0.12952	0.12759	0.12567	0.12377	0.12189	0.12002	0.11816	0.11633	0.11451	0.11271
1.6	0.11093	0.10916	0.10741	0.10568	0.10397	0.10227	0.10059	0.09893	0.09729	0.09567
1.7	0.09406	0.09247	0.09090	0.08934	0.08780	0.08629	0.08478	0.08330	0.08184	0.08039
1.8	0.07896	0.07755	0.07615	0.07477	0.07342	0.07207	0.07075	0.06944	0.06815	0.06688
1.9	0.06562	0.06439	0.06316	0.06196	0.06077	0.05960	0.05845	0.05731	0.05619	0.05509
2.0	0.05400	0.05293	0.05187	0.05083	0.04981	0.04880	0.04781	0.04683	0.04587	0.04492
2.1	0.04399	0.04307	0.04217	0.04129	0.04041	0.03956	0.03871	0.03788	0.03707	0.03627
2.2	0.03548	0.03471	0.03395	0.03320	0.03247	0.03175	0.03104	0.03034	0.02966	0.02899
2.3	0.02833	0.02769	0.02705	0.02643	0.02582	0.02522	0.02464	0.02406	0.02350	0.02294
2.4	0.02240	0.02187	0.02135	0.02083	0.02033	0.01984	0.01936	0.01889	0.01843	0.01798
2.5	0.01753	0.01710	0.01667	0.01626	0.01585	0.01545	0.01506	0.01468	0.01431	0.01394
2.6	0.01359	0.01324	0.01290	0.01256	0.01224	0.01192	0.01160	0.01130	0.01100	0.01071
2.7	0.01042	0.01015	0.00987	0.00961	0.00935	0.00910	0.00885	0.00861	0.00837	0.00814
2.8	0.00792	0.00770	0.00749	0.00728	0.00707	0.00688	0.00668	0.00649	0.00631	0.00613
2.9	0.00595	0.00578	0.00562	0.00546	0.00530	0.00514	0.00500	0.00485	0.00471	0.00457

z	0.00	0.10	0.20	0.30	0.40
3.0	0.0044318	0.0032668	0.0023841	0.0017226	0.0012322
3.5	0.00087269	0.00061191	0.00042479	0.00029195	0.00019866
4.0	0.00013383	0.000089264	0.000058945	0.000038536	0.000024943
4.5	0.000015984	0.000010141	0.0000063701	0.0000039615	0.0000024391
5.0	0.0000014868	0.00000089730	0.00000053614	0.00000031716	0.00000018575
5.5	0.00000010771	0.00000006183	0.00000003514	0.00000001978	0.00000001102

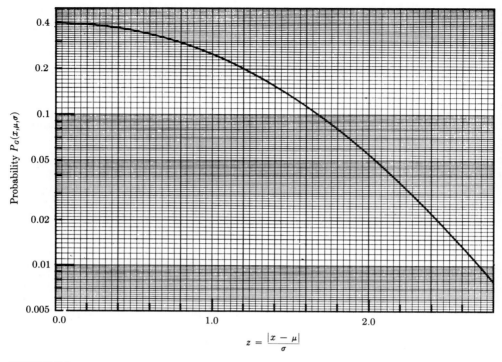

FIGURE C.1
The Gaussian probability function $P_G(x; \mu, \sigma)$ vs. $z = |x - \mu|/\sigma$.

of 0.1. This function is graphed on a semilogarithmic scale as a function of z in Figure C.1.

The function that is tabulated and graphed is $P_G(z; 0, 1)$, which gives the probability that $x = \mu \pm z\sigma$. It is the curve of Figure 2.5 tabulated only for positive values of z as indicated.

C.2   INTEGRAL OF GAUSSIAN DISTRIBUTION

The integral $A_G(x, \mu, \sigma)$ of the probability function $P_G(x, \mu, \sigma)$ for the Gaussian or normal error distribution is given by

$$A_G(x, \mu, \sigma) = \frac{1}{\sigma\sqrt{2\pi}} \int_{\mu - z\sigma}^{\mu + z\sigma} \exp\left[-\frac{1}{2}\left(\frac{x - \mu}{\sigma} \right)^2 \right] dx$$

$$z = \frac{|x - \mu|}{\sigma}$$

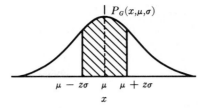

TABLE C.2
Integral of Gaussian distribution. The integral of the Gaussian probability
distribution $A_G(x; \mu, \sigma)$ **vs.** $z = |x - \mu| / \sigma$

z	0.00	0.01	0.02	0.03	0.04	0.05	0.06	0.07	0.08	0.09
0.0	0.0	0.00798	0.01596	0.02393	0.03191	0.03988	0.04784	0.05581	0.06376	0.07171
0.1	0.07966	0.08759	0.09552	0.10343	0.11134	0.11924	0.12712	0.13499	0.14285	0.15069
0.2	0.15852	0.16633	0.17413	0.18191	0.18967	0.19741	0.20514	0.21284	0.22052	0.22818
0.3	0.23582	0.24344	0.25103	0.25860	0.26614	0.27366	0.28115	0.28862	0.29605	0.30346
0.4	0.31084	0.31819	0.32551	0.33280	0.34006	0.34729	0.35448	0.36164	0.36877	0.37587
0.5	0.38292	0.38995	0.39694	0.40389	0.41080	0.41768	0.42452	0.43132	0.43809	0.44481
0.6	0.45149	0.45814	0.46474	0.47131	0.47783	0.48431	0.49075	0.49714	0.50350	0.50981
0.7	0.51607	0.52230	0.52847	0.53461	0.54070	0.54674	0.55274	0.55870	0.56461	0.57047
0.8	0.57629	0.58206	0.58778	0.59346	0.59909	0.60467	0.61021	0.61570	0.62114	0.62653
0.9	0.63188	0.63718	0.64243	0.64763	0.65278	0.65789	0.66294	0.66795	0.67291	0.67783
1.0	0.68269	0.68750	0.69227	0.69699	0.70166	0.70628	0.71085	0.71538	0.71985	0.72428
1.1	0.72866	0.73300	0.73728	0.74152	0.74571	0.74985	0.75395	0.75799	0.76199	0.76595
1.2	0.76985	0.77371	0.77753	0.78130	0.78502	0.78869	0.79232	0.79591	0.79945	0.80294
1.3	0.80639	0.80980	0.81316	0.81647	0.81975	0.82298	0.82616	0.82930	0.83240	0.83546
1.4	0.83848	0.84145	0.84438	0.84727	0.85012	0.85293	0.85570	0.85843	0.86112	0.86377
1.5	0.86638	0.86895	0.87148	0.87397	0.87643	0.87885	0.88123	0.88358	0.88588	0.88816
1.6	0.89039	0.89259	0.89476	0.89689	0.89898	0.90105	0.90308	0.90507	0.90703	0.90896
1.7	0.91086	0.91272	0.91456	0.91636	0.91813	0.91987	0.92158	0.92326	0.92491	0.92654
1.8	0.92813	0.92969	0.93123	0.93274	0.93422	0.93568	0.93711	0.93851	0.93988	0.94123
1.9	0.94256	0.94386	0.94513	0.94638	0.94761	0.94882	0.95000	0.95115	0.95229	0.95340
2.0	0.95449	0.95556	0.95661	0.95764	0.95864	0.95963	0.96059	0.96154	0.96247	0.96338
2.1	0.96426	0.96513	0.96599	0.96682	0.96764	0.96844	0.96922	0.96999	0.97074	0.97147
2.2	0.97219	0.97289	0.97358	0.97425	0.97490	0.97555	0.97617	0.97679	0.97739	0.97797
2.3	0.97855	0.97911	0.97965	0.98019	0.98071	0.98122	0.98172	0.98221	0.98268	0.98315
2.4	0.98360	0.98404	0.98448	0.98490	0.98531	0.98571	0.98610	0.98648	0.98686	0.98722
2.5	0.98758	0.98792	0.98826	0.98859	0.98891	0.98922	0.98953	0.98983	0.99012	0.99040
2.6	0.99067	0.99094	0.99120	0.99146	0.99171	0.99195	0.99218	0.99241	0.99264	0.99285
2.7	0.99306	0.99327	0.99347	0.99366	0.99385	0.99404	0.99422	0.99439	0.99456	0.99473
2.8	0.99489	0.99504	0.99520	0.99534	0.99549	0.99563	0.99576	0.99589	0.99602	0.99615
2.9	0.99627	0.99638	0.99650	0.99661	0.99672	0.99682	0.99692	0.99702	0.99712	0.99721

	0.00	0.10	0.20	0.30	0.40
3.0	0.9973002	0.9980648	0.9986257	0.99903315	0.99932614
3.5	0.99953474	0.99968178	0.99978440	0.99985530	0.999903805
4.0	0.999936656	0.999958684	0.999973308	0.999982920	0.999989174
4.5	0.9999932043	0.9999957748	0.9999973982	0.9999984132	0.99999904149
5.0	0.99999942657	0.99999966024	0.99999980061	0.99999988410	0.99999993327
5.5	0.99999996193	0.99999997847	0.99999998793	.99999999328	0.99999999627

If measurements of a quantity x are distributed according to the Gaussian distribution around a mean μ with a standard deviation σ, $A_G(x; \mu, \sigma)$ is equal to the probability for observing a value of x in a random sample measurement that is between $\mu - z\sigma$ and $\mu + z\sigma$; that is, it is the probability that $|x - \mu| < z\sigma$.

Values of the integral $A_G(x; \mu, \sigma)$ are tabulated in Table C.2 as a function of z for z ranging from 0.0 to 3.0 in increments of 0.01 and up to 5.9 in increments of 0.1. This function is graphed on a probability scale as a function of z in Figure C.2.

A related function is the error function erf Z:

$$\text{erf } Z = \frac{1}{\sqrt{\pi}} \int_{-Z}^{Z} e^{-z^2} dz = A_G(z\sqrt{2}, 0, 1)$$

The function that is tabulated and graphed is the shaded area between the limits $\mu \pm z\sigma$ as indicated.

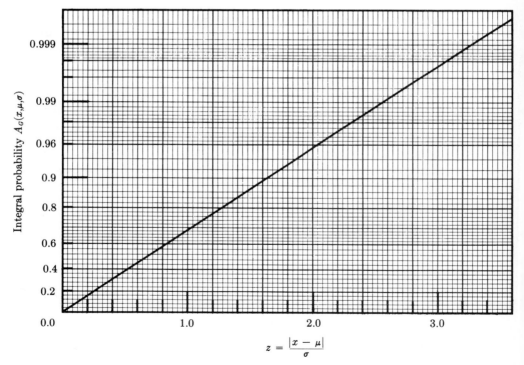

FIGURE C.2
The integral of the Gaussian probability distributions $A_G(x; \mu, \sigma)$ vs. $z = |x - \mu|/\sigma$.

C.3 LINEAR-CORRELATION COEFFICIENT

The probability distribution $P_r(r, \nu)$ for the linear-correlation coefficient r for ν degrees of freedom is given by

$$P_r(r;\nu) = \frac{1}{\sqrt{\pi}} \frac{\Gamma[(\nu + 1)/2]}{\Gamma(\nu/2)} (1 - r^2)^{(\nu-2)/2}$$

The probability of observing a value of the correlation coefficient larger than r for a random sample of N observations with ν degrees of freedom is the integral of this probability $P_c(r; N)$:

$$P_c(r; N) = \frac{1}{\sqrt{\pi}} \frac{\Gamma[(\nu + 1)/2]}{\Gamma(\nu/2)} \int_{|r|}^{1} (1 - x^2)^{(\nu-2)/2} \, dx \qquad \nu = N - 2$$

If two variables of a parent population are uncorrelated, the probability that a random sample of N observations will yield a correlation coefficient for those two variables greater in magnitude than $|r|$ is given by $P_c(r; N)$.

Values of the coefficient $|r|$ corresponding to various values of the probability $P_c(r; N)$ are tabulated in Table C.3 for N ranging from 3 to 100, and values of $P_c(r; N)$ ranging from 0.001 to 0.5. The functional dependence of r corresponding to representative values of $P_c(r, N)$ is graphed on a semilogarithmic scale as a smooth variation with the number of observations N in Figure C.3.

The function that is tabulated and graphed is the shaded area under the tails of the probability curve for values larger than $|r|$ as indicated.

C.4 χ^2 DISTRIBUTION

The probability distribution $P_\chi(x^2; \nu)$ for χ^2 is given by

$$P_\chi(x^2; \nu) = \frac{1}{2^{\nu/2}\Gamma(\nu/2)} (x^2)^{(\nu-2)/2} e^{-x^2/2}$$

The probability of observing a value of chi-square larger than χ^2 for a random sample of N observations with ν degrees of freedom is the integral of this

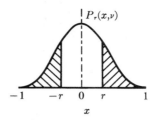

$P_r(x,\nu)$

x

TABLE C.3
Linear-correlation coefficient. The linear-correlation coefficient r vs. the number of observations N and the corresponding probability $P_c(r; N)$ of exceeding r in a random sample of observations taken from an uncorrelated parent population ($\rho = 0$)

N	P								
	0.50	0.20	0.10	0.050	0.020	0.010	0.005	0.002	0.001
3	0.707	0.951	0.988	0.997	1.000	1.000	1.000	1.000	1.000
4	0.500	0.800	0.900	0.950	0.980	0.990	0.995	0.998	0.999
5	0.404	0.687	0.805	0.878	0.934	0.959	0.974	0.986	0.991
6	0.347	0.608	0.729	0.811	0.882	0.917	0.942	0.963	0.974
7	0.309	0.551	0.669	0.754	0.833	0.875	0.906	0.935	0.951
8	0.281	0.507	0.621	0.707	0.789	0.834	0.870	0.905	0.925
9	0.260	0.472	0.582	0.666	0.750	0.798	0.836	0.875	0.898
10	0.242	0.443	0.549	0.632	0.715	0.765	0.805	0.847	0.872
11	0.228	0.419	0.521	0.602	0.685	0.735	0.776	0.820	0.847
12	0.216	0.398	0.497	0.576	0.658	0.708	0.750	0.795	0.823
13	0.206	0.380	0.476	0.553	0.634	0.684	0.726	0.772	0.801
14	0.197	0.365	0.458	0.532	0.612	0.661	0.703	0.750	0.780
15	0.189	0.351	0.441	0.514	0.592	0.641	0.683	0.730	0.760
16	0.182	0.338	0.426	0.497	0.574	0.623	0.664	0.711	0.742
17	0.176	0.327	0.412	0.482	0.558	0.606	0.647	0.694	0.725
18	0.170	0.317	0.400	0.468	0.543	0.590	0.631	0.678	0.708
19	0.165	0.308	0.389	0.456	0.529	0.575	0.616	0.662	0.693
20	0.160	0.299	0.378	0.444	0.516	0.561	0.602	0.648	0.679
22	0.152	0.284	0.360	0.423	0.492	0.537	0.576	0.622	0.652
24	0.145	0.271	0.344	0.404	0.472	0.515	0.554	0.599	0.629
26	0.138	0.260	0.330	0.388	0.453	0.496	0.534	0.578	0.607
28	0.133	0.250	0.317	0.374	0.437	0.479	0.515	0.559	0.588
30	0.128	0.241	0.306	0.361	0.423	0.463	0.499	0.541	0.570
32	0.124	0.233	0.296	0.349	0.409	0.449	0.484	0.526	0.554
34	0.120	0.225	0.287	0.339	0.397	0.436	0.470	0.511	0.539
36	0.116	0.219	0.279	0.329	0.386	0.424	0.458	0.498	0.525
38	0.113	0.213	0.271	0.320	0.376	0.413	0.446	0.486	0.513
40	0.110	0.207	0.264	0.312	0.367	0.403	0.435	0.474	0.501
42	0.107	0.202	0.257	0.304	0.358	0.393	0.425	0.463	0.490
44	0.104	0.197	0.251	0.297	0.350	0.384	0.416	0.453	0.479
46	0.102	0.192	0.246	0.291	0.342	0.376	0.407	0.444	0.469
48	0.100	0.188	0.240	0.285	0.335	0.368	0.399	0.435	0.460
50	0.098	0.184	0.235	0.279	0.328	0.361	0.391	0.427	0.451
60	0.089	0.168	0.214	0.254	0.300	0.330	0.358	0.391	0.414
70	0.082	0.155	0.198	0.235	0.278	0.306	0.332	0.363	0.385
80	0.077	0.145	0.185	0.220	0.260	0.286	0.311	0.340	0.361
90	0.072	0.136	0.174	0.207	0.245	0.270	0.293	0.322	0.341
100	0.068	0.129	0.165	0.197	0.232	0.256	0.279	0.305	0.324

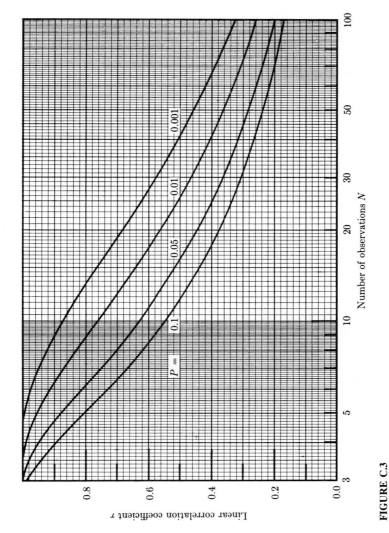

FIGURE C.3
The linear-correlation coefficient r vs. the number of observations N and the corresponding probability $P_c(r; N)$ that the variables are not correlated.

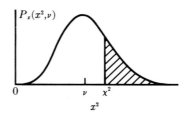

$$P_z(x^2,\nu)$$

0 $\quad \nu \quad \chi^2$

x^2

TABLE C.4
χ^2 **distribution. Values of the reduced chi-square** $\chi_\nu^2 = \chi^2 / \nu$ **corresponding to the probability** $P_\chi(\chi^2; \nu)$ **of exceeding** χ^2 **vs. the number of degrees of freedom** ν

					P			
ν	0.99	0.98	10.95	0.90	0.80	0.70	0.60	0.50
1	0.00016	0.00063	0.00393	0.0158	0.0642	0.148	0.275	0.455
2	0.0100	0.0202	0.0515	0.105	0.223	0.357	0.511	0.693
3	0.0383	0.0617	0.117	0.195	0.335	0.475	0.623	0.789
4	0.0742	0.107	0.178	0.266	0.412	0.549	0.688	0.839
5	0.111	0.150	0.229	0.322	0.469	0.600	0.731	0.870
6	0.145	0.189	0.273	0.367	0.512	0.638	0.762	0.891
7	0.177	0.223	0.310	0.405	0.546	0.667	0.785	0.907
8	0.206	0.254	0.342	0.436	0.574	0.691	0.803	0.918
9	0.232	0.281	0.369	0.463	0.598	0.710	0.817	0.927
10	0.256	0.306	0.394	0.487	0.618	0.727	0.830	0.934
11	0.278	0.328	0.416	0.507	0.635	0.741	0.840	0.940
12	0.298	0.348	0.436	0.525	0.651	0.753	0.848	0.945
13	0.316	0.367	0.453	0.542	0.664	0.764	0.856	0.949
14	0.333	0.383	0.469	0.556	0.676	0.773	0.863	0.953
15	0.349	0.399	0.484	0.570	0.687	0.781	0.869	0.956
16	0.363	0.413	0.498	0.582	0.697	0.789	0.874	0.959
17	0.377	0.427	0.510	0.593	0.706	0.796	0.879	0.961
18	0.390	0.439	0.522	0.604	0.714	0.802	0.883	0.963
19	0.402	0.451	0.532	0.613	0.722	0.808	0.887	0.965
20	0.413	0.462	0.543	0.622	0.729	0.813	0.890	0.967
22	0.434	0.482	0.561	0.638	0.742	0.823	0.897	0.970
24	0.452	0.500	0.577	0.652	0.753	0.831	0.902	0.972
26	0.469	0.516	0.592	0.665	0.762	0.838	0.907	0.974
28	0.484	0.530	0.605	0.676	0.771	0.845	0.911	0.976
30	0.498	0.544	0.616	0.687	0.779	0.850	0.915	0.978
32	0.511	0.556	0.627	0.696	0.786	0.855	0.918	0.979
34	0.523	0.567	0.637	0.704	0.792	0.860	0.921	0.980
36	0.534	0.577	0.646	0.712	0.798	0.864	0.924	0.982
38	0.545	0.587	0.655	0.720	0.804	0.868	0.926	0.983
40	0.554	0.596	0.663	0.726	0.809	0.872	0.928	0.983
42	0.563	0.604	0.670	0.733	0.813	0.875	0.930	0.984
44	0.572	0.612	0.677	0.738	0.818	0.878	0.932	0.985
46	0.580	0.620	0.683	0.744	0.822	0.881	0.934	0.986
48	0.587	0.627	0.690	0.749	0.825	0.884	0.936	0.986
50	0.594	0.633	0.695	0.754	0.829	0.886	0.937	0.987
60	0.625	0.662	0.720	0.774	0.844	0.897	0.944	0.989
70	0.649	0.684	0.739	0.790	0.856	0.905	0.949	0.990
80	0.669	0.703	0.755	0.803	0.865	0.911	0.952	0.992
90	0.686	0.718	0.768	0.814	0.873	0.917	0.955	0.993
100	0.701	0.731	0.779	0.824	0.879	0.921	0.958	0.993
120	0.724	0.753	0.798	0.839	0.890	0.928	0.962	0.994
140	0.743	0.770	0.812	0.850	0.898	0.934	0.965	0.995
160	0.758	0.784	0.823	0.860	0.905	0.938	0.968	0.996
180	0.771	0.796	0.833	0.868	0.910	0.942	0.970	0.996
200	0.782	0.806	0.841	0.874	0.915	0.945	0.972	0.997

TABLE C.4
χ^2 **distribution** (*continued*)

					P			
ν	0.40	0.30	0.20	0.10	0.05	0.02	0.01	0.001
1	0.708	1.074	1.642	2.706	3.841	5.412	6.635	10.827
2	0.916	1.204	1.609	2.303	2.996	3.912	4.605	6.908
3	0.982	1.222	1.547	2.084	2.605	3.279	3.780	5.423
4	1.011	1.220	1.497	1.945	2.372	2.917	3.319	4.617
5	1.026	1.213	1.458	1.847	2.214	2.678	3.017	4.102
6	1.035	1.205	1.426	1.774	2.099	2.506	2.802	3.743
7	1.040	1.198	1.400	1.717	2.010	2.375	2.639	3.475
8	1.044	1.191	1.379	1.670	1.938	2.271	2.511	3.266
9	1.046	1.184	1.360	1.632	1.880	2.187	2.407	3.097
10	1.047	1.178	1.344	1.599	1.831	2.116	2.321	2.959
11	1.048	1.173	1.330	1.570	1.789	2.056	2.248	2.842
12	1.049	1.168	1.318	1.546	1.752	2.004	2.185	2.742
13	1.049	1.163	1.307	1.524	1.720	1.959	2.130	2.656
14	1.049	1.159	1.296	1.505	1.692	1.919	2.082	2.580
15	1.049	1.155	1.287	1.487	1.666	1.884	2.039	2.513
16	1.049	1.151	1.279	1.471	1.644	1.852	2.000	2.453
17	1.048	1.148	1.271	1.457	1.623	1.823	1.965	2.399
18	1.048	1.145	1.264	1.444	1.604	1.797	1.934	2.351
19	1.048	1.142	1.258	1.432	1.586	1.773	1.905	2.307
20	1.048	1.139	1.252	1.421	1.571	1.751	1.878	2.266
22	1.047	1.134	1.241	1.401	1.542	1.712	1.831	2.194
24	1.046	1.129	1.231	1.383	1.517	1.678	1.791	2.132
26	1.045	1.125	1.223	1.368	1.496	1.648	1.755	2.079
28	1.045	1.121	1.215	1.354	1.476	1.622	1.724	2.032
30	1.044	1.118	1.208	1.342	1.459	1.599	1.696	1.990
32	1.043	1.115	1.202	1.331	1.444	1.578	1.671	1.953
34	1.042	1.112	1.196	1.321	1.429	1.559	1.649	1.919
36	1.042	1.109	1.191	1.311	1.417	1.541	1.628	1.888
38	1.041	1.106	1.186	1.303	1.405	1.525	1.610	1.861
40	1.041	1.104	1.182	1.295	1.394	1.511	1.592	1.835
42	1.040	1.102	1.178	1.288	1.384	1.497	1.576	1.812
44	1.039	1.100	1.174	1.281	1.375	1.485	1.562	1.790
46	1.039	1.098	1.170	1.275	1.366	1.473	1.548	1.770
48	1.038	1.096	1.167	1.269	1.358	1.462	1.535	1.751
50	1.038	1.094	1.163	1.263	1.350	1.452	1.523	1.733
60	1.036	1.087	1.150	1.240	1.318	1.410	1.473	1.660
70	1.034	1.081	1.139	1.222	1.293	1.377	1.435	1.605
80	1.032	1.076	1.130	1.207	1.273	1.351	1.404	1.560
90	1.031	1.072	1.123	1.195	1.257	1.329	1.379	1.525
100	1.029	1.069	1.117	1.185	1.243	1.311	1.358	1.494
120	1.027	1.063	1.107	1.169	1.221	1.283	1.325	1.446
140	1.026	1.059	1.099	1.156	1.204	1.261	1.299	1.410
160	1.024	1.055	1.093	1.146	1.191	1.243	1.278	1.381
180	1.023	1.052	1.087	1.137	1.179	1.228	1.261	1.358
200	1.022	1.050	1.083	1.130	1.170	1.216	1.247	1.338

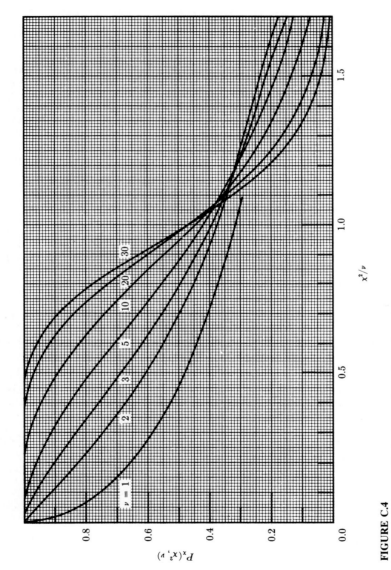

FIGURE C.4
The probability $P_x(\chi^2; \nu)$ of exceeding χ^2 vs. the reduced chi-square $\chi^2_\nu = \chi^2/\nu$ and the number of degrees of freedom ν.

probability $P_\chi(\chi^2; \nu)$:

$$P_\chi(\chi^2; \nu) = \frac{1}{2^{\nu/2}\Gamma(\nu/2)} \int_{\chi^2}^{\infty} (x^2)^{(\nu-2)/2} e^{-x^2/2} d(x^2)$$

Values of the reduced chi-square $\chi_\nu^2 = \chi^2/\nu$ corresponding to various values of the integral probability $P_\chi(\chi^2; \nu)$ of exceeding χ^2 in a measurement with ν degrees of freedom are tabulated in Table C.4 for ν ranging from 1 to 200. The functional dependence of $P_\chi(\chi^2; \nu)$ corresponding to representative values of ν is graphed in Figure C.4 as a smooth variation with the reduced chi-square χ_ν^2.

The function that is tabulated and graphed is the shaded area under the tail of the probability curve for values larger than χ^2 as indicated.

C.5 F DISTRIBUTION

The probability distribution for F is given by

$$P_f(f, \nu_1, \nu_2) = \frac{\Gamma[(\nu_1 + \nu_2)/2]}{\Gamma(\nu_1/2)\Gamma(\nu_2/2)} \left(\frac{\nu_1}{\nu_2}\right)^{\nu_1/2} \frac{f^{(\nu_1-2)/2}}{(1 + f\nu_1/\nu_2)^{1/2(\nu_1+\nu_2)}}$$

The probability of observing a value of F test larger than F for a random sample with ν_1 and ν_2 degrees of freedom is the integral of this probability:

$$P_F(F; \nu_1, \nu_2) = \int_F^{\infty} P_f(f; \nu_1, \nu_2) \, df$$

Values of F corresponding to various values of the integral probability $P_F(F; \nu_1, \nu_2)$ of exceeding F in a measurement are tabulated in Table C.5 for $\nu_1 = 1$ and graphed in Figure C.5 as a smooth variation with the probability. Values of F corresponding to various values of ν_1 and ν_2 ranging from 1 to ∞ are listed in Table C.6 and graphed in Figure C.6 for $P_F(F; \nu_1, \nu_2) = 0.05$ and in Table C.7 and Figure C.7 for $P_F(F; \nu_1, \nu_2) = 0.01$. These values were adapted by permission from Dixon and Massey.

The function that is tabulated and graphed is the shaded area under the tail of the probability curve for values larger than F as indicated.

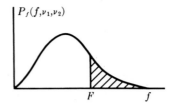

TABLE C.5
F distribution, $\nu = 1$. Values of F corresponding to the probability $P_F(F;1,\nu_2)$ of exceeding F (with $\nu_1 = 1$ degrees of freedom) vs. the larger number of degrees of freedom ν_2†

Degrees of freedom ν_2	Probability (P) of exceeding F							
	0.50	**0.25**	**0.10**	**0.05**	**0.025**	**0.01**	**0.005**	**0.001**
1	1.000	5.83	39.9	161	648	4050	16200	406000
2	0.667	2.57	8.53	18.5	38.5	98.5	198	998
3	0.585	2.02	5.54	10.1	17.4	34.1	55.6	167
4	0.549	1.81	4.54	7.71	12.2	21.2	31.3	74.1
5	0.528	1.69	4.06	6.61	10.0	16.3	22.8	47.2
6	0.515	1.62	3.78	5.99	8.81	13.7	18.6	35.5
7	0.506	1.57	3.59	5.59	8.07	12.2	16.2	29.2
8	0.499	1.54	3.46	5.32	7.57	11.3	14.7	25.4
9	0.494	1.51	3.36	5.12	7.21	10.6	13.6	22.9
10	0.490	1.49	3.28	4.96	6.94	10.0	12.8	21.0
11	0.486	1.47	3.23	4.84	6.72	9.65	12.2	19.7
12	0.484	1.46	3.18	4.75	6.55	9.33	11.8	18.6
15	0.478	1.43	3.07	4.54	6.20	8.68	10.8	16.6
20	0.472	1.40	2.97	4.35	5.87	8.10	9.94	14.8
24	0.469	1.39	2.93	4.26	5.72	7.82	9.55	14.0
30	0.466	1.38	2.88	4.17	5.57	7.56	9.18	13.3
40	0.463	1.36	2.84	4.08	5.42	7.31	8.83	12.6
60	0.461	1.35	2.79	4.00	5.29	7.08	8.49	12.0
120	0.458	1.34	2.75	3.92	5.15	6.85	8.18	11.4
∞	0.455	1.32	2.71	3.84	5.02	6.63	7.88	10.8

†For larger values of the probability P, the value of F is approximately $F \simeq [1.25(1-P)]^2$.

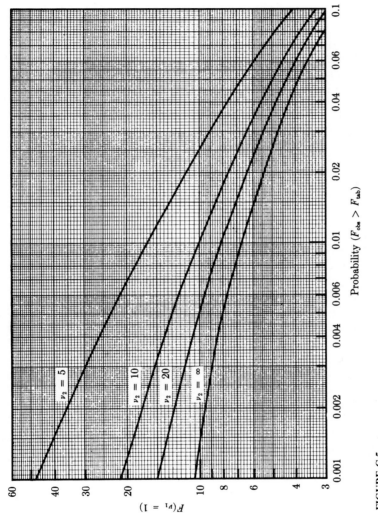

FIGURE C.5

The probability $P_F(F; 1, \nu_2)$ of exceeding F vs. F and ν_2 for $\nu_1 = 1$.

TABLE C.6
F distribution, 5%. Values of F corresponding to the probability $P_F(F; \nu_1, \nu_2) = 0.05$ of exceeding F for ν_1 vs. ν_2 degrees of freedom

Degrees of freedom ν_2	Degrees of freedom ν_1							
	2	**4**	**6**	**8**	**10**	**15**	**20**	**100**
1	200	225	234	239	242	246	248	253
2	19.0	19.2	19.3	19.4	19.4	19.4	19.4	19.5
3	9.55	9.12	8.94	8.85	8.79	8.70	8.66	8.55
4	6.94	6.39	6.16	6.04	5.96	5.86	5.80	5.66
5	5.79	5.19	4.95	4.82	4.73	4.62	4.56	4.41
6	5.14	4.53	4.28	4.15	4.60	3.94	3.87	3.71
7	4.74	4.12	3.87	3.73	3.64	3.51	3.44	3.27
8	4.46	3.84	3.58	3.44	3.35	3.22	3.15	2.97
9	4.26	3.63	3.37	3.23	3.14	3.01	2.94	2.76
10	4.10	3.48	3.22	3.07	2.98	2.85	2.77	2.59
11	3.98	3.36	3.09	2.95	2.85	2.72	2.65	2.46
12	3.89	3.26	3.00	2.85	2.75	2.62	2.54	2.35
15	3.68	3.06	2.79	2.64	2.54	2.40	2.33	2.12
20	3.49	2.87	2.60	2.45	2.35	2.20	2.12	1.91
24	3.40	2.78	2.51	2.36	2.25	2.11	2.03	1.80
30	3.32	2.69	2.42	2.27	2.16	2.01	1.93	1.70
40	3.23	2.61	2.34	2.18	1.08	1.92	1.84	1.59
60	3.15	2.53	2.25	2.10	1.99	1.84	1.75	1.48
120	3.07	2.45	2.18	2.02	1.91	1.75	1.66	1.37
∞	3.00	2.37	2.10	1.94	1.83	1.67	1.57	1.24

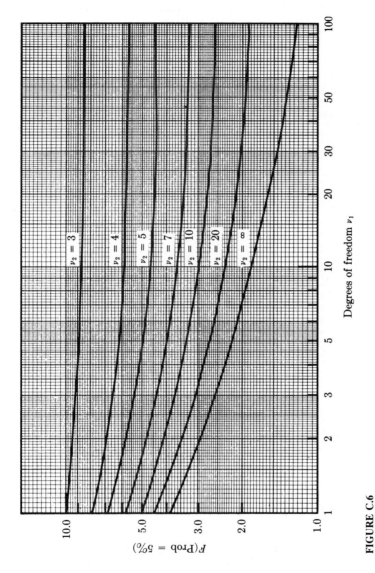

FIGURE C.6
Test values of $F(\nu_1, \nu_2)$ vs. the numbers of degrees of freedom ν_1 and ν_2 for a probability $P_F(F; \nu_1, \nu_2) = 0.05$ of exceeding F.

TABLE C.7
F distribution, 1%. Values of F corresponding to the probability
$P_F(F; \nu_1, \nu_2) = 0.01$ of exceeding F for ν_1 vs. ν_2 degrees
of freedom

Degrees of freedom ν_2	Degrees of freedom ν_1							
	2	4	6	8	10	15	20	100
1	5000	5620	5860	5980	6060	6160	6210	6330
2	99.0	99.2	99.3	99.4	99.4	99.4	99.4	99.5
3	30.8	28.7	27.9	27.5	27.2	26.9	26.7	26.2
4	18.0	16.0	15.2	14.8	14.5	14.2	14.0	13.6
5	13.3	11.4	10.7	10.3	10.1	9.72	9.55	9.13
6	10.9	9.15	8.47	8.10	7.87	7.56	7.40	6.99
7	9.55	7.85	7.19	6.84	6.62	6.31	6.16	5.75
8	8.65	7.01	6.37	6.03	5.81	5.52	5.36	4.96
9	8.02	6.42	5.80	5.47	5.26	4.96	4.81	4.42
10	7.56	5.99	5.39	5.06	4.85	4.56	4.41	4.01
11	7.21	5.67	5.07	4.74	4.54	4.25	4.10	3.71
12	6.93	5.41	4.82	4.50	4.30	4.01	3.86	3.47
15	6.36	4.89	4.32	4.00	3.80	3.52	3.37	2.98
20	5.85	4.43	3.87	3.56	3.37	3.09	2.94	2.54
24	5.61	4.22	3.67	3.36	3.17	2.89	2.74	2.33
30	5.39	4.02	3.47	3.17	2.98	2.70	2.55	2.13
40	5.18	3.83	3.29	2.99	2.80	2.52	2.37	1.94
60	4.98	3.65	3.12	2.82	2.63	2.35	2.20	1.75
120	4.79	3.48	2.96	2.66	2.47	2.19	2.03	1.56
∞	4.61	3.32	2.80	2.51	2.32	2.04	1.88	1.36

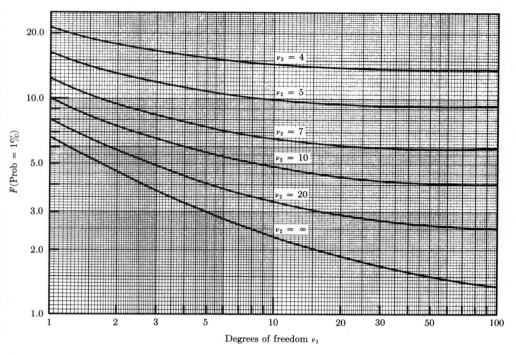

FIGURE C.7
Test values of $F(\nu_1, \nu_2)$ vs. the numbers of degrees of freedom ν_1 and ν_2 for a probability $P_F(F; \nu_1, \nu_2) = 0.01$ of exceeding F.

APPENDIX
D

HISTOGRAMS AND GRAPHS

Graphs of experimental data and of theoretical predictions have always been important tools for scientists, in both the actual performance of research and in presentations of results. In recent years we have seen a proliferation of graphics displays in all media as fast inexpensive computers have facilitated the display-making process. Scientists have benefitted from the new techniques and equipment, with many excellent commercial programs available for creating high quality scientific graphics suitable for publication. In this section, we shall limit our discussion to less ambitious plotting systems, computer routines that can often be written by the user and adapted to his or her particular needs. Graphs produced by such routines can be used by students and scientists in the data collection and analysis stages of an experiment, in discussions of results with colleagues, and in informal presentations and talks.

In science, the object is to present results in a straightforward manner so that relevant points are illustrated clearly but without bias. Bar graphs with suppressed zeros, which are common in advertisements, are not often seen in scientific papers. Bar graphs tend to be simple histograms rather than the multibar, brightly colored displays of the magazines. In fact, very few scientific preprints and published papers make use of color at all, although it is one of the clearest methods of presenting graphical information. Error bars, which are rare indeed in advertisements, are essential in a scientific presentation. Exaggerated perspective and distorted scales have very limited use in scientific work, whereas semilogarithmic plots that are often used in science are not often seen in business publications.

It is often convenient to have graphics routines that are part of a simulation or an analysis program, rather than to use free-standing programs. For example, in a Monte Carlo simulation, it is essential to be able to produce histograms and data graphs quickly at each stage of the study. Generations of scientists have made line-printer histograms of varying degrees of sophistication to display the distributions of generated or experimentally measured and analyzed data. We list some simple routines of this type that could be adapted and incorporated into a user's program.

Graphs that are more elegant and detailed than the line-printer plots can be created by making use of the graphics features of particular computers and languages. Some high level languages for personal computers include graphics commands to expedite plotting and drawing on the screen, and screen-dump programs are generally provided for transferring the screen plot to a dot matrix printer or even a laser printer. Hard copies of screen plots created by the screen-dump procedure generally have their resolution defined by the number of pixels on the screen, rather than by the generally higher resolution of the printer. For ordinary systems, this usually limits resolution to less than about 100 dots per inch. Such plots are often satisfactory for reports, overhead projections, and slides for talks, but not for publication. For highest resolution, plotting routines are available commercially that take full advantage of the resolution available on high quality printers.

D.1 MAKING A GRAPH

Whether a scientific graph is produced by hand or by computer, there are several basic principles that should be followed. The graph should be large enough to be read and understood easily, with appropriately proportioned abscissa and ordinate. Axes should be clearly and briefly labelled with large, clean letters, and the axes scales should be clearly marked. If more than a single function is displayed, or if both data and a curve are displayed, a box, or *legend*, may be superimposed on the graph to indicate the meaning of different symbols. In graphs published in scientific journals, a description of the graph is generally included as text below the abscissa label. In internal papers and preprints, these descriptions are often collected in a separate section of the paper. For visual presentation, some descriptive material may be included in a box on the graph, but it is important that text be sufficiently large so as to be legible to all viewers. One should avoid scattering too much material over any graph, which gives a busy appearance. A properly made graph should not require many words of explanation.

It is generally advisable to plot the independent variable as the abscissa and the dependent variable as ordinate. Reasonable, convenient values and intervals should be chosen for the scale marks on the two axes. For example, if x-axis values range from 0 to 400, it might be reasonable to divide the axis into eight parts and thus to mark the abscissa with *tick marks* at 50, 100, 150, and so on. Numbers might be placed on the axis at every second tick mark, 0, 100, 200, and so on. Dividing the x axis into six parts and putting ticks at 66.7, 133.3, and so forth, would make it difficult for a reader to interpret.

In general, error bars should be included for the ordinate data points, except for simple histograms where the text clearly specifies that the errors are Poisson and therefore given by the square root of the value of the coordinate. Unless otherwise noted, error bars generally indicate the standard deviation. Error bars may not be necessary on the abscissa. If appropriate, they may be drawn to indicate the resolution of the measurement or setting, or they may simply indicate the range of the variable over which data have been collected or grouped, as in the case of the width of a histogram bin. The text must explain the meaning of any error bar. If no error bar is shown for the abscissa, then it is useful to draw a circle or other symbol at each data point to indicate the position of the ordinate.

D.2   GRAPHICAL ESTIMATION OF PARAMETERS

A graph of y versus x often provides a convenient way of estimating parameters of the relation $y = y(x)$. The simplest example is the straight line

$$y = A + Bx \qquad (D.1)$$

where the slope and the intercept can be estimated by making a graph and drawing a straight line that relates y to x. Clearly the better way to handle this problem is by a least-squares fitting technique, but the graphical method can be useful in both research and instructional laboratories for obtaining quick preliminary estimates of experimental results.

If we wish to find from the graph the uncertainty in our estimate of the slope, for example, then we should attempt to draw two lines through the data, corresponding to reasonable estimates of the largest and smallest reasonable slopes, s_1 and s_2. We should take account in the uncertainties in the data points, if they are available and, because we are trying to estimate the uncertainty as a standard deviation, we should attempt to draw these two lines to bracket about 68% of the data points, not all the points. This is often difficult and subjective, especially if there are few points and they exhibit a lot of scatter. The mean slope s is just the average of our two slopes,

$$s = (s_1 + s_2)/2 \qquad (D.2)$$

and our estimate of the uncertainty is the difference

$$\sigma_s = (s_1 - s_2)/2. \qquad (D.3)$$

It is a worthwhile exercise to make estimates from a graph of the slope of a straight line and its uncertainty and to compare those estimates with the results of a least-squares fit to the data. We should note that the two lines selected to give a reasonable estimate of the uncertainty in the slope may not be the same two lines one would draw to obtain a reasonable estimate of the uncertainty in an intercept.

Semilogarithmic Graphs

When dealing with an exponential decay function, it is convenient to display the activity as a function of time on a semilogarithmic graph. That is, if the relation is

$$y(t) = y_0 e^{-at} \qquad (D.4)$$

we plot a graph of $\log(y)$ versus x. Fortunately semilogarithmic graph paper is readily available so that it is not necessary actually to calculate any logarithms to make this plot. We merely have to select paper with the appropriate number of powers of 10 for our plot, label the axes, and plot y versus x on the graph. Such a plot is illustrated in Figure 8.1 for Example 8.1.

Semilogarithmic graph paper comes in various *cycles*, corresponding to the number of *decades* or powers of 10 that can be plotted on a single sheet. Thus, for example, on three-cycle paper we can plot y values that range from 1 to 1000 (or from 0.01 to 10.0, etc.). Note that we can never plot y values that are zero or negative on semilogarithmic paper. This is a problem when dealing with subtracted distributions, such as the counting experiment of Example 8.1, where, if we wish to plot the number of counts remaining after we have subtracted the average background from cosmic rays, we discover that, at large times, some bins have negative net counts. Those points, of course, cannot be displayed on a semilogarithmic graph. A full, least-squares fit to the total, unsubtracted data sample is clearly the right way to solve this problem, but if we are to attempt a graphical solution, we should be aware of this limitation.

We can determine from our data the parameter a in Equation (D.4) by finding the slope of the straight line on the semilogarithmic graph just as we found the slope on ordinary graph paper for a simple linear plot. Note that when calculating the slope we must compute the logarithms of the y values. Thus, if the two ends of the straight line have coordinates (x_1, y_1) and (x_2, y_2), the slope is given by

$$s = \frac{\ln(y_2) - \ln(y_1)}{x_2 - x_1} = \frac{\ln(y_2/y_1)}{x_2 - x_1} \qquad (D.5)$$

The uncertainty in the slope can again be determined by drawing two straight lines that bracket the mean slope, although the logarithmic form of the plot increases the inaccuracy in this determination.

Full-Logarithmic Graphs

If we wish to display a power relation of the form $y = Ax^n$, we may make a plot of y versus x on full-logarithmic paper or *log-log* paper. The result will be a straight line with slope n and we can obtain the slope, and therefore the exponent n, from the graph. This technique could be used, for example, to check the $1/r^2$ law for radiation intensity as a function of distance, by plotting a graph of intensity versus distance on log-log paper.

In Section 7.4 we discuss variable transformation as a method of converting a nonlinear fitting problem to a linear problem, and the distortions that may be introduced into the uncertainties in the process. Plotting on semilogarithmic or full-logarithmic paper is equivalent to such a variable change and we should attempt to compensate for these distortions, if necessary.

D.3   HISTOGRAMS

If we wish to display the frequency distribution of a measured variable x, then a histogram is generally the simplest and clearest form of presentation. For example, we may have observed particles emitted in the decay of an unstable state and wish to present the number detected in successive time intervals as in Example 2.4. Alternatively, we may have measured secondary particles in a scattering experiment and wish to display the distribution of their energies. In both examples, we display the frequency distribution of the individual measurements, or *events*, as a histogram of $y(x)$ versus x, where $y(x)$ is the number of events that have values of x between $x + \Delta x$, and Δx is the histogram interval or *bin width*.

A convenient procedure for making a histogram of a continuous variable x is to label a bin with a tick mark at the lower limit x_i of the bin and to count within a bin those events for which $x_i \leq x < x_i + \Delta x$. This is convenient for most, but not all, data sets. Choice of the bin width depends on a number of factors. In the ideal situation with a large quantity of high precision data, the bin width could be chosen to be very small. However, in real experiments, the number of events may not be very large and each x coordinate will have some uncertainty. As a general rule, the bin width should not be less than the uncertainty in the measured variable x and one should be very wary of any data structure that is narrower than the uncertainty in x. (The exception is a study of uncertainties where the distribution of the measurements is the subject of interest.) If the number of events is relatively small, then even wider binning may be necessary. With such data, the competition between statistical significance and resolution of narrow effects in the histogram becomes important. A histogram with less than 10 events in its highest bin is not generally very informative because the uncertainty in the bin height will be over 30%.

A problem arises when the bin width of a histogram is close to or equal to the least count of the data. This can happen with integral data or with data that have been collected by a digital device. The previous suggestions that the histogram bins be labelled with the lower limit at the left of the bin may not be reasonable for such data, and it may be better to place tick marks at the center of the bins.

Example D.1. A student in an introductory physics laboratory attempts to measure the value of the acceleration of gravity by timing a ball that she drops 50 times from a height of 3 m. She uses an electronic timer with a least count of 0.01 s. The timer starts when the ball is released and stops when it hits the floor. Uncertainties in the measurements come mainly from variations in the starting and stopping times.

The student's measurements have been plotted in the histogram of Figure D.1 where the bin width is equal to the least count (0.01 s). We may assume that the digital clock truncates the measured times so each time measurement corresponds to the left-hand edge of a bin and the actual value of the time is somewhere within the bin limits. Thus, in this case it is appropriate to indicate the lower value of the bin limit at the left-hand edge of the bin.

The dashed Gaussian curve was calculated from the mean ($\bar{t} = 0.431$ s) and standard deviation ($s = 0.0184$) of the measurements. The curve clearly is shifted to the left relative to the data. The discrepancy is caused by the fact that we neglected to correct for the truncation of the data by the digital clock. To correct the mean we must add to it half the width of a bin to obtain $\bar{t} = 0.431 + 0.005 = 0.436$ s. The Gaussian curve, calculated from the corrected mean, is shown as a solid curve.

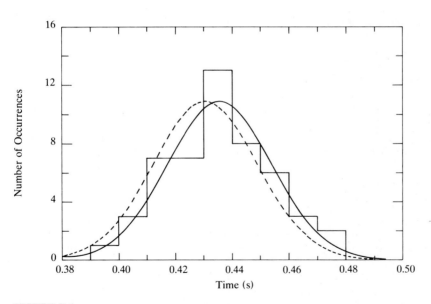

FIGURE D.1
Histogram of measured times plotted with the bin width equal to the least count of a digital clock. The numbers on the abscissa correspond to the lower time limit of the bin. The dashed Gaussian curve was calculated from the mean and standard deviation of the measurements. The solid curve was calculated with the mean increased by half the bin width to correct for the truncation of the data.

Normalized Curves on Histograms

When superimposing a theoretical curve on a data histogram, we often want to scale the area of the curve to that of the histogram, or to *normalize* the curve. The required scale factor can be determined in the following way. We assume that the curve has been calculated from a probability distribution function that is normalized to unit area, such as the Gaussian probability function of Equation (2.23). The area of one event on the histogram is equal to the bin width Δx multiplied by a unit interval on the ordinate. Thus, the total area of the histogram is equal to the bin width multiplied by the total number of events ($A = N\Delta x$). To scale the curve to the area of the histogram, we multiply the values $p(x_i)$, calculated from the equation of the probability distribution, by the product of the number of events on the plot and the bin width, so that the plotted curve becomes

$$y(x_i) = p(x_i) \times N\Delta x. \tag{D.6}$$

Program unit D.1. QuikHist Simple routines that can be included in Monte Carlo or analysis programs to make histograms of selected variables are listed in Appendix E. The routines produce plots either on the computer monitor or the printer. Screen displays can be either block characters or true graphics displays. The routines are listed in Appendix E and their use is illustrated by Program 5.2. The following routines are called.

HistInit(outPut) Selects the output device. The argument may be a printer or a disk file. The screen is the default output device for outPut = ´ ´. The routine is called once in a program.

HistSetUp(hNum, bot, int, top, hLabel) Defines each histogram. The routine is called once for each histogram with the histogram number, the lower limit of the histogram, the histogramming interval, the upper limit, and an identifying title or label as arguments.

Histogram(hNum,X) Increments the appropriate bin for variable X. The routine is called once for each variable that is to be tested for entry into a histogram. The histogram includes in a given bin all events for which the variable x is greater than or equal to the lower limit of the bin and less than the upper limit.

HistStats(hNum) Displays histogram statistics—the numbers of events that fell below, within, and above the range of the histogram and the mean, standard deviation, and the error in the mean.

HistDisplay(hNum) Displays each histogram as block characters, or as digits, with the bins running across the screen or page. This is the simplest method of display, and requires no special graphics features. The routine HistStats is called to display the statistics for each histogram. Output is to the monitor, the printer, or a disk file as specified in the call to Histinit. The internal variable printCode can be set to the ASCII value of the desired display character. If it is set to ´N ´, a digit or character corresponding to the number of events represented by each character will be displayed.

GHistDisplay(hNum) Displays a histogram on the monitor screen in graphics mode. The graphics routines in QuikPlot are called. (See the following section.) HistStats is not called, but may be called separately. Note that with separate calls to HistDisplay and GHistDisploay, a histogram can be both displayed in graphics mode on the screen and stored as numbers in a disk file for later examination or analysis.

HistDisplayAll(graph) Displays all histograms sequentially, in graphics mode if the Boolean variable graph is true, or as characters to the file defined in HistInit (which may be the screen or printer), if graph is false. The computer pauses after each histogram is displayed until the keyboard is struck.

HistFetch(hNum, hBin) Returns number of events recorded in a specified bin of a selected histogram.

D.4 GRAPHICS ROUTINES

Program unit D.2. QuikPlot A set of routines for making graphs and histogram is listed in Appendix E. All graphs and histograms in the main sections of this book were produced with slightly modified versions of these routines. Hard copy can be produced by replacing the unit Quickplot with the unit QuikScrp. Insructions are provided in the Read.Me file on the accompanying diskette. The sample program SamplPlt illustrates the use of the QuikPlot routines. The following calls extracted from this program illustrate the basic steps in making a graph with the QuikPlot procedures.

1. InitGraphics(setLog, 'D:\TP\') The first argument is a Boolean variable and if true, specifies that a semilogarithmic plot is to be made. The second argument must give the full path to the graphics driver.
2. Border(xLo, yLo, xHi, yHi) The four arguments are the ranges of the x and y variables.
3. Xgrid(FullXgrid,5,0); Ygrid(FALSE,5,1) These two optional calls place tick marks or grids on the plots. In each case the first argument is a Boolean variable that selects tick marks if FALSE and a full dashed-line grid if TRUE.
4. Xlabel('This is the X AXIS '); Ylabel('This is the Y AXIS ')-These two optional calls label the axes of the graph.
5. OpenWindow This function opens the window for plotting. The function must be called after the graph has been set up but before the actual plotting is done.
6. Calculations and Plotting The sample program calculates and plots a simple exponential function. Three types of plots are illustrated in both linear and semilogarithmic mode. Note that variable arrays are only required for the spline plots.
   (a) Rhist(xPrev,yPrev,xx[i] + dx,yy[i]) Plots a histogram of the data. On first entry to the routine, the variable xPrev should be set to the left-hand edge of the first bin and the variable yPrev to 0. These values are updated for subsequent calls and need not be reset by the user. On all entries,

the third and fourth variables are set to the value at the right-hand side of the current histogram bin.

(b) `RDataCross(xx[i],yy[i],5,sqrt(yy[i]))` Plots data crosses centered at the first and second arguments with width and height given by twice the third and fourth arguments, respectively.

(c-1) `SplineMake(nPts,0,0,xx,yy)` Calculates a table of second derivatives from the coordinate pairs in arrays `xx` and `yy`. The second and third arguments are the second derivatives at the ends of the plots. The values used correspond to the *natural spline* convention (i.e., second derivatives set to 0) and are not the actual second derivatives. The arrays `xx` and `yy` are declared in the program unit `FitVars`.

(d-2) `Scurve(1,nPts,5,0,xx)` Plots a spline curve with five intervals between each tabulated coordinate pair. The first and second arguments are the indices of the first and last points to be plotted; the third argument is the number of interpolated points between each pair of data points; the fourth coordinate is an offset that allows a curve to be shifted relative to the *x* coordinates (useful for plotting at the center of histogram bins). The last argument is the data array, which must be declared as `xx:rDataArray`, where the type `rDataArray` is defined in the program unit `QuikPlots` or in the unit `FitVars` (see comment at beginning of `QuikPlots`).

7. `CloseGraphics` Exits graphics mode and returns to text mode. Several other routines are available in `QuikPlot` for users. A list of these routines appears in the "Interface" section of the Program unit.

APPENDIX

E

COMPUTER ROUTINES IN PASCAL

Many of the techniques discussed in the book are illustrated by computer routines, which are described briefly in the text and listed in this appendix. For some routines that are not directly related to examples in the text, such as graphics and histogramming procedures, only the introductory, or *Interface*, sections are listed, although complete source code is available on the accompanying computer diskette.

Routines are written in Turbo Pascal, version 4.0, and are collected into *units* of that programming language. For readers who are not familiar with the notation, we note that units can be considered as convenient collections of related routines. The Interface section of each unit lists those routines that are accessible to external programs and serves as a directory of the unit. Users should be aware that the Turbo Pascal compiler includes in the executable file only those routines that are referenced by the program. Pascal routines may be divided into *programs*, *procedures*, and *functions*. We generally refer to these as *routines* in the text.

We have tried to keep the routines as simple as possible, trading efficiency for clarity when necessary. Familiarity with basic programming techniques is assumed. The precision of *Real* variables in Turbo Pascal is 11–12 digits, so double or extended precision was not required for any of the sample calculations. However, higher precision is strongly recommended for research calculations. For convenient reference, arguments of procedures and functions are listed both in the declaration of routines in the Interface section of the program units and in the header to the individual routines, although the second listing is not required.

Computer Disk

Complete source code for the computer routines is listed on the IBM-compatible computer diskette that accompanies this book. The disk also includes an executable program, Qdisplay, which can produce high resolution graphics plots on a dot-matrix or laser printer. Drivers are included for several popular types of printers. The routine communicates with a user's program through a script file, and can be linked through the Pascal Exec instruction, or can be run separately. The routine can also be used with script files produced by other programs to produce screen and hard-copy graphics plots. Instructions for use are provided in the READ.ME file on the diskette. With the exception of the program Qdisplay, which includes proprietary routines of GRAF/DRIVE PLUS, source code is provided for all routines.

PROGRAM LISTINGS

E.1 ROUTINES FROM CHAPTER 5

```
{Program 5.1: Simulated variation of temperature along a metal rod}
Program HotRod;
{ 10 cm rod-Temperature is zero at one end, 100 degrees C at other }
Uses CRT, MonteLib;
var
    i ,s1,s2,s3   : integer;    x, T, Tprime  : real;
const
    n      =  10;         {generate 10 points at 1 cm intervals}
    sigmaT =  1.0;        {with an uncertainty of +-1 degree}
begin
    s1 := 1171;
    s2 := 343;
    s3 := 1322;
    SetRandomDeviateSeed(s1, s2, s3);
    clrScr;
    writeln('       Hot Rod Test Data, sigma=',sigmaT:10:1);
    x := -0.5;
    for i := 1 to n do
    begin
        x       := x + 1;                {position along rod}
        T       := 10*x;                 {calculate mean temperature at point}
        Tprime := GaussSmear(T,sigmaT);  {smear it}
        writeln(i:10, x:10:1, T:10:2, Tprime:10:2);
    end;
    repeat until KeyPressed;
END.
```

```
{Program 5.2: Simulated decay of an unstable state}
Program PoisDcay;  { Generate a 200-event Poisson  histogram }
Uses  CRT, QuikHist, MonteLib;
const
    lo = 0; int = 1; hi = 22;
    mu = 8.4;   nEvents = 200;  nbins =  10;
var s1 ,s2 ,s3 ,i ,k : integer;    x, mean : real;
begin
    clrScr;
    s1 := 1171;  s2 := 343;  s3 := 1322;
    SetRandomDeviateSeed(s1, s2, s3);
    HistInit('');                        {output to monitor}
    HistSetup(1,lo,int,hi,'Poisson - Counts/10 sec');
    k:=PoissonDeviate(mu,TRUE);         {Initialize - make the table}
    for i := 1 to nEvents do
    begin
       k := PoissonDeviate(mu,FALSE);
       x := k;
       Histogram(1,x);
    end;
    HistDisplayAll(FALSE);  {block characters - no graphics}
    writeln('Seeds at end = ', s1:6, s2:6, s3:6);
    readln;
END.
```

```
{Program 5.3: Monte Carlo library routines}
Unit MonteLib;
Interface
Uses GenUtil;
procedure SetRandomDeviateSeed(sa,sb,sc:integer);
procedure GetRandomDeviateSeed(var sa,sb,sc:integer); {Wichmann & Hill}
function  RandomDeviate:real;
function  RandomGaussDeviate:real;  {Box-Muller -- find a random
                    variable drawn from the Gaussian distribution --}
function  GaussSmear(x,dx:real):real;
function  PoissonRecur(j:integer; m:real):real;{-- recursion method
         for Poisson Probability (P(n;m). To find P(n;m) must call
         with successive arguments j=0,1,...n. Max mu=85,no limit on x}
function  PoissonDeviate(mu:real; INIT:boolean):integer;{-- find a
                 random variable drawn from the Poisson distribution --}
Implementation
const  maxBins = 100;
var  seed1, seed2, seed3 : integer;     x2RanGauss, poiss  : real;
        Ptable     : array [0..maxBins] of real;
```

```
procedure SetRandomDeviateSeed(sa,sb,sc:integer);
begin
   seed1 := sa;   seed2 := sb;   seed3 := sc;
end;

procedure GetRandomDeviateSeed(var sa,sb,sc:integer);
begin
   sa := seed1;   sb := seed2;   sc := seed3;
end;

function RandomDeviate:real;
var  temp:real;
begin
   seed1 := 171*(seed1 mod 177) - 2*(seed1 div 177);
      if seed1 < 0 then seed1 := seed1 + 30269;
   seed2 := 172*(seed2 mod 176) - 35*(seed2 div 176);
      if seed2 < 0 then seed2 := seed2 + 30307;
   seed3 := 170*(seed3 mod 178) - 63*(seed3 div 178);
      if seed3 < 0 then seed3 := seed3 + 30323;
   temp := seed1/30269. + seed2/30307. + seed3/30323.;
   randomDeviate := temp-trunc(temp);
end;

function RandomGaussDeviate:real;
const  nextVar                   : boolean = false;
var  r, f, z1, z2, x1RanGauss  : real;
begin
   if nextVar = true then
   begin
      nextVar := false;
      RandomGaussDeviate := X2RanGauss;
   end
   else
   begin
      repeat
         z1 := -1 + 2*randomDeviate;
         z2 := -1 + 2*randomDeviate;
         r  := z1*z1 + z2*z2;
      until r < 1;
      f := sqrt(-2*Ln(R)/R);
      X1RanGauss := z1*f;
      X2RanGauss := z2*f;
      RandomGaussDeviate := X1RanGauss;
      nextVar := true;
   end;
end;
```

```pascal
function GaussSmear(x,dx:real):real;
begin
   GaussSmear := x + RandomGaussDeviate * dx;
end;


function PoissonRecur(j:integer; m:real):real;
begin
   if j=0 then poiss := exp(-m)
             else poiss := poiss*m/j; { poiss =  (m^j)exp(-mu/j!) }
   PoissonRecur := poiss;
end;


function PoissonDeviate(mu:real; init:boolean):integer;
var  i,x,n  :integer;  p,r     :real;
begin
   if init then
   begin                          {--- Make table of sums ---}
      n := trunc(mu + 8* sqrt(mu));   (ie., 8*sigma)
      if n > maxBins then ErrorAbort('Overflow err in Poisson dev');
      Ptable[0] := PoissonRecur(0,mu);
      for i := 1 to n-1 do
      begin
         p := PoissonRecur(i,mu);
         Ptable[i] := Ptable[i-1]+p;
      end;
      Ptable[n] := 1;              {--- Assure unit probability ---}
   end  else
   begin                          {--- Generate an event ---}
      x := -1;
      r := RandomDeviate;
      repeat
         inc(x);
      until Ptable[x] > r;
      PoissonDeviate := x;
   end;
end;
END.
```


E.2 ROUTINES FROM CHAPTER 6

```pascal
{Program 6.1: Least-squares fit to a straight
              line (Y=a[1]+a[2]*X )by the method of determinants  }
Program FitLine;
Uses  CRT, FitVars, FitUtil, GenUtil;


procedure CalculateY;           {fills array Ycalc}
```

```
var  i   : integer;
begin
    for i:= 1 to nPts do   yCalc[i] := a[1] + a[2]*x[i];
end;


function CalcChiSq:real; {assumes array yCalc has been filled}
var  i          :integer;    chi2     :real;
begin
    chi2:=0.;
    for i := 1 to nPts do
           chi2 := chi2 + sqr( (y[i]-yCalc[i]) / sigY[i]);
    CalcChiSq := chi2;
end;


procedure LineFit(var det :real);
var i        : integer;         dev     : text;
      sumWt, sumX, sumY, sumX2, sumY2, sumXY, weight, variance   : real;
begin
    sumWt := 0;   sumX   := 0;  sumY  := 0;
    sumX2 := 0;   sumY2 := 0;   sumXY := 0;
{--------- accumulate weighted sums -----------}
    for i:= 1 to nPts do
    begin
      weight  := 1/sqr(sigY[i]);
      sumWt   := sumWt  + weight;
      sumX    := sumX   + weight * x[i];
      sumY    := sumY   + weight * y[i];
      sumX2   := sumX2  + weight * sqr(x[i]);
      sumY2   := sumY2  + weight * sqr(y[i]);
      sumXY   := sumXY  + weight * x[i]*y[i];
  end;
{-calculate the parameters - cut out if determinant is not > 0 -}
    det :=  sumWt * sumX2 - sumX * sumX;
    if det > 0 then
    begin
        a[1]      := (sumx2*sumY  - sumX*sumXY)/det;
        a[2]      := (sumXY*sumWt - sumX*sumY) /det;
        sigA[1]   := sqrt(sumX2/det);
        sigA[2]   := sqrt(sumWt/det);
    end else
    ErrorAbort('Determinant <= 0 in LineFit');
end;
{-------------------------MAIN  ROUTINE----------------------}
var  title : string;    VorG : char;  i : integer; det, chi2 : real;
```

```
begin
   clrscr;
   m := 2;        {find 2 parameters}
   CenterWrite('(V)olts or (G)eiger? ',10);
   VorG := ReadChar;
   Case VorG of
     'V':  FetchData('VOLTS.Dat' ,title);                {Example 6.1}
     'G':begin
               FetchData('Geiger.Dat',title);     {table63.datExample 6.2}
               for i := 1 to nPts do x[i]  := 1/sqr(x[i]); {fitting 1/r^2}
          end;
   end;
   LineFit(det);
   CalculateY;      {fill array Ycalc for CalcChiSq and PlotIt}
   chi2 := CalcChiSq;
   OutPut(FALSE, 'CON', chi2, title); {FALSE for no error matrix}
   case VorG of
     'V':PlotIt(FALSE,FALSE,'C',abs(x[2]-x[1])/20,
                      {NoLog,Spline,style=circles,circle  radius}
             0, 0, 100, 3.0,                     {x1,y1,x2,y2}
             5, 6,                               {# x-div, # y-div}
             'X (cm)', 'Potential Difference (Volts)');
     'G':PlotIt(FALSE, FALSE, 'C',abs(x[2]-x[1])/50,
             0, 0.0, 30, 1000,
             6, 5,'Squared Inverse Distance (1/m^2)',
             'Number of Counts per sec');
   end;
   readln;
END.



{Program 6.2: Variable arrays for least-squares fitting}
Unit FitVars;
{Constants, variables and arrays for least-squares fits}
Interface
{-All global Types, Constants and Variables are declared here.---}
{ Global variables for linear regression        }
{ The array limits maxData and MaxParam can be set as required}
{                 for particular problems.  }
const maxData          =    100;
      maxParam         =     10;
      type rParamArray = array[1..maxParam]     of real;
      dataArray        = array[1..maxData]      of real;
      sqArray          = array[1..maxParam, 1..maxParam] of real;
      iparamArray      = array[1..maxParam]      of integer;
```

```
var    nPts, m, nFree    : integer;
       x, y, sigY, yCalc : dataArray;
       a, sigA, beta     : rParamArray;
       alpha             : sqArray;
{-- Variables used by non-linear methods --}
type   dblDataArray      = array[1..maxData, 1..maxParam] of Real;
var    y0                : dataArray;
       deltaA, dA        : rParamArray;
       dYda              : dblDataArray;
       chiSq1, chiSq2, chiSq3, chiOld, chiCut, stepSize : real;


Implementation
END.



{Program 6.3: Utility programs for least-squares fitting}
Unit FitUtil;   {Functions used in least-squares fits}
Interface
Uses CRT, Graph, GenUtil, QuikPlot, FitVars;
function  ChiProb(nFree:integer; chi2:real):real;
function  ChiProbDens(x:real; nFree:integer):real;
{ --- Unlisted routines- available on diskette -----------}
procedure FetchData(Fname:string; var title:string);{from file}
procedure FetchParameters;                      {from the keyboard}
procedure OutPut(errMatrix:boolean;OutDev:string;chi2:real;
                                                title:string);
procedure PlotIt(ifLogPlot, ifSpline:boolean;
       dataStyle:char;dx:real; {(C)ircle or (H)ist; dx for circle}
       xLo,yLo,xHi,yHi: real;nxGrid, nyGrid:integer;
       xTitle, yTitle: string);
var  outFile    : text;
Implementation


function ChiX(x:real):real; { Used by ChiProb for speed  }
begin                       { glSimps (global) = nFree/2 }
   if x = 0 then ChiX:= 0 else
   begin
      ChiX := exp((glSimps-1)*ln(x) - x/2); {  x^(h-1) e^(-x/2)}
   end;
end;


function ChiProb(nFree:integer; chi2:real):real; {max nFree = 56}
const    cLim      = 2;           {expansion limit for nFree = 1}
         intFromLim = 0.157;      {integral from cLim to infinity}
         dx        = 0.2;     {determines accuracy of integration}
var      nInt : integer;
```

```
    begin
        nInt := trunc(chi2/dx);
        if chi2 > 15*sqrt(nFree) then     {quick cutout}
        ChiProb := 0
    else
    begin
        glSimps := nFree/2;  {glSimps is global for ChiX}
        if nFree = 1 then
        begin                             { Integrate expansion }
            if chi2 < cLim then
            begin
                ChiProb := 1-sqrt(chi2/2/pi)*
                (2 - chi2*(1/3 - chi2*(1/20 - chi2*(1/168 - chi2/1728))));
            end else
            ChiProb    := intFromLim - Simpson(ChiX,nInt,cLim,chi2)
                                    /Gamma(nFree/2)/exp(nFree/2*ln(2))
        end else
        if nFree = 2 then ChiProb := exp(-chi2/2)  { Integrable }
        else
            ChiProb :=  1 - Simpson(ChiX, nInt, 0, chi2)
                            /Gamma(nFree/2)/exp(nFree/2*ln(2));
    end;
end;


function ChiProbDens(x:real; nFree:integer):real;
var num, den, h : real;
begin
    h             := nFree/2;
    num           := exp((h-1)*ln(x) - x/2);  { x^(h-1) e^(-x/2) }
    den           := exp(h*ln(2))*Gamma(h);   { 2^h * gamma(h)   }
    ChiProbDens := num/den;
end;


{Program 6.4:  General utilities routines}
Unit GenUtil; {A collection of general utility routines}
Interface
Uses  Dos, CRT, GRAPH;
type   SimpsFun = function(x:real):real;
                                        {functions for Simpson ->1 arg}
var  glSimps:real; {global parameter for Simpson's rule functions}
function  Gamma(h:real):real;
                        {approx Gamma functn by Stirling's approx}
function  Simpson(Funct:SimpsFun; nInt:integer;
                                        loLim,hiLim:real):real;
{ ------ Unlisted routines- available on diskette ----------}
```

```
function  Ask(s:string; yLoc:integer): boolean;
procedure CenterWrite(s:string; y : integer); {screen message}
procedure ErrorAbort(s:string);
function ReadChar:char;     {read a character, convert to upper case}
function ReadDigit:integer; {read a single digit from the keyboard}
function IntToStr(i:integer):string;  {convert interger to a string}
function StrFromNum(num:real;totalLen,aftDec:integer):string;
Implementation
function Gamma(h:real):real;              {Approximate Gamma function }
begin                            {   for integers and half-integers}
{gamma  = sqrt(2*pi)*[h^(h-1/2)]*[1 + 0.0833/h]  with h = nFree/2}
  Gamma := sqrt(2*pi)*exp(-h+(h-0.5)*ln(h))*(1 + 0.0833/h);
end;
{---Simpson's rule for functions of the form "Funct(x:real):real"--}
{If Funct has other parameters, they must be global, e.g., glSimps }
function Simpson(Funct:SimpsFun; nInt:integer;

                                        loLim, hiLim:real):real;
var sum,x,dx:real; i:integer;            {2 calcs/interval}
begin
    x     := loLim;
    dx    := (hiLim - loLim)/(2*nInt);
    sum   :=Funct(x);
    sum   := sum - Funct(hiLim);
    for i:=1 to nInt do
    begin
       x:=x+2*dx;
       sum:=sum + 4*Funct(x-dx) + 2*Funct(x);
    end;
    sum   := sum;
    Simpson := sum*dx/3;
end;
END.
```

E.3 ROUTINES FROM CHAPTER 7

```
{Program 7.1: Least-squares fit to a power
                        series and to Legendre polynomials}
Program MultRegr;
{   m=num of parameters, nPts=number of data pairs,
            arrays x,y,dy are data and errors.              }
Uses CRT, GenUtil, FitVars, FitUtil, Matrix, MakeAB7, FitFunc7;
var i     : integer; det, chi2   : real;   spl  :  boolean;
      ch      : char;   title        : string;
```

```pascal
begin
  clrscr;
  CenterWrite('(P)ower series, (A)ll Legendre terms to L=4,',10);
  CenterWrite('or (E)ven Legendre terms(L=0,2,4).',11);
  CenterWrite('Type P,A or E ',12);
  PAE := ReadChar;
  case PAE of
    'P'     :begin
               FetchData ('ThermCou.Dat', title);
               CenterWrite('Type number of parameters ',13);
               m := ReadDigit; writeln;
             end;
    'A','E' :begin
               FetchData('LegeAngl.Dat',title);
               if PAE= 'E' then m := 3 else m := 5 ;
             end;
  end; {case}
  MakeBeta;                          {set up the linear beta matrix}
  MakeAlpha;                         {set up the square alpha matrix}
  Matinv(m, Alpha, det);            {invert alph to get beta matrix}
  LinearBySquare(m,Beta,Alpha,a);   {beta x epsilon=parameter matrix}
  CalculateY;
  chi2 := CalcChiSq;
  for i := 1 to m do sigA[i] := sqrt(alpha[i,i]);
  OutPut(TRUE, 'CON.out', chi2, Title);  {TRUE to print error matrix}
  if m > 2 then spl := TRUE else spl := FALSE;
  case PAE of
    'P':PlotIt(FALSE,spl,'C',(x[2]-x[1])/12,{log?,splin?,Circl,radius}
              -10, -2, 110, 4,                    {x1,y1, x2,y2}
              6, 6,                              {x,y grid marks}
              'Temperature (degrees Celsius)','Voltage (mV)');
    'A','E':PlotIt(FALSE, TRUE, 'C', (x[2]-x[1])/10,
              0, 0, 180, 1500, 6, 6,
              'Theta(degrees)', 'Number of Counts');
  end;
END.


{Program 7.2: Fitting functions - Power series and Legendre polynomials}
Unit FitFunc7;
Interface
Uses FitVars;
var   PAE : char; {'P'-power series,
                          'A'-all Legendre terms to order m,
                          'E'-even Legendre terms}
function  Funct(k:integer; x:real):real;
```

```pascal
procedure CalculateY;
function  CalcChiSq:real; {assumes array yCalc has been filled}
Implementation
function PowerFunc(k:integer; x:real):real;
var  y:real;
         i:integer;
begin
   y:= 1;
   if k > 1 then
   begin
      for i:= 2 to k do
      begin
         y := x * y;
      end;
   end;
   PowerFunc := y;
end;


function LegFunc(k:integer; x:real):real;
{--Define separate terms in a series, y=a0*L0(x) + a1*L1(x) + ..
    Note k = 1 corresponds to zeroth order.}
var  LegPoly:array[1..11] of real; {i.e., 0th thru 10th order}
        i,kk:integer;         c:real;
begin
   case PAE of
      'E' : kk := 2*k-1;
      'A' : kk := k;
   end;
   c := cos(pi*x/180);
   LegPoly[1] := 1;{for better efficiency, could calc once and save}
   if kk > 1 then
   begin
      LegPoly[2] := c;
      if kk > 2 then
      begin
         for i := 3 to kk do
         begin
            LegPoly[i]:=((2*i-1)*c*LegPoly[i-1]-(i-1)*LegPoly[i-2])/i;
         end;
      end;
   end;
   LegFunc := LegPoly[kk];
end;


function Funct(k:integer; x:real):real;
```

```
begin
   case PAE of
      'A','E' : Funct := LegFunc(k, x);
      'P'     : Funct := PowerFunc(k , x);
   end;
end;

procedure CalculateY;
var  i, k : integer;      y  :  real;
begin
   for i:=1 to nPts do
   begin
      y := 0;
      for k := 1 to m do
      begin
         y := y + a[k] * Funct(k,x[i]);
      end;
      yCalc[i] := y;
   end;
end;

function CalcChiSq:real; {assumes array yCalc has been filled}
var  i  :  integer;     chi2  :  real;
begin
   chi2:=0.;
   for i := 1 to nPts do
   begin
      chi2 := chi2 + sqr( (y[i]-yCalc[i]) / sigY[i]);
   end;
   CalcChiSq := chi2;
end;
END.
```

{Program 7.3: Set up the alpha and beta matrices for linear
                                 least-squares fitting}

```
Unit MakeAB7;
Interface
Uses   FitVars, Matrix, FitFunc7;
procedure MakeBeta;
procedure MakeAlpha;
Implementation
procedure MakeBeta;  { Make the beta matrices }
var  i,k:integer;   tem : real;
begin
   for k:=1 to m do
```

```
    begin
        Beta[k]:=0;
        for i:=1 to nPts do
        begin
            Beta[k]:=Beta[k] + y[i]*Funct(k, x[i])/sqr(sigY[i]);
        end;
    end;
end;

procedure MakeAlpha;                    { Make the alpha matrices }
var  i,j,k : integer;     tem : real;
begin
    for j:=1 to m do
    begin
        for k:=1 to m do
        begin
            Alpha[j,k]:=0;
            for i:=1 to nPts do
            begin
                Alpha[j,k]  := Alpha[j,k]+Funct(j, x[i])
                            *Funct(k, x[i])/sqr(sigY[i]);
            end;
        end;
    end;
end;
END.

{Program 7.4: Matrix operations}
Unit Matrix;
Interface
Uses FitVars;
procedure MatInv(m:integer; var mArray:sqArray; var Det:real);
procedure LinearBySquare(m:integer; a:rParamArray;b:sqArray;
                                    var c:rParamArray);

Implementation

procedure Matinv;          {Invert a square matrix}
var ik,jk:iParamarray; i,j,k,L:integer; aMax,save:real;
label  100;
begin
    Det:=0;
{--- find largest element  ---}
    for k := 1 to m do
    begin
        aMax:=0;
```

```
100:for i := k to m do
       begin
           for j := k to m do
           begin
               if  Abs(mArray[i,j]) > Abs(aMax) then
               begin
                   aMax := mArray[i,j];
                   ik[k] := i;
                   jk[k] := j;
               end   {if}
           end; {for j}
       end; {for i}
       if aMax = 0 then EXIT;  {with 0 determinant as signal}
       Det := 1;
{---interchange rows and columns to put aMax in mArray[k,k]---}
       i := ik[k];
       if i < k then goto 100 else
       if i > k then
       begin
           for j := 1 to m do
           begin
               save := mArray[k,j];
               mArray[k,j] := mArray[i,j];
               mArray[i,j] := -save;
           end; {for j}
       end; {else if i}
       j := jk[k];
       if j < k then goto 100 else
       if j > k then
       begin
          for i := 1 to m do
          begin
              save := mArray[i,k];
              mArray[i,k] := mArray[i,j];
              mArray[i,j] := -save;
          end; {for i}
       end;   {if j}
{--- accumulate elements of inverse matrix ---}
       for i := 1 to m do
       begin
           if i <> k then
                 mArray[i,k] := -mArray[i,k]/aMax;
       end; {for i}
       for i := 1 to m do
```

```
        begin
            for j := 1 to m do
            begin
                if (i <> k) and (j <> k) then
                        mArray[i,j] := mArray[i,j] + mArray[i,k]*mArray[k,j];
            end; {for j}
        end;    {for i}
        for j := 1 to m do
        begin
            if j <> k then
                    mArray[k,j] := mArray[k,j]/aMax;
        end; {for j}
        mArray[k,k] := 1/aMax;
        Det := Det * aMax;
    end;    {for k}
{--- restore ordering of matrix ---}
    for L := 1 to m do
    begin
        k := m + 1 - L;
        j := ik[k];
        if j > k then
        begin
            for i := 1 to m do
            begin
                save := mArray[i,k];
                mArray[i,k] := -mArray[i,j];
                mArray[i,j] := save;
            end;  {for i}
        end;  {if}
        i := jk[k];
        if i > k then
        begin
            for j := 1 to m do
            begin
                save := mArray[k,j];
                mArray[k,j] := -mArray[i,j];
                mArray[i,j] := save;
            end; {for j}
        end; {if i}
    end; {for L}
end;


procedure LinearBySquare(m:integer; a:rParamArray;b:sqArray;
                    var c:rParamArray);  {matrix product}
var  i,j : integer;
```

```
begin
   for i:= 1  to m do
   begin
      c[i]:=0;
      for j:= 1 to m do
      begin
         c[i]:=c[i] +a[j]*b[i,j];
      end;
   end;
end;
END.
```

Routines listed in other sections: The units FitVars, FitUtil, and GenUtil are listed in Section E.2 (for Chapter 6).

E.4: ROUTINES FROM CHAPTER 8

```
{Program 8.0: Main calling routine for non-linear fitting methods}
program NonLinFt;
Uses CRT, GridSear, GradSear, ExpndFit, MarqFit, FitFunc8,
   MakeAB8, NumDeriv, Matrix, GenUtil,  FitVars,  FitUtil;
var  trial, j, method : integer;
        stepDown, chiSqr, lambda : real; title : string;
const stepScale:array[1..4] of real=(0.49999,0.99999,0.001,0.001);
begin
   clrscr;
   CenterWrite('     (1)Grid Search,      (2)Gradient Search',10);
   CenterWrite('     (3)ChiSq expansion, (4)Function expansion',11);
   CenterWrite('Type 1, 2, 3, or 4  --- ',12);
   method := ReadDigit;  writeln;
   chiCut    := 0.01;
   stepDown  := 0.1;            {step down the gradient in Gradls}
   lambda    := 0.001;          {for Marquardt method only}
   stepSize  := stepScale[method]; {scales deltaA[j]}
   FetchData('RadioDk.hst',title);
   FetchParameters;             {uses nPts, must follow FetchDAta}
   trial     := 0;
   chiSqr    := CalcChisq;
   repeat
      chiOld := chiSqr;
      writeln('Trial #',trial:4,' chiSq=',chiSqr:10:2);
      for j:= 1   to m do write(a[j]:12:4);
      writeln;
```

```
     case method of
          1: Gridls(chiSqr);
          2: Gradls(chiSqr, stepDown);
          3: ChiFit(chiSqr);
          4: Marquardt(chiSqr, chiCut, lambda);
     end;
     inc(trial);
until (abs(chiOld - chiSqr) < chiCut)  or  keypressed;
if keyPressed then readln;
CalculateY;
case method of
 1,2: begin
         for j := 1  to m do SigA[j] := SigParab(j);  {dChi2=1}
         OutPut(FALSE, 'CON', chiSqr, title); {no error matrix}
      end;
 3,4: begin
         if method =  4   then
         begin
           Marquardt(chiSqr,chiCut,0);  {get error matrix}
         end;
         for j := 1 to m do SigA[j]:=SigMatrx(j); {error matrix}
         OutPut(TRUE, 'CON', chiSqr, title); {with error matrix}
      end;
end;  {case}
PlotIt(TRUE,TRUE,'C',(x[2]-x[1])/5,{Log?,spline?,Circle,radius}
              0, 1, 900, 1000,                {ranges-x1,y1,x2,y2}
              6, 6,                  {num x-axis div, num y-axis div}
              'Time (sec)','Number of Counts');        {axis labels}
END.



{Program 8.1:Non-linear least-squares fit by the grid-search method}
Unit GridSear;
Interface
Uses      FitVars, FitFunc8, FitUtil;
procedure Gridls(var chiSqr:real);
Implementation
var    delta : real;

procedure Gridls(var chiSqr:real);
var
    save,delta1,del1,del2,aa,bb,cc,disc,alpha,x1,x2:real;
    j : integer;
begin
    chiSq2:= CalcChiSq;
```

```
{ -find local minimum for each parameter- }
   for j := 1 to m do
   begin
       delta    := deltaA[j];
       a[j]     := a[j] + delta;
       chiSq3   := CalcChiSq;
       if chiSq3 > chiSq2 then
       begin                       {started in wrong direction}
          delta  := -delta;
          a[j]   := a[j] + delta;
          save   := chiSq2;    {interchange 2 and 3 so 3 is lower}
          chiSq2 := ChiSq3;
          chiSq3 := save;
       end;
{ -Increment or decrement a[j] until chi squared increases- }
       repeat
          chiSq1 := chiSq2; {move back to prepare for quad fit}
          chiSq2 := chiSq3;
          a[j]   := a[j] + delta;
          chiSq3 := CalcChiSq;
       until chiSq3 > chiSq2;
{ -Find minimum of parabola defined by last three points  -}
       del1 := chiSq2 - chiSq1;
       del2 := chiSq3 - 2*chiSq2 + chiSq1;
       delta1 := delta * (del1/del2 + 1.5);
       a[j] := a[j]  - delta1;
       chiSq2 := CalcChiSq;     {at new local minimum}
{ -Adjust delta for change of 2 from chiSq at minimum  -}
       aa := del2/2;       {chiSq = aa*sqr(a[j] + bb*a[j] + cc}
       bb := del1 - del2/2;
       cc := chiSq1-chiSq2;
       disc := sqr(bb) -4*aa*(cc-2); {chiSqr difference=2}
       if disc > 0 then {if not true, then probably not parabolic yet}
       begin
          disc :=sqrt(disc) ;
          alpha:= (-bb - disc)/(2*aa);
          x1 := alpha*delta + a[1] - 2*delta;{a[j] at chiSq minimum+2}
          disc := sqr(bb) - 4*aa*cc;
          if disc>0 then disc:=sqrt(disc) else disc:=0;{elim round err}
          alpha:= (-bb - disc)/(2*aa);
          x2 := alpha*delta + a[1] - 2*delta;  {a[j] at chiSq minimum}
          delta := x1 - x2;
          deltaA[j] := delta;
       end;
   end;    { for j := 1 to m}
```

```
      chiSqr := chiSq2;
  end;
  END.




{Program 8.2 Non-linear least-squares fit by gradient-search method}
Unit GradSear;
Interface
Uses          FitVars, FitFunc8, FitUtil;
procedure Gradls(var chiSqr, stepDown:real);
Implementation
var grad:rParamArray; sum,stepSum,step1:real; j:integer;
procedure CalcGrad;
const
    fract = 0.001;
var
    j  :integer;
    dA :real;
begin
    sum := 0;
    for j := 1 to m do
    begin
       chiSq2 := CalcChiSq;
       dA   := fract * deltaA[j];{differential element for gradent}
       a[j]    := a[j] + dA;
       chiSq1  := CalcChiSq;
       a[j]    := a[j] - dA;
       grad[j] := chiSq2 - ChiSq1;    {2*da*grad}
       sum     := sum + sqr(grad[j]);
    end;
    for j := 1 to m do
    grad[j] :=  deltaA[j]*grad[j]/sqrt(sum); {step * grad}
end;


procedure Gradls(var chiSqr, stepDown:real);
label  5;
begin
    CalcGrad;   {calculate the gradient}
{-Evaluate chiSqr at new point and make sure chiSqr decreases-}
    repeat
       for j := 1 to m
                   do a[j] := a[j] + stepDown * grad[j]; {slide down}
       chiSq3 := CalcChiSq;
       if chiSq3 >= chiSq2 then
```

```
      begin                           {must have overshot minimum}
         for j := 1 to m do
                  a[j] := a[j] - stepDown * grad[j]; {restore}
         stepDown := stepDown/2;              {decrease stepSize}
      end;
   until chiSq3 <= chiSq2;
   stepSum := 0;
{ -- Increment parameters until chiSqr starts to increase -- }
   repeat
      stepSum := stepSum + stepDown;    {counts total increment}
      chiSq1 := chiSq2;
      chiSq2 := chiSq3;
      for j := 1 to m do a[j] := a[j] + stepDown * grad[j];
      chiSq3 := CalcChiSq;
   until chiSq3 > chiSq2;
{ -- Find minimum of parabola defined by last three points -- }
   step1:=stepDown*((chiSq3-chiSq2)/(chiSq1-2*chiSq2+chiSq3)+0.5);
   for j := 1 to m do
   begin
      a[j] := a[j] - step1 * grad[j];      {move to minimum}
   end;
   chiSqr := CalcChiSq;
   stepDown := stepSum;                 {start with this next time}
end;
END.




{Program  8.3: Non-linear least-squares fit by expansion of the
                        fitting function}
Unit  ExpndFit;
Interface
Uses       FitVars, FitFunc8, MakeAB8, Matrix;
procedure ChiFit(var chiSqr:real);
Implementation
var      j : integer;      det, chiSq1  : real;
procedure ChiFit(var chiSqr:real);
begin
   MakeBeta;
   MakeAlpha;
   MatInv(m, alpha, det);                {Invert matrix}
   LinearBySquare(m,Beta,Alpha,dA); {Evalulate parameter increments}
   for j := 1 to m do
   begin
      a[j] := a[j] + dA[j];  { Increment to next solution. }
   end;
```

```
      chiSqr:= CalcChiSq;
end;
END.
```

```
{Program 8.4: Non-linear least-squares fit by
                        the gradient-expansion (Marquardt) method}
Unit MarqFit;    {Marquardt method}
Interface
Uses      FitVars, FitFunc8, MakeAB8, Matrix;
procedure Marquardt(var chiSqr:real; chiCut, lambda:real);
Implementation
var  j   : integer;      det, chiSq1 : real;

procedure Marquardt(var chiSqr:real; chiCut, lambda:real);
label 1;
begin
1:MakeBeta;
   MakeAlpha;
   for j := 1 to m do  alpha[j,j] := (1 + lambda) * alpha[j,j];
   MatInv(m, alpha, det);            { Invert matrix }
   if lambda > 0 then {On final call, we just want the error matrix.}
   begin
      LinearBySquare(m,Beta,Alpha,dA);{Evaluate parameter increments }
      chiSq1 := chiSqr;
      for j := 1 to m do  a[j] := a[j] + dA[j];{Incr to next solution}
      ChiSqr := CalcChiSq;
      if chiSqr > chiSq1 + chiCut   then
      begin
         for j := 1 to m do  a[j]:=a[j]-dA[j];{Return to prev solution}
         chiSqr := CalcChiSq;
         lambda := 10*lambda; {and repeat the calc, with larger lambda}
         goto 1;
      end;
      lambda := 0.1 * lambda;
      end;
   end;
   END.
```

```
{Program 8.5: Sample fitting functions for non-linear fits}
Unit FitFunc8; { User supplied function for non-linear fitting }
     { Example is  sum of 2 exponentials on a constant background }
Interface
Uses FitVars;
function  yFunction(x:real):real;
procedure CalculateY;
```

```pascal
function  CalcChiSq:real;
function  SigParab(j:integer):real;
function  SigMatrx(j:integer):real;
Implementation
var i        : integer;    yy , arg  : real;
{|---------- Example 8.1 ------- linear bkgnd + 2 exponentials -|}
function  ExpF(a,x:real):real;
begin
   arg := abs(x/a);
   if arg > 60 then yy := 0 else  yy := Exp(-arg);
   ExpF := yy;
end;


function  yFunction(x:real):real;
var y : real;
begin
   Yfunction := a[1] + a[2]*ExpF(a[4],x) + a[3]*ExpF(a[5],x);
end;


{-- The following routines are general for fitting any function --}
procedure CalculateY;
  begin
    for i:=1 to nPts do  yCalc[i] := Yfunction(x[i]);
end;


function CalcChiSq:real;
var  chi2     :real;
begin
   chi2:=0.;
   for i := 1 to nPts do
        chi2 := chi2 + sqr( (y[i]-Yfunction(x[i]))/sigY[i]);
   CalcChiSq := chi2;
end;


{ Standard deviation calc'd from chiSq change of 1 (parabola fit)}
function  SigParab(j:integer):real;
begin
   chiSq2 := CalcChiSq;
   a[j]   := a[j] + deltaA[j];
   chiSq3 := CalcChiSq;
   a[j]   := a[j] - 2*deltaA[j];
   chiSq1 := CalcChiSq;
   a[j]   := a[j] + deltaA[j];
   SigParab := deltaA[j]*sqrt(2/(chiSq1-2*chiSq2+chiSq3));
end;
```

```
{Standard deviation calc'd from diagonal terms in error matrix}
function SigMatrx(j:integer):real;
var  sig : real;
begin
    sig := sqrt(abs(alpha[j,j]));
    if alpha[j,j] < 0 then sig := - sig;  {note- an error}
    SigMatrx := sig;
end;
END.


{Program 8.6: Matrix set-up for non-linear fits}
Unit MakeAB8;
Interface
Uses FitVars, Fitfunc8, NumDeriv;
procedure MakeBeta;
procedure MakeAlpha;
Implementation


procedure MakeBeta;   {Make beta matrices for non-linear fitting}
var  j   : integer;
begin
    for j := 1 to m do
    begin
       beta[j] := -0.5*dXiSq_da(j);
  end;
end;


procedure MakeAlpha; {Make alpha matrices for non-linear fitting}
var   j, k :integer;
begin
    for j := 1 to m do
begin
    alpha[j,j] :=  0.5 * d2XiSq_da2(j);
    if alpha[j,j] = 0 then
    begin
       writeln('Diagonal element is zero, j=',j:3);
       halt;
    end;
    if j > 1 then
    begin
       for k := 1 to j-1 do
       begin
          alpha[j,k] :=  0.5*d2XiSq_dajk(j,k);
          alpha[k,j] :=  alpha[j,k];
       end;  {for k}
```

```
            end;    {if j}
       end;       {for j}
       for j := 1 to m do
       begin
           if alpha[j,j] < 0 then
           begin
               alpha[j,j] := -alpha[j,j];
               if j > 1 then
               begin
                   for k := 1 to j-1 do
                   begin
                       alpha[j,k] := 0;
                       alpha[k,j] := 0;
                   end; {for k}
               end;   {if j}
           end;       {if alpha}
       end;          {for j}
end;
END.
```

```
{Program 8.7: Numerical derivatives for non-linear fits}
Unit NumDeriv;{Can be replaced by analytic derivatives, if they}
Interface    {can be calculated. However, numerical calculation}
Uses FitVars, FitFunc8;     { is general, and convenient}
function dXiSq_da(j:integer):real;
function d2XiSq_da2(j:integer):real;
function d2XiSq_dajk(j,k:integer):real;
Implementation
var  i : integer;     XiSq0, tem  : real;
function dXiSq_da(j:integer):real;
var  XiSqPlus : real;
begin
    if j=1 then  XiSq0:=CalcChiSq;  {starting point-calculate it once}
    a[j]          := a[j] + deltaA[j];
    XiSqPlus      := CalcChiSq;
    a[j]          := a[j] - deltaA[j];  {restore}
    tem           := (XiSqPlus - XiSq0)/(deltaA[j]);
    dXiSq_da      := tem;
end;

function d2XiSq_da2(j:integer):real;
var tem1:real;
begin
    if j = 1 then
```

```
   for i := 1 to nPts do
      y0[i] := yFunction(x[i]);{Starting point-calculate it once}
   a[j] := a[j] + deltaA[j];
   tem := 0;
   for i := 1 to nPts do
   begin
      tem1      := yFunction(x[i]);
      dYda[i,j] := (yFunction(x[i]) - y0[i])/deltaA[j]/sigY[i];
      tem       := tem + sqr(dYda[i,j]);
   end;
   a[j]       := a[j] - deltaA[j];
   d2XiSq_da2 := 2*tem;
end;


function d2XiSq_dajk(j,k:integer):real;
begin
   tem := 0;
   for i := 1 to nPts do
   begin
      tem := tem + dYda[i,j]*dYda[i,k];
   end;
   d2XiSq_dajk := 2*tem;
end;
END.
```

Previously defined program units: The units FitVars, FitUtil
and Gen Util were listed in Section E.2 (Chapter6).
The unit Matrix was listed in Section E.3 (Chapter 7).

E.5 ROUTINES FROM CHAPTER 9

```
{Program 9.1: Main calling routine for fit to Lorentzian
                                     + polynomial}
{- Note-must use  FitFUNCS8 from this directory -}
program LorenFit;
uses
   CRT, GRAPH, FitVars, Matrix, FitUtil, FitFunc8
   NumDeriv, MakeAB8, MarqFit, QuikPlot, GenUtil;
var title                           : string;
       trial, j                     : integer;
       xShift, dx, chiSqr, lambda : real;
const
   stepScale : array[1..4] of real = (0.49999, 0.99999, 0.001, 0.001);
begin
   clrscr;
   chiCut     := 0.01;
```

```
lambda     := 0.001;          {for Marquardt method only}
stepSize   := stepScale[4];   {scales deltaA[j]}
FetchData('single.hst',title);
xShift     := (x[2]- x[1])/2;
for j := 1 to nPts do x[j] := x[j] + xShift;  {move to bin center}
FetchParameters; {uses nPts, must follow FetchData}
trial      := 0;
chiSqr     := CalcChisq;
repeat
   chiOld := chiSqr;
   writeln('Trial #',trial:4,' chiSq=',chiSqr:10:2);
   for j   := 1 to m do write(a[j]:12:4);
   writeln;
   Marquardt(chiSqr, chiCut, lambda);
   inc(trial);
until (abs(chiOld - chiSqr) < chiCut)  or  keypressed;
if keyPressed then readln;
CalculateY;
Marquardt(chiSqr,chiCut,0);  {get error matrix}
for j     := 1 to m do SigA[j] := SigMatrx(j);{error matrix}
OutPut(TRUE,'CON',chiSqr,title); {with error matrix}
for j     := 1 to nPts do x[j] := x[j] - xShift;{restore to left edge}
PlotIt(FALSE,TRUE,'H', 0, {log?,spline?,Hist,0-not used}
            0.0, 0, 3.0, 90,  {x1,y1,x2,y2 for plot}
            6, 6,             {num grid marks x,y}
            'E (GeV)','Number of Counts'); {labels}
{ Plot the background }
   a[4]  := 0;    a[7] := 0;
   for j := 1 to nPts do  yCalc[j] := yFunction(x[j]);
   SplineMake(nPts,0,0,x,yCalc);
   Scurve(1, 40, 5, 0.025, x);  {spline curve}
   SetLineStyle(dashedLn,  0, ThickWidth);
   readln;
   CloseGraph;
END.


{Program 9.2 - Lorentzian peak on a quadratic background}
Unit FitFunc8; {Must be named FitFunc8 to use Chapter 8 routines}
Interface
uses FitVars;
function  yFunction(x:real):real;
{ -------- Unlisted routines- available on diskette ----------}
{ These are identical to the corresponding Chapter 8 routines}
procedure CalculateY;
function  CalcChiSq:real;
```

```
function  SigParab(j:integer):real;
function  SigMatrx(j:integer):real;

Implementation
var  i          : integer;
       yy , arg : real;
{|--------- Example 9.1 ------quadratic bkgnd + Lorentzian -|}
function  yFunction(x:real):real;
var y : real;
begin
    Y          := a[1] + a[2]*x + a[3]*sqr(x) +
                     a[4]*a[6]/(2*pi)/(sqr(x-a[5]) + sqr(a[6]/2));
    Yfunction := y;
end;
END.
```

E.6 ROUTINES FROM CHAPTER 10

```
{Program 10.1: Direct maximum likelihood example}
Program MaxLike;
Uses GRAPH, CRT, PRINTER, FitVars, FitUtil, GenUtil;
const maxEvents=2000;s{max number of events} c=3.0;{vel of light}
var
{-------------- Set or used in StartUp --------------------}
    inF : text;  nTrials  : integer;
    loSearch , hiSearch, tauStep,   {search range and step}
    xLo, xhi, yLo, yHi: real;  {plotting range}
{------ Input data from file -------------See Fig 10.1--
    for d1 and d2, the short and long fiducial volume cutoffs;
    xProduction and xDecay correspond to vertices V1 and V2.
------------------------------------------------------}
    nEvents:integer; mass,d1,d2,pLab:real; title:string;
    xProduction, xDecay:array [1..maxEvents] of real;
{------------------ Data arrays for search --------------}
    times, LtoTscale             : array [1..maxEvents] of real;
{ ------ Unlisted routines- available on diskette ---------}
procedure StartUp(inFile: string);chData;
procedure OutPut(var dTau:real; tauAtMax, maxM: real);
procedure PlotLikeCurve(tau, dtau, maxM:real);{M vs tau +-nSig st devs}

function LogProb(k:integer; tau:real):real;
var    loTlim, hiTlim, a, b : real;
begin                     {proper time/length}
{ d1, d2 are beginning and end of the fiducial region.  Must cvt
   to loTlim and hiTlim which are integration limits in proper time
   measured from production vertex.}
```

```pascal
      loTlim := (d1-xProduction[k]) * LtoTscale[k];
      hiTlim := (d2-xProduction[k]) * LtoTscale[k];
{ -------Now, calc probability------- }
      if hiTlim > 50 then b := 0
                          else b := exp(-hiTlim/tau);
      a := exp(-loTlim/tau);
{ prob      := exp(-times[k]/tau)/(tau*(a - b)); }
      LogProb  := -times[k]/tau - ln(tau*(a - b));  {log likelihood}
end;


function LogLike(t:real):real;
var  i  : integer;     M, prob : real;
begin
    M := 0;
    for i    := 1 to nEvents do
    begin
        prob  := LogProb(i,t);
        M     := prob + M;
    end;
    LogLike := M;
end;


procedure Search(var   tauAtMax, maxM : real);
var trial : integer; M1,M2,M3,del1,del2,delta1,tau,MLikeLi : real;
begin
    M2       :=   -1000;
    maxM     :=   -1.0e20;
    tau      :=   loSearch;
    for trial := 0 to nTrials do
    begin
        MLikeLi := logLike(tau);
        writeln('trial',trial:3, ' tau=', tau:8:4,
                                       ' Log Likelihood=',MLikeli:12);

        M3 := MLikeLi;
        if M3 > M2 then              {Remember, these are negative}
        begin                        {Still heading to maximum}
           M1 := M2;
           M2 := M3;
        end else
        begin    {leaving maximum.}
{  -Find maximum of parabola defined by last three points-  }
             del1     := M2 - M1;
             del2     := M3 - 2*M2 + M1;
             delta1   := tauStep * (del1/del2 + 1.5);
             tau      := tau - delta1;
```

```
            tauAtMax := tau;
            maxM     := LogLike(tau);    {at  maximum of parabola}
            writeln('Max from parabola, tau=', tau:8:4,
                                  ' Log Likelihood=',maxM:10);
            exit;
        end;
        tau := tau + tauStep;
      end;
end;


function Error(t, dt:real):real; {1/sqrt[-2nd derivative of log(L)]}
var
   t1, t2, d2Ydt2, err : real;
begin
   t1 := t - dt;
   t2 := t + dt;
   d2Ydt2 := (LogLike(t2) - 2*LogLike(t) + LogLike(t1))/sqr(dt);
   err    := 1/sqrt(-d2Ydt2);
   Error  := err;
end;
{===================== MAIN ROUTINE =======================}
var  sigTau, tauAtMax, maxM : real;{M is log of likelihood function}
begin
    Startup('Da50.dat');           {assign file, set search limits}
    FetchData;                     {read data file}
    Search(tauAtMax, maxM);        {search for max of Log L}
    dTau := Error(tauAtMax, tauStep); {find uncertainty in tauAtMax}
    writeln('N-events', nEvents:5,' max log likelihood=',maxM:10,
          ' tau at max =',tauAtMax:10:6, ' +-', sigTau:7:4);
    PlotLikeCurve(tauAtMax, sigTau, maxM);{plot the likelihood curve}
    repeat until keypressed;
    CloseGraph;
END.
```

Routines listed in other sections: The program units FitUtil
and GenUtil are listed in Section E.2 (Chapter 6).

E.7 ROUTINES FROM CHAPTER 11

```
{Program 11.1: Calculate the chi-square probability integral}
Program Chi2Prob;
Uses FitUtil;
var  chi2:real;  nFree:integer;
label 1;
begin
   writeln('Test Integral of Chi2 probability');
```

```
 1:write('type nDof, chi2 ');
   readln(nFree, chi2);
   writeln(' new ', ChiProb(nFree, chi2):10:3);
   goto 1;
END.
```

```
{Program 11.2: Calculate linear correlation probability integral}
Program LcorProb;
Uses LcorLate;
var nObserv:integer;  rCorr:real;
label 1;
begin
   writeln('Test Integral of Linear Correlation Function');
 1:write('Type-# observations, linear correlation coefficient : ');
   readln(nObserv, rCorr);
   writeln('Integral Corr fn=',LinCorProb(nObserv-2, rCorr):10:3);
   goto 1;
END.
```

```
{Program 11.3: Linear-correlation probability function & integral}
Unit LcorLate;
Interface
Uses FitUtil, GenUtil;
function  LinCorProb(nFree:integer; hiLim: real):real;
function  LinCorrel(r:real):real;
Implementation

function LinCorProb(nFree:integer; hiLim: real):real;
const  dx = 0.01;    loLim = 0.0;
var  nInt : integer;
begin
   glSimps    := nFree;  {global for LinCorrel when called by Simpson}
   nInt       := trunc((hiLim - loLim)/dx);
   LinCorProb := 1-2*Simpson(LinCorrel, nInt, loLim,  hiLim);
end;

function LinCorrel(r:real):real;{Set glSimps=nFree, global for
                       "Simpsons" which allows only 1 arg.}
const sqrtPi = 1.7724539;
begin
   LinCorrel := Gamma((glSimps+1)/2)/Gamma(glSimps/2)
             *exp( (glSimps-2)/2 * ln(1 - sqr(r)))/sqrtPi;
end;
END.
```

Routines listed in other sections: The program units FitUtil
and GenUtil are listed in Section E.2 (Chapter 6).
The routine ChiProb is part of the unit FitUtil.

E.8 ROUTINES FROM APPENDIX A
{Program A.1 Routines for cubic spline interpolation. Constant
   intervals in the independent variable are assumed. These
   routines are included in QuikPlots (Appendix D)   }

```
Unit Splines;
Interface
Uses FitVars;
procedure SplineMake(nn: integer; d2ydx2A, d2ydx2B: real;
  xIn, yIn:dataArray);   { Set up table of numerical 2nd derivs }
function SplineInt(x :real):real;                  { Interpolate }
Implementation

var           {variables set in SplineMake, used in SplineInt}
   xx,yy, d2ydx2 : dataArray;              {defined in FitVars}
   n             : integer;
   h             : real;
procedure SplineMake(nn: integer; d2ydx2A, d2ydx2B: real;
                                    xIn, yIn:dataArray);
var
   i                  : integer;
   a, delt1, delt2, b  : dataArray;
begin
   n := nn;           { This makes n available to procedure SplineInt}
   h := (xIn[n] - xIn[1])/(n-1);
   for i := 1 to n do
   begin
      xx[i] := xIn[i];
      yy[i] := yIn[i];
   end;
   d2ydx2[1]:=d2ydx2A;{End values of 2nd derivatives are input vars}
   d2ydx2[n] := d2ydx2B;
   a[2] := 4;
   for i := 3 to n-1 do
                a[i] := 4-1/a[i-1];                { coefficients    }
   for i := 2 to n  do
                delt1[i] := yIn[i] - yIn[i-1];    { 1st differences  }
   for i :=  2 to n-1 do
                delt2[i] := 6*(delt1[i+1] - delt1[i])/sqr(h);{ 2nd dif x 6 }
   b[2]  := delt2[2] - d2ydx2[1];                { b coefficients   }
   for i := 3 to n-1 do
                b[i] := delt2[i] - b[i-1]/a[i-1];
```

```
      b[n-1] := b[n-1] - d2ydx2[n];
      d2ydx2[n-1] := b[n-1]/a[n-1];
      for i := n-2 downto 2 do
              d2ydx2[i] := (b[i] - d2ydx2[i+1])/a[i]; { 2nd derivatives }
end;


function SplineInt(x:real):real;        { perform the interpolation }
   function DyDx(i : integer):real;              { 1st derivative }
   begin
      DyDx := (yy[i+1]-yy[i])/h - h*(d2ydx2[i]/3+d2ydx2[i+1]/6);
   end;

   function D3yDx3(i:integer):real;              { 3rd derivative }
   begin
      D3yDx3 := (d2ydx2[i+1] - d2ydx2[i])/h;
   end;
var dx   :   real;   i        :   integer;
begin
   i := trunc((x-xx[1])/h)+1;
   if i < 1   then i := 1;
   if i > n-1 then i := n-1;
   dx := x -xx[i] ;
   { interpolate x }
   if i = n then SplineInt := yy[i]
      else SplineInt := yy[i] + (DyDx(i) + (d2ydx2[i]/2 +
                          D3yDx3(i)/6*dx)*dx)*dx;

end;
END.
```

Routines listed in other sections: The program unit FitVars
is listed in Section E.2 (Chapter 6).

E.9 ROUTINES FROM APPENDIX D

```
{----Program D.1 Sample plotting program - uses QuikPlot----}
Program SampPlt;
Uses CRT, GRAPH, QuikPlot, GenUtil;
const setLog :   boolean = FALSE;   fullXgrid : boolean = FALSE;
{----------------------plot ranges----------------------}
   xLo  = 0;    ylo: real = 0;   xHi = 500;   yHi = 100;
{----------------------------------------------------------}
   nPts = 11;   dx = (xHi-xLo)/(nPts-1);
var x, y, xPrev, yPrev           :  real;        i : integer;
        semiLog, monoPrint, plotType : char;   xx, yy : dataArray;
begin
```

```
{------------------------- MAIN BODY ----------------------}
   ClrScr;
   If ASK('Do you want a semi-log plot? ',5) then
   begin
      yLo      := 1.0;
      setLog   := TRUE;      FullXGrid := TRUE;
   end;
   if ASK('Do you want mono for screen dump (Y or N)?  ',6)
            then SetMono(0,15,1,0); {forg, bkg pal; forg, bkg col}
   CenterWrite('(H)istogram (D)ata crosses, or (S)pline curve? ',7);
   plotType := ReadChar;
{---------------------- Calls to Plotting Routines ----------------}
   InitGraphics(setLog,'D:\TP\Graphics\');{Set path to graphics driver}
   Border(xLo, yLo, xHi, yHi);
   Xgrid(FullXGrid,5,0);    {FullXGrid TRUE if LOG plot,otherwise FALSE}
   Ygrid(FALSE,5,1);        {FALSE gives short ticks at edges}
   XLabel('This is the X AXIS'); {max = 50 chars in VGA}
   YLabel('This is the Y AXIS'); {max = 35 chars in VGA}
   OpenWindow;                      {define the plotting window}
   x := dx/2-dx;
   xPrev := x + dx;    yPrev := 0;  {yPrev is used by RHist}
   for i:=1 to nPts do
   begin
      x := x + dx;
      y := 90*exp(-x/130);        {calculate a simple exponential example}
      case plotType of          {histogram, data circles, or spline curve}
        'H': RHist(xPrev,yPrev, x+dx, y);
        'D': RDataCircle(x, y, 2, sqrt(y));               {X,Y,dX,dY}
        'S': begin                        {Splines requires arrays}
                xx[i] := x;
                yy[i] := y;
             end;
      end;
   end;
   if plotType = 'S' then                             {Spline curve}
   begin
      Setlinestyle(solidln,0,thickWidth);
      SplineMake(nPts, 0, 0, xx, yy);
      Scurve(1,nPts,5,0,xx); {1st pt,last pt,# interp pts,x-offset,x-array}
      Setlinestyle(solidLn,0,normWidth);
   end;
   readln;
   CloseGraphics;
{-------------------------END MAIN BODY ----------------------}
END.
```

```
{Program D.2 Sample histogramming program}
Program SampHist;
uses  CRT, QuikHist, MonteLib, GenUtil;
var i : word;
      r : real;
      which : char;
const
   nEvents = 500;
   lo=0; int=1; hi=10;  {histogram range and interval}
   mean = (hi+lo)/2; sigma = (hi+lo)/6;
                 {Gaussian mean and standard deviation}
begin
   clrScr;
   CenterWrite('Screen Display - (G)raphics  or (C)haracters? ',10);
   which := ReadChar; writeln(which);
   SetRandomDeviateSeed(317,1375,1237);
   HistInit('');
   HistSetup(1, lo, int, hi,'Gaussian Deviate');
   clrScr;
   for i:= 1 to nEvents do
   begin
      gotoxy(57,25);
      write('event #',i:6);
      r:=sigma*RandomGaussDeviate + mean;
      Histogram(1,r);
   end;
   HistDisplayAll(which='G');
END.
```

```
{Program D.3: Printer or screen histograms}
{Interface section of unlisted routines- available on diskette}
Unit QuikHist;  {Printer or screen histograms}
Interface
Uses  CRT, GRAPH, QuikPlot, GenUtil;
procedure GHistDisplay(hNum:integer);{graphics display}
procedure HistDisplay(hNum:integer); {block character display}
procedure HistDisplayAll(graph:boolean);{TRUE for graphics output}
function  HistFetch(hNum,hBin:integer):integer;{fetch one bin}
procedure HistInit(outPut:string);{select output device, ''=screen}
procedure Histogram(hNum:integer;X:real);{build the histogram}
procedure HistSetBorder(grid:boolean;xLow,xHigh:real;nxticks,
    nxdecPt:integer) {for graphics, can change the default border}
procedure HistSetup(hNum:integer;bot,int,top:real;hLabel:string);
procedure HistStats(hNum:integer);{display mean, st.dev.,etc.}
```

Implementation
```
----------------See diskette for remainder of program--------------

{Program D.4: Printer or screen graphs}
{ --- Unlisted routines- available on diskette ----------}
Unit QuikPlot;{Collection of graphics plotting routines}
Interface
uses  CRT, GRAPH, GenUtil;
{NOTE:When QuikPlots is used with other programs in this book,
remove the following "type" statement and add "FitVars" to
the "uses" statement.}
type dataArray = array[1..100] of real;
{----------------------INITIALIZATION ROUTINES--------------------}
procedure Border(x1,y1,x2,y2:real);
procedure InitGraphics(logOn:boolean ; graphDriverPath:string);
{--set path to graphics drivers---}
procedure SetDefaultColors(c0,c1,c2,c3,c4,c5,c6,c7:integer);
procedure SetMono(fgPal, bkgPal, fgCol, bkgCol:integer);
{--to make a monochrome plot for a screen dump to a printer,
    enter with fgPal and bkgPal and fgCol and bkgCol set to the
    appropriate palettes for a sceen dump on your system. Systems
    vary, e.g., VGA screen to HP laser printer requires fgPal=0
    and bkgPal=15. Set fgCol and bkgCol to the desired screen colors}
procedure OpenWindow;
{-----------------------TERMINATION ROUTINES--------------------}
procedure CloseGraphics;    {for compatibility with QuikScrp}
procedure CloseWindow;
{-----------------------POINT SCALING ROUTINES------------------}
function  XScale(x:real):Integer;{change data to screen coordinates}
function  YScale(y:real):Integer;
procedure SetScreenRange(ix1, iy1, ix2, iy2:integer);
{-set pixel count of screen size if default is not satisfactory}
{--------------------------GRID ROUTINES--------------------------}
procedure Xgrid(fullGrid:boolean;nTics,nDec:integer);
procedure Ygrid(fullGrid:boolean;nTics,nDec:integer);
procedure SetXYformat(xForm, yForm: word);
{ set number of digits for numbers on X and Y axes, or use default.}
{-----------------------LABELLING ROUTINES-----------------------}
procedure XLabel(w:String); {-- max = 50 characters in VGA --}
procedure YLabel(w:String); {-- max = 35 characters in VGA --}
procedure WriteTextXY(x,y:real;s:string);{write string s at x,y}
{-----------------DATA AND CURVE PLOTTING ROUTINES----------------}
procedure RCurve(i:integer; x,y:real);{Line segments,i=0 on 1st call}
procedure RDataCircle(x,y,dr,dy:real);{Vertical line through circle}
procedure RDataCross(x,y,dx,dy:real);
```

```
procedure RHist(var x1, y1:real; x2, y2:real);
{-plot a histogram.  On 1st call, enter with limits of first
  bin. Routine updates x1 and y1 on later calls.}
procedure RLine(x1,y1,x2,y2:real);
{-------------------------SPLINE ROUTINES------------------------}
procedure SplineMake(nn:integer;d2ydx2A,d2ydx2B:real;
                                 xIn,yIn:dataArray); {N equal steps of h}
function  SplineInt(x :real):real;{spline interpolation}
{Spline curve-pt1, ptn are 1st and last points to plot from array x,
   nInt = # spline intervals; xOffset=x bin plotting offset.}
procedure SCurve(pt1, ptn, nInt:integer; XoffSet:real; x:dataArray);
Implementation
-----------------See diskette for remainder of program---------------
```

```
{Program D.5: Printer or screen graphs by a script file}
program SmpScrip;         {--Sample program to produce a script file for
                   Qdisplay and to run QDISPLAY.EXE via EXEC statement.
   See Interface section of QUIKSCRP.PAS for available instructions.--}
{$M $4000,0,0}  {16k stack, no heap}
Uses DOS, CRT, QuikScrp, GenUtil;
const setLog :    boolean = FALSE;  fullXgrid : boolean = FALSE;
{-----------------------plot ranges-----------------------}
   xLo = 0;   ylo: real = 0;  xHi = 500;  yHi = 100;
{----------------------------------------------------------}
   nPts = 11;   dx = (xHi-xLo)/(nPts-1);
var  x, y, xPrev, yPrev             : real;      i : integer;
        semiLog, monoPrint, plotType : char;  xx, yy : dataArray;
const  {------------- Change to match user's system ----------------}
        scriptF   = 'test.scr';         {script file for graphics commands}
        workDrive : char   = ' ';  {' ' gives current drive for work files}
        workMem:  word   = 0;        {0 gives 32k work space for hard copy}
        printer   : string[4] = 'FX';                     {see list below}
        printPort : string[4] = 'LPT1';     {'LPT1','PT2','COM1',or 'COM2'}
        pageSize  : char     = 'H';       {'H' or 'F' for Half or Full;}
        resolution: char     = 'M';    {'L','M','H' for Low, Medium, High}
        baudRate  : string[4] = '2400';{or'300','600','1200','4800','9600'}
        parity    : char     = 'N';            {'E'ven, 'N'o, or 'O'dd}
        nDataBits : char     = '8';                        {'7' or '8'}
        nStopBits : char     = '1';                        {'1' or '2'}
        XON       : char     = 'Y';               {'Y'es or 'N'o}
{ printers      Epson compatible  9-pin dot matrix 'FX'
                          Epson compatible 24-pin dot matrix 'LQ'
                          IBM Proprinter X24               'PP24'
                          IBM Quietwriter                  'IBMQ'
```

```
                        Toshiba 24-pin dot matrix         'TSH'
                        LaserJet                          'LJ'
                or   PaintJet                           'PJET'  }
{---------------------------------------------------------------}
begin
   ClrScr;
   printIt := ASK('Do you want to print your graph? (Y or N) ',4);
   InitDisplay(printIt, scriptF,workDrive,workMem);
{------------------Sample printer setup ------------------}
   if printIt then
 begin
 begin
     SelectPrinter(printer, printPort, pageSize, resolution);
{ The following is needed only for a serial port           }
{ SetSerialPort(baudRate,parity,nDataBits,nStopBits,XON);}
   end;
{---------------------------------------------------------------}
          See MAIN BODY of preceding program SampPlt.
          Note that SetLine Style is replaced by QsetLineStyle.
{---------------------------------------------------------------}
{-Sample execution from user's program - note $M compiler directive.
   If there isn't enough memory, can always run QDISPLAY separately.-}
   SWAPVECTORS; {Otherwise, from DOS type QDisplay test.scr}
   Exec('C:\bev\chapters\pubpgms\QDisplay.exe',scriptF);
   SWAPVECTORS;
{ --------------------------------------- }
   if DosError <> 0 then
   begin
     Writeln('DosError #',DosError);
     readln;
   end;
END.



{Program D.6: Routine to write a script file for graphs}
{ --- Unlisted routines- available on diskette -----------}
Unit QuikScrp; {Write script file for input to Qdisplay}
Interface
type dataArray = array[1..100] of real;
{----------------------INITIALIZATION ROUTINES--------------------}
procedure InitDisplay(printPic:boolean; scriptFile:string;
                                          workDrive:char; workSpace:integer);
          {printPic = TRUE for hardCopy, FALSE for screen;
            workDrive = work space for temporary files
            (e.g., 'A' or 'B' or 'C', etc. ' ' for current directory)
```

```
         workSpace : allocate memory for drivers. Min = 16k, defaults
         to 32k if set to 0 }
procedure InitGraphics(logOn:boolean ; graphDriverPath:string);
{--graphDriverPath to graphics drivers--}
procedure SetDefaultColors(c0,c1,c2,c3,c4,c5,c6,c7:integer);
procedure SetMono(fgPal, bkgPal, fgCol, bkgCol:integer);
{--to make a monochrome plot for a screen dump to a printer,
     enter with fgPal and bkgPal and fgCol and bkgCol set to the
     appropriate palettes for a sceen dump on your system. Systems
     vary, e.g., VGA screen to HP laser printer requires fgPal=0
     and bkgPal=15. Set fgCol and bkgCol to the desired screen colors}
procedure OpenWindow;
procedure Border(x1,y1,x2,y2:real);
{------------------------TERMINATION ROUTINES---------------------}
procedure CloseGraphics;
procedure CloseWindow;
{------------------------POINT SCALING ROUTINES-------------------}
function  XScale(x:real):Integer;{change data to screen coordinates}
function  YScale(y:real):Integer;
{Set pixel count of screen size if default is not satisfactory}
procedure SetScreenRange(ix1, iy1, ix2, iy2:integer);
{-set pixel count of screen size if default is not satisfactory}
{-------------------------GRID ROUTINES---------------------------}
procedure Xgrid(fullGrid:boolean;nTics,nDec:integer);
procedure Ygrid(fullGrid:boolean;nTics,nDec:integer);
procedure SetXYformat(xForm, yForm: word);
{ set number of digits for numbers on X and Y axes, or use default.}
{------------------------LABELLING ROUTINES-----------------------}
procedure XLabel(w:String); {-- max = 50 characters in VGA --}
procedure YLabel(w:String); {-- max = 35 characters in VGA --}
procedure WriteTextXY(x,y:real;s:string);{write string s at x,y}
{-----------------DATA AND CURVE PLOTTING ROUTINES----------------}
procedure RCurve(i:integer; x,y:real);{Line segments,i=0 on 1st call}
procedure RDataCircle(x,y,dr,dy:real);{Vertical line through circle}
procedure RDataCross(x,y,dx,dy:real);
procedure RHist(var x1, y1:real; x2, y2:real);
{-plot a histogram.  On 1st call, enter with limits of first
  bin. Routine updates x1 and y1 on later calls.}
procedure RLine(x1,y1,x2,y2:real);
{------------------------SPLINE ROUTINES--------------------------}
procedure SplineMake(nn:integer;d2ydx2A,d2ydx2B:real;
                                 xIn,yIn:dataArray); {N equal steps of h}
procedure SCurve(pt1, ptn, nInt:integer; XoffSet:real; x:dataArray);
{------------------------GRAPHICS CONTROLS------------------------}
procedure qSetLineStyle(style, pattern, thickness:integer);
```

```
const {style} SolidLn   = 0; DottedLn   = 1; CenterLn = 2; DashedLn = 3;
  {thickness} NormWidth = 1; ThickWidth = 3;
procedure qSetTextJustify(horiz, vert : integer);
const {horiz} LeftText   = 0;  CenterText = 1; RightText = 2;
      { vert} BottomText = 0;  TopText    = 2;
procedure qSetTextStyle(font, dir, size:integer);
const { font} DefaultFont  = 0; TriplexFont = 1; SmallFont=2;
              SanSerifFont = 3; GothicFont  = 4;
      { dir}  HorizDir     = 1; VertDir     = 2;
          {size  1,2,3,  etc. }
{---------------------------Hard Copy Functions----------------------}
type  qString4 = string[4];     {calling programs must use this type}
procedure SelectPrinter(driverC, portC:qString4; qSize, qResolution:char);
{ driverC   = 'FX', 'LQ','PP24','IBMQ', 'TSH', 'LJ','PJET'
      for  Epson compatible  9-pin dot matrix (FX)
              Epson compatible 24-pin dot matrix (LQ)
              IBM Proprinter X24            (PP24)
              IBM Quietwriter               (IBMQ)
              Toshiba 24-pin dot matrix     (TSH)
              LaserJet                      (LJ)
              PaintJet                      (PJET)         }
{ portC = 'LPT1','LPT2','COM1', or 'COM2'              }
{ sizeC      = 'H' or 'F' for half or full page        }
{ resolutionC = 'L','M' or 'H'for low, medium or high  }


procedure SetSerialPort(baudC:qString4; parityC, dataBitsC,
                                    stopBitsC, XonC: char);
{ baudC   ='300','600','1200','2400','4800' or '9600'
   parityC   = 'E','N',or 'O' for Even, None, or Odd
   dataBitsC = '7 'or '8' for 7 or 8 bits
   stopBitsC = '1' or '2' for 1 or 2 stop bits
   Xon       = 'Y' or 'N'                              }

const
   HalfHi = 1;  {etc}
   printit: boolean = FALSE;


Implementation
----------------See diskette for remainder of program----------------
```

Routines listed in other sections: The program units FitVars and GenUtil are listed in Section E.2 (Chapter 6). The unit Splines is listed in Section E.7 (Appendix A). The units CRT, GRAPH, and PRINTER are part of the Turbo Pascal operating system.

REFERENCES

Anderson, R. L. and E. E. Houseman, *Tables of Orthogonal Polynomial Values Extended to N = 104*, Research Bulletin 297, Agricultural Experimental Station, Iowa State University (April, 1942).

Arndt, R. A. and M. H. MacGregor, Nucleon-Nucleon Phase Shift Analysis by Chi-Squared Minimization, in *Methods in Computational Physics*, vol. 6, pp. 253–296, Academic Press, New York (1966).

Bajpai, A. C., I. M. Calus, and J. A. Fairley, *Numerical Methods for Engineers and Scientists*, Wiley, Chichester (1977).

Baird, D. C., *Experimentation: An Introduction to Measurement Theory and Experiment Design*, Prentice-Hall, Englewood Cliffs, N.J. (1988).

Beers, Y., *Introduction to the Theory of Error*, Addison-Wesley, Reading, Mass. (1957).

Box, G. E. P. and M. E. Müller, A Note on the Generation of Random Normal Deviates, *Ann. Math. Statist.*, vol. 29, pp. 610–611 (1958).

David, F. N., *Tables of the Correlation Coefficients*, (Cambridge University Press, London (1938).

Dixon, W. J. and F. J. Massey, Jr., *Introduction to Statistical Analysis*, McGraw-Hill, New York (1969).

Eadie, W. T., D. Drijard, F. E. James, M. Roos, and B. Sadoulet, *Statistical Methods in Experimental Physics*, North-Holland, Amsterdam (1971).

Hamilton, W. C., *Statistics in Physical Science*, Ronald Press, New York (1964).

Hamming, R. W., *Numerical Methods for Scientists and Engineers*, McGraw-Hill, New York (1962).

Handbook of Chemistry and Physics, Chemical Rubber Co., Cleveland, Ohio (1973).

Hildebrand, F. B., *Introduction to Numerical Analysis*, McGraw-Hill, New York (1956).

Hoel, P. G., *Introduction to Mathematical Statistics*, Wiley, New York (1954).

IBM, *System/360 Scientific Subroutine Package*, *Programmer's Manual* (360A-CM-03X).

Knuth, D. E., Seminumerical Algorithms, in *The Art of Computer Programming*, vol. 2, pp. 29ff., Addison-Wesley, Reading, Mass. (1981).

Marquardt, D. W., An Algorithm for Least-Squares Estimation of Nonlinear Parameters, *J. Soc. Ind. Appl. Math.*, vol. II, no. 2, pp. 431–441 (1963).

Melkanoff, M. A., T. Sawada, and J. Raynal, Nuclear Optical Model Calculations, in *Methods in Computational Physics*, vol. 6, pp. 2–80, Academic Press, New York (1966).

Merrington, M. and C. M. Thompson, Tables of Percentage Points of the Inverted Beta (F) Distribution, *Biometrica*, vol. 33, pt. 1, pp. 74–87 (1943).

Orear, J., Notes on Statistics for Physicists, UCRL-8417, University of California Radiation Laboratory, Berkeley, Calif. (1958).

Ostle, B., *Statistics in Research*, Iowa State College Press, Ames, Iowa (1963).

Pearson, K., *Tables for Statisticians and Biometricians*, Cambridge University Press, London (1924).

Press, W. H., B. P. Flannery, S. A. Teukolsky, and W. T. Vetterling, *Numerical Recipes, The Art of Scientific Computing*, Cambridge University Press, New York (1986).

Pugh, E. M. and G. H. Winslow, *The Analysis of Physical Measurements*, Addison-Wesley, Reading, Mass. (1966).

Review of Particle Properties, *Phys. Lett.*, vol. 170B, p. 53 (1986). Published in even-numbered years alternately in *Physics Letters* and *Reviews of Modern Physics*. Includes discussions of probability and statistics and the properties of particle detectors.

Taylor, J. R., *An Introduction to Error Analysis*, University Science Books, Mill Valley, Calif. (1982).

Thompson, W. J., *Computing in Applied Science*, Wiley, New York (1984).

Turbo Pascal Reference Guide, Version 5.0, Borland International, Scotts Valley, Calif. (1989).

Wichmann, B. and D. Hill, Building a Random Number Generator, *Byte Magazine*, March, p. 127 (1987). *Applied Statistics*, vol. 31, pp. 188–190 (1982).

Young, H. D., *Statistical Treatment of Experimental Data*, McGraw-Hill, New York (1962).

Zerby, C. D., Monte Carlo Calculation of the Response of Gamma-Ray Scintillation Counters, in *Methods in Computational Physics*, vol. 1, pp. 90–133, Academic Press, New York (1963).

ANSWERS TO SELECTED EXERCISES

CHAPTER 1

1.1. (*a*) 5   (*b*) 2   (*c*) 2   (*d*) 5   (*e*) 4
    (*f*) 1   (*g*) 3   (*h*) 3   (*i*) 3   (*j*) 4

1.3. (*a*) 980.   (*b*) 84,000   (*c*) 0.0094   (*d*) 3.0×10^2
    (*e*) 4.0   (*f*) NA   (*g*) 5300   (*h*) 4.0×10^2
    (*i*) 4.0×10^2 (*j*) 3.0×10^4

1.5. Mean = 73.48; median = 73; most probable value = 70

1.7. Standard deviation = 15.52

CHAPTER 2

2.1. (*a*) 20   (*b*) 6   (*c*) 120   (*d*) 270, 725

2.2. For $p = 1/6$, 0.015625, 0.093750, 0.234375, 0.31250, 0.234375, 0.093750, 0.015625

2.5. 4.1 for one lemon; 37 for two lemons; 1000 for three lemons

2.8. (*a*) $2.3 \simeq 2$ students   (*b*) 8%

2.12. (*a*) 0.0011   (*b*) $\sim 5 \times 10^{-17}$

2.14. Mean number in the time interval: $\bar{x} = \sum_{x=0}^{\infty} x P_P(x; \mu)$; mean number recorded $= \sum_{x=1}^{\infty} 1 P_P(x; \mu) = 1 - P_P(0; \mu) = 1 - e^{-\mu}$; efficiency $= \mu/e^{-\mu}$

2.16. $C = 4/R^2$

CHAPTER 3

3.3. The relative uncertainty in r should be one-half the relative uncertainty in L

3.5. 1.503 ± 0.024

3.7. (a) 15300 ± 6700 (b) 165 ± 11
3.9. $\bar{n} = 3.61$; $s = 1.88$

CHAPTER 4

4.1. $\sigma = 1.49$; $\sigma_\mu = 0.30$
4.3. Fig. 2.3: $\chi^2 = 1.39$ for 5 bins; $\chi_\nu^2 = 0.35$
Fig. 2.4: $\chi^2 = 4.88$ for 7 bins; $\chi_\nu^2 = 0.81$
4.7. Mean total counts in 1-min interval = 123.2; $\sigma = 9.4$; $\sigma_\mu = 3.0$
Background counts in 1-min interval = 11.6; $\sigma = 1.5$
Difference = 111.6 ± 3.3 counts per minute from the source
4.9. 32.81 ± 0.46

CHAPTER 5

5.10. For 6 rows: (b) $8, 48, 120, 160, 120, 48, 8$ (c) $\sigma = 1.22$

CHAPTER 6

6.1. $a = 114.3 \pm 9.6$; $b = 9.58 + 0.89$; $\chi^2 = 10.1$
6.4. $b = 3.60 \pm 0.03$; $\chi^2 = 11.9$

CHAPTER 7

7.2. $a_1 = 512.0 \pm 45.9$; $a_2 = 348.3 \pm 21.8$; $\chi^2 = 13.2$
$\alpha_{11} = 21.09.$; $\alpha_{12} = \alpha_{21} = -147.1$; $\alpha_{22} = 476.1$
7.4. All terms: $\chi^2 = 17.21$ for 12 degrees of freedom
Even terms: $\chi^2 = 17.59$ for 14 degrees of freedom
$a_1 = (849.6 \pm 15.4) - (335.5 \pm 85.7)x^2 + (847.3 \pm 87.8)x^4$
with $x = \cos(\theta)$
7.10. $a_1 = 0.0001 \pm 0.0009$; $a_2 = v_0 = 0.871 \pm 0.018$
$a_1 = g/2 = 4.870 \pm 0.057$ (after iterating)

CHAPTER 8

8.3. (a) $\mu = 1.8741 \pm 0.0005$; $\chi^2 = 13.70$
(b) $\mu = 1.8471 \pm 0.0005$; $\Gamma = 0.0555 \pm 0.0008$; $\chi^2 = 13.3$
8.4. $a_1 = 148.6 \pm 31.0$; $a_2 = 31.0 \pm 1.1$; $\chi^2 = 13.0$
$\epsilon_{11} = 65.6$; $\epsilon_{12} = \epsilon_{21} = -6.26$; $\epsilon_{22} = 1.156$

CHAPTER 9

9.3. (b) $\chi^2 = 21.9$ for 24 degrees of freedom
Fitted parameters a_1 through a_6:

$$3.0 \qquad 86.8 \qquad -25.3 \qquad 26.8 \qquad 0.989 \qquad 0.168$$

Uncertainties σ_1 through σ_6:

$$2.3 \qquad 5.7 \qquad 2.1 \qquad 4.5 \qquad 0.011 \qquad 0.046$$

CHAPTER 10

10.1. $a_1 = 4.16$; $a_2 = 22.8$ at the maximum of the likelihood function

CHAPTER 11

11.4. Approximately 10% probability
11.5. Approximately 0.1% probability; not a very good fit
11.9. 0.9985
11.10. 0.9729
11.12. 0.9997
11.14. $F \simeq 10$ for $\nu_1 = 1$; $F \simeq 5$ for $\nu_1 \leq \nu_2$
11.20. $\Delta\chi^2 = 2.7$; $a_4 = 3.4^{+4.5}_{-3.1}$; $a_5 = 205^{+70}_{-30}$

INDEX

BESSIE SMITH

BESSIE SMITH

JACKIE KAY

faber

This new edition first published in 2021
by Faber & Faber Limited
Bloomsbury House
74–77 Great Russell Street
London WC1B 3DA
First published in 1997 by Absolute Press

Typeset by Faber & Faber Limited
Printed and bound by CPI Group (UK) Ltd, Croydon, CR0 4YY

The right of Jackie Kay to be identified as author of this work
has been asserted in accordance with Section 77 of the Copyright,
Designs and Patents Act 1988

A CIP record for this book
is available from the British Library

ISBN 978–0–571–36292–9

2 4 6 8 10 9 7 5 3 1

For my dear dad, John Kay (1925–2019),
who passed his love of Bessie on to me

Show me a hero and I will write you a tragedy.

F. Scott Fitzgerald

And freedom had a name. It was called the blues.

Walter Mosley

CONTENTS

THE RED GRAVEYARD

There are some stones that open in the night like flowers.
Down in the red graveyard where Bessie haunts her lovers.
There are stones that shake and weep in the heart of night.
Down in the red graveyard where Bessie haunts her lovers.

Why do I remember the blues?
I am five or six or seven in the back garden;
the window is wide open;
her voice is slow motion through the heavy summer air.
Jelly roll. Kitchen man. Sausage roll. Frying pan.

Inside the house where I used to be myself,
her voice claims the rooms. In the best room even,
something has changed the shape of my silence.
Why do I remember her voice and not my own mother's?
Why do I remember the blues?

My mother's voice. What was it like?
A flat stone for skitting. An old rock.
Long long grass. Asphalt. Wind. Hail.
Cotton. Linen. Salt. Treacle.
I think it was a peach.
I heard it down to the ribbed stone.

I am coming down the stairs in my parents' house.
I am five or six or seven. There is fat thick wallpaper
I always caress, bumping flower into flower.

She is singing. (Did they play anyone else ever?)
My father's feet tap a shiny beat on the floor.

Christ, my father says, that's some voice she's got.
I pick up the record cover. And now. This is slow motion.
My hand swoops, glides, swoops again.
I pick up the cover and my fingers are all over her face.
Her black face. Her magnificent black face.
That's some voice. His shoes dancing on the floor.

There are some stones that open in the night like flowers.
Down in the red graveyard where Bessie haunts her lovers.
There are stones that shake and weep in the heart of night.
Down in the red graveyard where Bessie haunts her lovers.

INTRODUCTION

There are some people whose voices ring out across the centuries, who, even after they have gone, possess a strange ability to still be effortlessly here. Bessie's voice has that quality. Unsettled most of her life, she still unsettles. Try to imagine asking her about anything that is going on today, from the floods, to the climate crisis, to the coronavirus, to the Black Lives Matter movement, to the Me Too movement, to the refugee crisis, and you would find an answer in her rich and resonant blues narratives. We could match any of today's troubles and anxieties to her music. The blues are not past. Bessie's blues are current.

Her narratives are even eerily prescient – she sang about floods, about sexual abuse, about financial crashes, about sudden changes in circumstances, changes in love. There isn't anything that life could currently throw at her that would surprise her. Her blues sought the truth – the truth in all its multiplicity; the hard truth, the strangest truth, the supernatural truth. The whole truth has a different ring to it in the world of Bessie's blues. In these surreal times, where distinguishing the truth is a challenge, Bessie's voice has a pure and true ring. She is telling it like it is. There's nothing fake about her. And because she was not afraid to bear witness to her times, to rising racism and the Ku Klux Klan, to inequalities and class differences, to hypocrisy and the dangers of celebrity, she also manages to bear witness to our times. Pioneers don't just lead the way in their own time; they continue to refract and reflect our time. Pioneers can perform the magic trick of being contemporary in any time.

For my twelfth birthday, my dad bought me my first double album. It was Bessie Smith's *Any Woman's Blues*. I was drawn to the two-sided picture of her on the cover, the smiling Bessie and the sorrowful one. It wasn't long before I made her part of my extended imaginary black family, before I felt not just as if she belonged to me, but as if I belonged to her. She felt like kith and kin. She felt like kindred. There was something in her that seemed to recognise something in me. Well, that's what we think with people whose writing or music or art we love – it is not so much that we see something in their work, but that we are deluded enough to imagine they might understand us, comprehend the complex workings of our minds. They feel like soul mates. We feel known, intimately known.

Now, more than twenty years since I first wrote about Bessie, that feeling is so ingrained in me that it feels a little awkward to state it. It feels like stating the obvious. I'm older now than she ever got to be, twenty years older, and yet trying to imagine her at fifty-seven or at seventy-seven is not all that difficult. Not difficult because the people you choose to accompany you don't die, they hold up one of life's oddly glinting mirrors. You're not the young girl that loved Bessie Smith any more and danced to the blues in your living room. You are a fifty-seven-year-old woman whose odd reflection in the waiting mirror perhaps Bessie would catch? You're not the same any more. You've changed physically, emotionally; you've learnt and unlearnt different things. But you still love Bessie. She is one of those folks you go to for comfort and understanding when the going gets rough, the rough gets going.

Bessie Smith is the perfect antidote to these times. She tells no lies. Her voice is still authentic. Her stories seem ever more urgent. She's still troubled. Her eyebrows still furrowed. Loving her blues, the exact timbre of her voice, no longer even feels like taste or choice. It goes deeper than that. I don't know what gave me the idea back in 1996 when I first wrote *Bessie* to write about my life and write about her life together. How odd to try and do both at the same time. I wasn't interested in writing a standard kind of biography. I think I was interested in how much our interests and passions form part of our own identity – how we beg and borrow and become ourselves, how much of a big mixture we are. I was interested in the point of intersection.

I was working on my novel *Trumpet* at the time when Nick Drake asked me if I would put on hold what I was writing and choose a gay icon to write about instead. It was strange. I was having trouble with *Trumpet* in trying to find the right tone to tell the story – and, strangely, returning to the blues and immersing myself in Bessie and in her contemporaries clarified the voice of *Trumpet*. I started to see the style of the book as a piece of music. The whole chapter called 'Music' in *Trumpet* was directly inspired by thinking about how the blues journeyed into jazz. I was trying to find a metaphor for that fluidity in our own gendered identities. I was thinking about how we imagine states of identity to be static when they are in fact fluid. I had arrived at the conclusion that my hero, Joss Moody, would be called 'he' after he made the decision to present himself to the world as a man, and that to refer to him as 'she' would be a kind of an affront. Writing about Bessie and her blues, about her very fluid identity, how she was as at home in pearls and plumes as in a man's suit, allowed me to create Joss Moody. The two books seem twinned. Writing about Bessie unleashed something and *Trumpet* kind of sprang to life.

Twenty-odd years on it is amazing how much has changed in a relatively short period of time. The shift in attitudes to gay and trans people has probably been the biggest social change of our lifetime. It is not to say that prejudice does not still exist – but still it would have been impossible to imagine all the terms that have so quickly become part of our new vocabulary; we have walked into a changing language about ourselves, which is still shifting, still open to question, still partly greeted with derision in places, but which none the less has reshaped the gendered landscape that we lived in just twenty years ago. I want every bit of it, Bessie sang.

I can see my dad, who died last year, sitting in his armchair singing 'Nobody Knows You When You're Down and

Out'. I can see him relishing the words – the notion that people are fickle and that capitalism is a sham. I can see the enjoyment of the philosophy that is in the blues. There is something egalitarian and equalising about the blues. There are salutary lessons to be found. There's a terrible reckoning in the space between 'once I lived' and now – and there's nobody to save you. Falling so low is something that might have been predicted. No place to go. It seemed to me, in the way that my dad sang Bessie, that everyone who had ever been poor without ever being rich could yet understand the falling from grace, and could empathise. Everyone might intuit the shallowness of the well-to-do. Might sympathise with the trauma of being left on your tod to cope with whatever after having had 'friends'. The story of the blues – Bessie's blues, going from something to nothing, from having happiness to being wretched, from having love to losing it, from being feted to being ignored – those terrible trajectories were all around us. They were real life. The reason that I loved the blues was because they didn't appear to me to be made up in any way. They sprang from life's source, life's true well and, well, they tapped into the well of loneliness on the way, allowed for a kind of transformation, a becoming. If you can recognise the other in you, the other side, then perhaps your life can be meaningful in some way you hadn't yet imagined.

Blues led the way. It's hard to think of so much music – jazz, rap, house – without the blues coming first. We can trace the etymology of the blues, the brilliant blues' bloodline and find it riveting, fascinating, like going onto a genealogy site and finding ancestors. The blues singers are the ancestral voice, the one that can be heard still – those hauntings, those defiant mournful calls. Bessie's blues are bang up to the minute. She calls and, across the years and miles, we respond.

IN THE HOUSE OF THE BLUES

I was adopted in 1961 and brought up in a suburban house in a suburban street in the north of Glasgow. A small semi-detached Wimpey house. Outside our house is a cherry-blossom tree that is as old as me. It doesn't seem the most likely place to be introduced to the blues, but then blues travel to wherever the blues lovers go. In my street and in the neighbouring streets to Brackenbrae Avenue, I never saw another black person. There was my brother and me. That was it. The butcher, the baker and the candlestick maker were all white. (Although I never actually met a candlestick maker – has anyone?)

So the first time I saw Bessie Smith, it really was like finding a friend. I saw her before I heard her. My father – a Scottish communist who loved the blues – bought me my first double album. I was twelve. The album was called *Bessie Smith: Any Woman's Blues* and produced by CBS Records (John Hammond and Chris Albertson; Albertson went on to write her biography). I remember taking the album off him and poring over it, examining it for every detail. Her image on the cover captivated me. She looked so familiar. She looked like somebody I already knew in my heart of hearts. I stared at the image of her, trying to recall who it was she reminded me of.

She looked so sorrowful and so strong. She would stand

up for herself, so she would. She wouldn't take any insults from people. I could see from her eyes that she was a fighter. I put her down and I picked her up. I stroked her proud, defiant cheeks. I ran my fingers across her angry eyebrows. I soothed her. Sometimes I felt shy staring at her, as if she was somehow able to see me looking. On the front cover she was smiling. Every feature of her face lit up by a huge grin bursting with personality. Her eyes full of hilarity. Her wide mouth full of laughing teeth. On the back she was sad. Her mouth shut. Eyes closed. Eyebrows furrowed. The album cover was like a strange two-sided coin. The two faces of Bessie Smith. I knew from that first album that I had made a friend for life. I would never forget her.

The names of the blues songs transported me places, created scenes and visions. 'Jail House Blues', 'Haunted House Blues', 'Eavesdropper's Blues', 'Graveyard Dream Blues', 'Whoa, Tillie, Take Your Time', 'St Louis Gal', 'New Orleans Hop Scop Blues', 'Kitchen Man', 'Chicago Bound Blues', 'Worn Out Papa Blues'. Each name was enough to make up a story. That's what I liked about the blues, they told stories. The opposite of fairy tales; these were grimy, real, appalling tragedies. There were people dying in the blues; people coming back to haunt the people who were living in the blues; there were bad men in the blues; there were wild women in the blues. People travelled places, or wished they were someplace else in the blues. Could I be a St Louis Gal? Or could I be Tillie? Might Chicago be a place I would go when I grew up? Was the New Orleans Hop Scop like hopscotch? There were Daddys and Mamas galore in the blues. Every next person was a Mama or a Daddy but they didn't sound like my mum and dad. 'Mistreating Daddy, mistreating Mama all the time.' People got drunk and ate pigs' feet in the blues. It was totally wild.

We had an old-fashioned record player that looked as if it was pretending to be a bit of furniture until you lifted its lid and saw its strange black arm and needle. I'd place the record down into the player, lift the arm of the needle and try and get it on the exact line of my favourite track. Sometimes I'd miss and hear the end of another unfamiliar track that would, for a moment, make me want to listen to it. I liked the way the piano and the horn sounded comical, as if they weren't taking any of this so seriously. Especially in songs like 'Cemetery Blues' and 'Graveyard Dream Blues'; the musical accompaniment sounded like it was all one big joke, like the funny music in silent movies.

My favourite of the lot was 'Dirty No-Gooder's Blues'. It sounded so bad. The very name made you think things you weren't supposed to be thinking at that age. A Dirty No-Gooder? What was he? Some man that said filthy words and had no morals. Some man who could not be good; some man who dedicated his whole life to being bad. I found the blues so exciting because the characters were real 'characters'; people who behaved badly, who did terrible things, who killed and murdered and went to prison and missed Chicago and wanted to 'Take it right back to the place where you got it', whatever the 'it' was. The men in the blues were lowdown and rotten, cheating and lying and lazy.

There's nineteen men living in my neighbourhood.
There's nineteen men living in my neighbourhood.
Eighteen of them are fools and the one ain't no doggone good.

'Dirty No-Gooder's Blues'

At the age of twelve in my small suburban neighbourhood, the idea of describing men in this way was scandalous, hilarious. Mr Aird, Mr Tweedie, Mr Dunsmore, Mr Macintosh, Mr Murray, Mr Kerr, Mr Cochrane. Which one was no doggone good? Whenever I listened to the lyrics of the blues they made me feel I was being entrusted with a secret. I was being told a secret story about some no-good, no-count man. The story was for me and not for him. The story was a joke on him. Bessie Smith was singing for women. It was women who were singing those blues and the lyrics were mainly about the 101 ways a man could let you down. I guess I took the warning. Every woman could understand the blues. That's why this album was called *Any Woman's Blues*.

I realised that I could choose always to have Bessie Smith in my life, that she was not going to betray me or go off with another friend, or move to Kirkintilloch, or suddenly turn nasty. Nobody could take her away from me. And when I grew up and went away, I could take her with me. And I did. Hours and hours looking at her face on the cover and many hours memorising the lyrics to the songs. They weren't all blues; some were vaudeville, some were Tin Pan Alley. But everything Bessie Smith sang sounded like a blues song. I had my favourite songs that I would play over and over again. 'Kitchen Man', 'Dirty No-Gooder's Blues', 'I'm Wild About That Thing', 'He's Got Me Goin''.

Got me goin'
He's got me goin'
But I don't know where I'm headed for!

When I listened to the songs on this first album, I always assumed that they were about Bessie Smith's own life. (It didn't occur to me to think this of folk, soul or rock singers.) The blues sounded like autobiography, like ordinary people telling the story of their lives. There was always an I in the blues. It was all first person. When I listened to 'Nobody Knows You When You're Down and Out' and 'Wasted Life Blues', I felt so sad for Bessie Smith. I could picture her whole life suddenly changing. (Her whole life did change, strangely mirroring 'Nobody Knows You'; she might as well have been singing about herself.) I felt so passionate about her, it made me angry. I'd listen to the lyrics of her songs and that voice of hers that no other voice can ever come near:

Nobody knows you
When you're down and out.
In my pocket, not one penny
And my friends, I haven't any,
But if I ever get on my feet again
Then I'll meet my long-lost friend.
It's mighty strange without a doubt
Nobody knows you when you're down and out,
I mean when you're down and out.

I'd imagine her, wandering through the streets of segregated America, penniless, friendless. *Down and out.* There were many down-and-outs in Glasgow on Sauchiehall Street on a Saturday night. If nobody loved Bessie Smith when she was down and out, then nobody would love me. All that was left then would be to dream of death. Graveyard Dream Blues. Cemetery Blues. Wasted Life Blues. Somehow being down and out in America seemed to be inextricably linked with her colour. I could not separate them. I could not separate myself. I am the same colour as she is, I thought to myself, electrified. I am the same colour as Bessie Smith. I am not the same colour as my mother, my father, my grandmother, my grandfather, my friends, my doctor, my dentist, my butcher, my teacher, my headmaster, my next-door neighbour, my aunt, my uncle, my mother's friends, my father's friends. The shock of not being like everyone else; the shock of my own reflection came with the blues. My own face in the mirror was not the face I had in my head.

There were cheeky songs too on this double album. 'I'm Wild About That Thing' – which contained the enigmatic line 'I like your ting-a-ling'. (I couldn't understand what was going on but it sounded sexual. Children approaching adolescence are adept at picking up on any hint of sex anywhere. And in the blues, it wasn't so much a hint as a wallop.) There were the songs of defiance, of not putting up with it: 'Take it right back to the place where you got it. I don't want a bit of it in here.' I feasted off those songs for years. It was ages before I got another album so this one was it. It is still my favourite. The first love. The first time I heard her voice it made me wonder what I was remembering. Her voice was so raw and fresh and different from any other singing voice I'd ever heard. It was not like the voice of Gallagher or Lyle. It was not like Lena Zavaroni or Joni Mitchell, Rod Stewart or the Bay City Rollers. It was not Simon or Garfunkel. It was not Donny Osmond or David Cassidy. It was not Lulu or Elton John. It wasn't even like the voice of Ella Fitzgerald, who I heard live at the Kelvin Hall in Glasgow when I was fourteen with Count Basie accompanying her. No, Ella Fitzgerald's voice had a girlish chuckle in it. Bessie Smith's raw unplugged voice dragged you right down to a place you had never been. It seemed to drag you down to the depths of yourself. Her voice carried a kind of knowing that made you feel this woman knew everything about life and was not frightened of any of it. It made me stop reading *Anne of Green Gables* and look up sharp and listen.

13

What was it I was remembering? Her voice made me want to be her. I could never listen to her as background music; I still can't. In the days when she performed live, people were mesmerised by her. They would not be distracted or try to do anything else while Bessie Smith was on. She captivated them. She had them under her spell. It is still like that even just listening to her. You listen so hard she practically inhabits you; her very soul seems to find a way inside you. I'd try and move my lips the way I imagined she moved her lips. I'd wonder if she was ever called 'Rubber lips' like I was. What would she have said? She would have fought anybody who called her names. I was certain of this. She wouldn't have stood for it. 'I won't have a bit of it in here.'

My mind would wander off to her life. I'd picture her on the road travelling from small Southern town to small Southern town. I had read all about segregation, how black people were not allowed to sit on the same bit of the bus as white people, how they had to use different roadside restaurants, how they couldn't sit down in ice-cream parlours. When I listened to Bessie singing 'Kitchen Man', I pictured her living in a small wooden house with a porch. I could see the Kitchen Man himself, a black man with a white hat. One of those tall hats that scare small children. He would be out the back baking, flour on his fists. He'd bring her plate after plate of those sweet jelly rolls. She just lay on her bed, dressed up in her feathers and plumes, and said, 'Thank you Honey,' and tucked in to those goodies. I was a bit nonplussed when I discovered that all those jelly rolls and sugar rolls in those songs had nothing to do with food.

What was it she reminded me of? Whenever I impersonated her in front of my mirror with my hairbrush microphone, I had a sense of something, at the edge of myself, that I mostly ignored; the first awareness of myself being black. I'd only ever think about it if something reminded me. Bessie Smith always reminded me. I am the same colour as Bessie Smith. I'd look at her hands then I'd look at my own hands. I'd look at her nose then I'd look at my own nose. Perhaps she is related after all. Maybe my great-great-grandmother was a blues singer. Who knows? The great thing about being adopted was that you could invent your family all the time. You could make them up and invent yourself in the process. At one point, every time I saw Shirley Bassey on the telly singing 'Goldfinger', I was convinced she was my birth mother. Any time I came across a black person, usually on TV, or in a book, or on a political poster that my father brought home, I tried to work out their relationship to me. I concocted an imaginary black family for myself through images that I had available to me. There were a lot of images in this politically international-ist household. Nelson Mandela, Angela Davis, the Soledad Brothers, Cassius Clay, Count Basie, Duke Ellington. I liked trying to connect one black person to another, a racial jigsaw puzzle. I was aware that racial discrimination existed all over the world. South Africa had sixteen legal defin-itions of black; South Africa had borrowed its system of apartheid from the American South. At Christmas time, I would sit down with my mum and send cards to the political prisoners in South Africa. All black people could at some point in their life face racism or racialism (I could never understand the difference) therefore all black people had a common bond. It was like sharing blood.

I did not think that Bessie Smith only belonged to African Americans or that Nelson Mandela belonged to South Africans. I could not think like that because I knew then of no black Scottish heroes that I could claim for my own. I reached out and claimed Bessie.

When I was a young girl, Bessie Smith comforted me, told me I was not alone, kept me company. I could imagine her life as I invented my own; I would not have grown up in the same way without her. Just at that crucial moment, before my periods, after my first bra, before I had any big romance, before I went to secondary school, there she was. It was perfect timing. I was just the right age for her to become my lifelong friend, my beautiful flame, my brave heart-throb, my paragon of virtue. My libidinous, raunchy, fearless blueswoman. My righteous, courageous, wild blueswoman. I am still full of her passion. I have her spark in my eye. I can burn with love, burn with the blues. I am still totally, utterly, Wild About That Thing.

I heard the singing of the Mississippi when Abe Lincoln went down to New Orleans, and I've seen its muddy bosom turn all golden in the sunset.

I've known rivers:
Ancient, dusky rivers.

My soul has grown deep like the rivers.

Langston Hughes, 'The Negro Speaks of Rivers'

CHATTANOOGA, 1894

According to the date on her marriage certificate, Bessie Smith was born in Chattanooga, Tennessee on 15 April 1894. But we cannot be sure. Chris Albertson tells us that 'Southern bureaucracy made little distinction between its black population and its dogs; such official records as a birth certificate were not always considered necessary.'[1] But Chattanooga it was, definitely. Not any of the other towns close by: Knoxville, Nashville, Birmingham, Atlanta, Oak Ridge, Blue Ridge, Scottsboro.

What does a girl from Bishopbriggs near Glasgow know about Chattanooga? She might have heard that song: *Pardon me boy, is that the Chattanooga Choo-choo? / Track 29, well you can give them a shine*. But that's about it. I looked up Chattanooga in my atlas. Chattanooga, Hamilton County, Tennessee. It made me feel even closer to her being able to find it in the atlas, to rest my finger on the town; it covered the whole of it. There's a certain satisfaction to be had from finding a place in an atlas. The next best thing to going there. With my finger pressing into the name Chattanooga, I tried to picture it. Well, it wasn't like Glasgow. It wouldn't be like anywhere I had been. It might look like a town in a Western. Maybe. It would have a small railway station where the trains coming in and out of Chattanooga were a major event, a happening. The train would make a huge

19

boasting noise as it came into the station, announcing its arrival. People with their collars turned up would be waiting in the cold wind. *There's gonna be a certain party at the station/ Satin and lace – I used to call funny face.* What else? A sheriff. A sheriff's hat. Dust. Horses. Poor black people living in shacks. A river. A long, long river that passes through practically every town in Tennessee. The Tennessee River. I traced it with my finger. The shape of a U. Parsons, Savannah, Decatur, Guntersville, Jasper, Chattanooga, Cleveland, Charleston, right up to Oak Ridge. *Dinner in the diner. Nothing could be finer / Than to have your ham and eggs in Carolina.* It was a bigger river than I could imagine. It was a dangerous river that could cause massive floods. A picture of people floating with furniture in the water rose up in my mind. A giant of a river.

Chattanooga, 1894. The sheriff would have a gun and a gun holster, wouldn't he? He would be white, definitely. He would be able to arrest a black man for just looking at him straight in the eye. A black man would have to keep his eyes down on the dusty street if he wanted to be safe. Someone new in town would be noticed. Chattanooga, 1894. The town would be packed with preachers. Many big brothers and big sisters. Large and lean religious people, black and white, the faces of people captured by religion. The white faces – pointed, pinched, spiteful. The black faces animated, arrested, actorish. Hymns, psalms, prayers, spirituals. 'Go Down, Moses', 'Swing Low, Sweet Chariot'. Bessie Smith would have gone to church as a young girl. (Church continued to play an important role in her life. She would often go after a heavy night's drinking!) The emotion of those songs – 'O Mary, Don't You Weep', 'Rough Rocky Road' – might have influenced the way she sang the blues.

Rough, rocky road, I'se most done suffering.
Rough, rocky road, I'se most done suffering.
Rough, rocky road, I'se most done suffering.
I'm bound to carry my soul to de Lord.
I'm bound to carry my soul to de Lord.

The preacher man would be the town's salvation. He would wear a white shirt and a black hat. In the black church the preacher man would get his congregation worked up till they were almost crazy with love for God. They would sing those spirituals the way Bessie Smith sang the blues, with total and utter belief and conviction. The Lord could lift them up and carry them away from the terrors and indignities of racism.

Bessie Smith was born into a place and a time where racist violence was in the air she breathed. In the name of a town. All over the South that Bessie loved were towns plagued by racism. The names of those places still ring with trauma. (We have our own traumatic town names in Scotland; we know the impact of just hearing the word Dunblane or Lockerbie.) The South is full of names of towns that make you sink and reel with disbelief. Birmingham, Alabama, where the black children were bombed in church. Scottsboro where nine innocent black men were accused of raping a white woman; eight of them were sent to the electric chair.

We have been sentenced to die for something we ain't never done. Us poor boys been sentenced to burn up on the electric chair for the reason that we is workers – and the color of our skin is black.

The Scottsboro boys

All over the South that Bessie Smith so loved are the graves of people murdered for being black. The blues never shied away from singing about tragedy. Later, Billie Holiday would sing about a lynching in 'Strange Fruit':

Southern trees bear a strange fruit,
Blood on the leaves and blood at the root,
Black bodies swaying in the Southern breeze,
Strange fruit hanging from the poplar trees.

The poet who wrote 'Strange Fruit', Lewis Allan, took it to Billie Holiday in 1939, two years after the death of Bessie Smith. When Lady Day first read the words of 'Strange Fruit', she didn't know what to make of it. But the song made such an impact on people that she later claimed to have written it herself. It became her song. 'Strange Fruit' provoked a huge reaction from its audience. Song and memory intertwined like branch and leaf. People in the audience would weep when they heard it, some would remember having witnessed lynchings themselves. Donald Clarke, in his biography of Billie,[2] tells us that:

> One evening . . . a woman followed Lady into the powder room crying, 'Don't you dare ever sing that song again. Don't you dare . . .' When she was seven or eight years old she had witnessed a lynching in the South.

Blues songs and real life weren't all that separate. These songs were not simple entertainment. They told terrible true stories. In the collection *American Negro Folk Songs* there is a disturbing little song about Chattanooga:

> Chattanooga, chickamauga,
> Tobe Domingus,
> Kill a nigger;
> I'm Alabama bound.

Reported in 1915 from Auburn, Alabama. Tobe Domingus was the name of a policeman from the nearby city of Dothan who killed a black man. Bessie Smith would have been twenty-one around this time. This song actually records that incident. Songs like this give us a kind of history we don't find in the books. One way of dealing with racism was to write and sing songs about it. There is so much healthy irony in these songs. Bessie would have grown up hearing those songs.

Chattanooga, 1894 – 138 miles south-east of Nashville. The town grew with the coming of the railroads in the middle of the nineteenth century. By 1900, Chattanooga had a population of thirty thousand, over half of it black. Unemployment was high. Opportunities for black workers were few. In fact there were two. You could either be an ill-paid manual or domestic labourer or you could join a travelling show.

Bessie Smith was one of eight children. Her father, William Smith, was a part-time Baptist preacher who ran a small mission. He died soon after Bessie was born. One of her brothers had died in infancy before she was born. Her family lived in crushing poverty. Bessie lived in what she later described as 'a little ramshackle cabin', where the rats outnumbered the children. A one-room wooden shack on Charles Street in the Blue Goose Hollow section of Chattanooga. Her mother, Laura Smith, was dead by the time Bessie was eight. Another brother died around the same time as her mother. With no adults to earn money or care for them, the Smith children had to do it for themselves. Viola, Bessie's elder sister, took on the main responsibility to make ends meet, taking in laundry which she boiled on top of an outdoor coal stove. She looked after the surviving Smiths – Bessie, Tinnie, Lulu, Andrew and Clarence. From the age of nine Bessie sang for nickels – maybe a dime once in a blue moon – on Chattanooga's Ninth Street, a stretch along which the city's black night-life centred. Her brother Andrew accompanied her on the

guitar. Sometimes they would stay in their own neighbourhood, in front of the White Elephant Saloon on Thirteenth
and Elm. From a very early age, singing was literally her
survival. It helped to feed her and her family. When anyone threw a coin, she would say, 'That's right, give to the
church.'

I can just about see her at the age of nine standing on a
street corner singing for a nickel or a dime. She is tall for her
age, and big-boned. She already has a mature sense of herself
and a deep voice. Any song suits her. She can sing anything.
What did she sing? A spiritual? Something like this:

Oh! brudder, can't you hol' out yo' light,
Oh! brudder, can't you hol' out yo' light,
Oh! brudder, can't you hol' out yo' light,
An' let yo' light shine ober de world.

She would have stood out in the cold, inadequately
clothed, singing whatever it was she sang. Perhaps it was a
spiritual, perhaps it was a work song.

Born in no'th Carlina,
Raised in Tennessee,
Worked like hell in Georgia,
Died in Germinee.

Whatever it was she sang, her voice even then would have
attracted attention. You've got to be born with a voice like

25

that. A voice like that doesn't come along often. Passers-by must have noticed. She made more nickels than her brothers or her sisters. Even at this early age, she was the main breadwinner.

One thing I notice about maself – I can sing when the sky is a certain way, hanging down low and mean and grey, looking as threatening as Ma could look sometimes. When the sky is like that, the songs come out more gutsy. I am a gutsy girl ain't nobody ever gonna run me down. Ma sistah Viola wants some meat this Satday. We gets meat once in a blue moon. When the sky is all moody blues, I holler ma songs here on the corner like I'm warning people about themselves or something. People stop and listen. I know people are always gonna stop and listen to me. I got that sense about maself. Ma mama done dead. Ma papa done dead. He was always saying, 'If it's the last thing I do before I die.' Ma mama said he said it so much, he made himself die. The last thing he did, she said was try and fix the door of our cabin 'cause the big storm had done damaged it and the wind was coming into our dreams at night. We was always cold and hungry. Ma father was good at catching rats and killing them. Viola says he would hold the dead rat up in the air by its tail and say, 'Gotcha.' I ain't nevah known nothin' else but being po'. I nevah owned a pair of ma own shoes. I seen girls on the other side with dresses on, but I ain't nevah had a new dress. All my clothes are nothing but rags. But that's not bothering

26

me. I know I'm gonna have me a nice dress one day. I just knows it. Viola laughs violent when I say such things. She tells me I'm strange and not like the rest of them. 'What you talking about I'm not the same!' I want to hit Viola when she says shit like that. I feel like I'm a grown woman already, but I'm no different from the rest of them. I want to be the same as the rest of my family. Maybe I is different. I know I sing different. Do singing different make the whole of me different? I know I gets these moods – coming in at me just like weather. They catch me and they knock me right down, my whole face like the devil himself is sitting inside me cooking up some pigs' feet. Ma eyebrows arch themselves like bows and arrows. I talk dirty and I'm only nine. But I don't really think of myself as only nine. I nevah like to laugh when Viola or Andrew tease me. I feel like dying when I hear the sound of mocking laughter. It eats on my insides and makes me evil. I go straight up and I grab hold of them. I knock them down, batter them with my fists till I hear them whimper. Don't care who they are. Ain't nobody gonna run me down. When my temper is hot, not even a whole ocean of cold water gonna cool that temper down. I could catch a chicken and wring its scrawny neck. If they say sorry enough times to me, I just might calm down. Viola say, 'Bessie, that temper of yours is gonna get you into trouble someday.' I know that's true but there ain't a thing I can do about it.

Mama died because she was poor. She didn't get enough. She didn't get enough of anything and she gone and died.

I gonna help my family if it's the last thing I do before I die. When the sky is all hung down low, like a bony dog, slithering and insinuating itself about the neighbourhood, I can hear something come out of me, something that's inside me come out. And the more the roar comes out of me, the more folk gather round me on the corner. I heard someone say, Listen to that girl sing. Listen to her. I was standing there listening to myself. I could hear it myself. I ain't stupid. I know I can sing. I know I sing like nobody else in the whole goddam world. And that's the truth. I just knows myself. One day I'm gonna be somebody big. And that sister of mine better bow her sweet head. I'm gonna have so much fun telling her I told you so. I'm gonna get me a fine dress and a pair of shoes. Just imagine these dirty dusty feet of mine going into a pair of sweet little shoes. Little black shiny shoes. I'm sure a shop in the town might have a pair when I'm ready. I nevah go that way. I nevah really been out of this here Blue Goose Hollow.

Goose Hollow is called Goose Hollow because everybody who lives here is hungry, hollow bellies. Can't get no bird or animal to eat. No goose, no chicken, no pig, no cow. We eat the scraps here. The bits that other neighbourhoods throw away. That's what Andrew tole me. But Tinnie say that's a lie. Tinnie say it's called Goose Hollow because one day after the big rains a whole flock of geese descended in the hollow there near the water. They all marched along following their mama like they had some purpose in life and then suddenly stopped. Well those geese nevah flew after

that day, nevah went anywhere. They just stayed in that
there Goose Hollow like they'd found their Mecca. So say
Tinnie. But Viola, now she say, that's just about as much
rubbish as she ever heard. It's called Goose Hollow because
old Sam's grandfather was asked by the bossman where he
lived and his mind went blank because Goose Hollow use
ta be called the Belly, but he didn't want ta say 'I live in
the Belly' to his bossman. Sam looked up to the sky and
saw a goose flapping its fat wings, a big one, so he just come
up with Goose Hollow. And when he came back that night
after working in that white man's yard he laughed and
told everyone the name of where they lived. And everyone
thought it was funny. Goose Hollow. It stuck. It stuck
'cause the folk are stuck and ain't flying no place. I believe
Viola. I've told Viola I'm going. I'm gonna be one bird
that does fly out of here even if it means making my own
wings. I'll be sure not to fly too close to the sun.

The blues began in the South and spread their wings.
Soon Bessie Smith's taxing tours could make a blues map
of the country. There were favourite spots all over America
where blues lovers, a lot of them migrant workers from
the South, congregated. Hearing the blues in New York
reminded the black Southerners of back home tongue.
Smith would travel all over America, to Philadelphia,
New York, Detroit, Cleveland, Memphis, New Orleans,
Washington, Pittsburgh, Cincinnati, Kansas City. It was
impossible to be a blues singer in the America of the 1920s

without travel being in the blood. She never spent any of her money on making a home. She had no real concept of an adult home, which is probably why her childhood one continued to mean so much to her. Her home as an adult was on the road, travelling all over America, becoming particularly intimate with the backstreet bars in different towns. She always knew a place to go where she could get wildly drunk.

But Chattanooga tugged at the heart, the way the place you are born in always does. Just seeing the landscape around there made her feel at home. She liked travelling in the South. She liked the folk of the South. She didn't like those 'uppity northern bitches'. The people in the North were all pretence. The people in the South were real, genuine. It was all in the cooking. The people in the North couldn't cook worth a damn. She had to return home to get a decent meal. She always preferred to do her own cooking even on the road; perhaps this gave her some sense of 'home'. If she could take her favourite foods and her favourite drinks around with her wherever she went, then she could be at home anywhere. She liked cooking up big stews. Being from the South, she loved Coca-Cola. (She might have been addicted to the sugar in Coca-Cola to substitute for corn liquor.) There was something about the South, about coming from the South, that filled her with longing. Nostalgia, homesickness. Good chicken, chitterlings, boiled with scallions, dipped in flour and fried. Cornbread, collards, rice and peas, you could fill the hole of homesickness with a good gravy. Bessie Smith never forgot the girl she had been even when she became rich and famous. She never turned her back on her past and pretended that the girl standing on Chattanooga's Ninth Street singing for a nickel and a dime had never existed. Her voice kept its incredible power to haunt and disturb us because she never denied that girl. Even when she was rich and famous and dressed in ermine coats, she disliked black people who put on airs and graces, who were what she called 'dicty' blacks. She never changed. She stayed faithful to herself, to the girl she was who didn't have enough money for food or clothes. Maybe if she hadn't been born with that big voice of hers, she would have stayed in Chattanooga and barely ever moved from that town on the river. *She's gonna cry until I tell her that I'll never roam, so Chattanooga Choo-choo, won't you Choo-choo me home!*

I was thinking about Ma Rainey
Wonder where could Ma Rainey be.
I've been looking for her
Even been in old Tennessee.

Memphis Minnie, 'Ma Rainey'

IN THE MA'S FOOTSTEPS

She was the voice of the south, singing of the south,
to the south.

Edward Lucie-Smith, of Ma Rainey

Bessie was seventeen when she joined the Moses Stokes Travelling Show in 1912. The show was playing in a Chattanooga storefront. Her brother Clarence, who had been with travelling troupes since 1904, got her her first audition. It is possible she would have never been in show business if it hadn't been for Clarence.

Performing represented virtually the only alternative to sharecropping and manual labour for the poor and the black. (Will Friedwald tells us that Ethel Waters 'came to showbiz only as an escape from the grimiest of scullery work'.[3]) W. C. Handy, the so-called 'Father of the Blues' (because he was the first to write and popularise a blues composition – 'Memphis Blues', 1912), snapped up the opportunity to join a travelling minstrel show:

It goes without saying that minstrels were a
disreputable lot in the eyes of a large section of upper-
crust Negroes . . . but it was also true that all the
best talent of that generation came down the same
drain. The composers, the singers, the musicians, the

speakers, the stage-performers – the minstrel shows got them all. For my part there wasn't a moment's hesitation . . . I took the break for what it was. The cards were running up my way at last.[4]

The very origin of minstrel shows is inextricably connected with racism. They began around the 1830s with white people wearing 'blackface' make-up and went on to become the most popular form of American entertainment through most of the nineteenth century. Burnt cork was applied to actors' faces to make them into caricature black people. Thomas Rice was the first white person to copy a black act; he'd seen an old black man in Cincinnati performing alone for his own amusement in an awkward dance (he was disabled) and singing a strange song:

Weel about, and turn about
And do jis so.
Ebry time I weel about
I jump Jim Crow.[5]

'Jim Crow' later became the term used for the laws of the segregating South.

Minstrelsy was a crude copying game. White actors created black caricatures and played them to the hilt, the Sambos and Zip Coons and Jim Dandys. All grotesque characters and simpletons that didn't possess a brain in their curly head. Minstrelsy distorted authentic

African-American folk music, dance, speech and style. Yet it was the black people's entertainment the whites were copying. W. C. Handy maintained that every plantation had its talented band that could crack jokes, and sing and dance to the accompaniment of banjo and bones. It is ironic that when black people themselves wanted to enter the world of showbiz towards the end of the nineteenth century, they often wore the burnt cork on their faces too. Black on black. As if it wasn't possible to be a genuine black face in entertainment; you had to wear some kind of mask.

Real black faces were not wanted for a very long time, and even today black actors get very few parts – there are few black faces to be seen on television and film apart from a handful of stars. It is possible to watch one Woody Allen film after another and not see a single black face. Irving C. Miller, who threw Bessie out of his chorus line in 1912 for being 'too black', was a product of his time as well. Black-skinned girls were not considered as beautiful as brown-skinned girls. In fact black girls weren't considered beautiful at all.

Like W. C. Handy, Bessie wouldn't have hesitated to join the Stokes show. The opportunity was a godsend. It offered her freedom from stultifying poverty. It gave her a chance to grow. The troupe was managed by Lonnie and Cora Fisher, and Bessie later cited Cora Fisher as a major influence, though there is no information about exactly how she influenced her. Perhaps she was the first person to give Bessie a belief in her own performing abilities. Perhaps she taught her essential performing tips that she never forgot.

Or she was simply kind to her, one of the first adults who was ever kind to her. Perhaps she was her salvation, saving her from a grim life. Maybe when Bessie looked back on her life, she always saw Cora Fisher at the helm, the woman who launched, wittingly or unwittingly, one of the biggest careers in blues history.

Although Bessie was initially taken on by Moses Stokes as a dancer, she soon began to sing the blues. She left Chattanooga for the first time to travel the South. With her in the troupe were Ma and Pa Rainey, who were dubbed 'the Assassinators of the Blues'. She moved on from Moses Stokes to join Irving C. Miller's tent show in 1913, until she was thrown out for being too dark. Later that same year, she was performing at the 81 Theatre in Atlanta, where actor Leigh Whipper saw her: 'She was just a teenager, and she obviously didn't know the artist she was. She didn't know how to dress – she just sang in her street clothes – but she was such a natural, she could wreck anybody's show.'[6] As a singer and dancer, she teamed up with Buzzin' Burton (as 'Smith & Burton') in Park's Big Revue at the Dixie Theatre, Atlanta, in 1914. And in 1915 she joined another travelling show, again with Ma Rainey – Fat Chappelle's Rabbit Foot Minstrels. She toured with the Rabbit Foots through the South in 1915 and also with Pete Werley's Florida Cotton Blossoms Minstrel Show, Silas Green's Minstrel Show and others in 1916.

The Rabbit Foot company was particularly popular in the Mississippi Valley, from where it drew many of its cast. Paul Oliver in his book *Songsters and Saints* quotes an advertisement:

> '100 Performers and Musicians WANTED. Both ladies and gentlemen for my 2 shows under canvas. A Rabbit's Foot Comedy & Funny Folks Comedy. 40 weeks engagement for the right parties . . .' These shows were on the road in 1929–30 when the Alabama Minstrels, Jordan's Swiftfoot Minstrels, Richard's and Pringle's Georgia Minstrels, J. F. Murphy's Georgia Minstrels, The Fashion Plate Minstrels, Warner and Moorman's Famous Brown Derby Minstrels and John Van Arnam's Minstrels were all playing the South and Midwest . . . Many featured celebrated show singers: Lizzie Miles with the Alabama Minstrels, Bessie Smith with Irving C. Miller's show, Butterbeans and Susie with the Cole Brothers' Carnival.[7]

The minstrel troupes and blues travelling shows were often booked by TOBA (the Theatre Owners' Booking Association, aka Tough on Black Asses). They sang comedy songs and ballads and performed dramatic routines in addition to singing the blues. They mixed the rural blues tradition with the more sophisticated fare of the vaudeville stage. Humphrey Lyttelton in *The Best of Jazz* points out: 'Few "classic" blues singers of note became famous without serving a tough apprenticeship in the tent shows, barn-storming from settlement to township to plantation.'

Ida Cox (born Ida Prather) ran away from home to tour with White and Clark's Black and Tan Minstrels (as a blackface 'Topsy'). And she, along with Bessie Smith, Ma Rainey and many others, toured the theatre circuits from 1910 with the Rabbit Foot Minstrels, Silas Green from New Orleans and the Florida Cotton Blossoms. Ida Cox was often billed as 'The Sepia Mae West' – an unimaginable billing for Smith, who would be known as the Empress.

For Bessie, the difference between singing on Ninth Street with her brother accompanying her on guitar and travelling the South with a whole entourage of talented people must have been mind-blowing. Even later in her life when she could have afforded not to travel all over the place, she continued to do so. The exhilarating freedom of being on the road and arriving at completely new places with a whole group of people who knew how to have a good time made her into a lifelong travel addict. Travel was in her bones and in her blues. Places influenced the songs she wrote: 'Backwater Blues', 'Hot Springs Blues', 'Long Old Road', 'Lonesome Desert Blues', 'Dixie Flyer Blues'.

Hold that engine, let sweet Mama get on board
'Cause my home ain't here, it's a long way down the road
On that choo-choo, Mama's gonna find a berth
Going to Dixieland, it's the grandest place on earth
Dixie flyer, come on and let your drivers roll
Wouldn't see us now to save nobody's doggone soul.
'Dixie Flyer Blues'

This song has the sound of the train comically all the time in the background, the whistle and the chug-chugging of the engine. Travel in her blues, as in her real life, was a metaphor for getting away from men. Many of the blues she sang contained this restless desire to be on the road, going to someplace else. Either it's the man she loved who has just left this town, or it's Bessie herself who is leaving to get away from him. 'Florida Bound Blues', 'Nashville Women's Blues', 'Chicago Bound Blues', 'Far Away Blues', 'Gulf Coast Blues'. Going places, going anywhere was freedom. Bessie Smith grew up on the road, became intimate with places she passed through. It was all happening on the road. Sally Placksin, in her book *Jazzwomen*, says:

Travelling by Pullman car, bus, or automobile, the most famous of the minstrel troupes – the Rabbit Foots, Florida Cotton Blossoms, Georgia Smart Set, Silas Green of New Orleans – would hit the cane brakes and the factory towns, the theaters of the bigger cities, laying down a bedrock of blues among the people. They followed the tobacco crops in the spring, the cotton crops in the fall, hit the coal mines and the Sea Islands off Georgia, arriving in each town at the season when the people had the most money to spend and were ready to go out and party. When the troupes arrived, the tents would go up, the bands would march through town, flyers would be handed out, and soon the show would be on . . .[8]

The company on the troupes was as exciting as the towns themselves. Ma Rainey was on that very first troupe Bessie ever joined, the Moses Stokes Travelling Show. Ma Rainey is now known as the 'Mother of the Blues'. She was the first of the blueswomen; all the rest followed her. She is the earliest link between the male country blues artists who wandered the streets of the South and the 'classic blues' women singers. Bessie Smith, Clara Smith, Mamie Smith and Laura Smith, Ida Cox, Victoria Spivey, Alberta Hunter, Sippie Wallace, Lizzie Miles, Bertha 'Chippie' Hill, Memphis Minnie Douglas – all owed a debt to Ma. Ma Rainey was born Gertrude Pridgett on 26 April 1886. Her father, Thomas Pridgett, and her mother, Ella Allen, were both minstrel troupers. Gertrude was one of five children baptised in the First African Baptist Church, Columbus, Georgia. She was already working as early as 1900 in a talent show, *A Bunch of Blackberries*, at Springer Opera House, and in 1902 she was singing blues songs on the road, long, long before W. C. Handy wrote his famous 'St Louis Blues' in 1914. The gold-toothed Ma Rainey was there before him (Humphrey Lyttelton describes her mouth as 'revealing in a broad grin a veritable Fort Knox of gold fillings') and it was a woman singing on the road that inspired 'St Louis Blues'. Ma Rainey, along with Bessie Smith and most of the other classic blues singers, also wrote songs, such as 'Bo Weevil Blues', 'Don't Fish in My Sea', 'Louisiana Hoo Doo Blues', 'Prove It on Me Blues', 'Rough and Tumble Blues', 'Titanic Man Blues', 'Weepin' Woman Blues'.

Ma Rainey sang, 'I can't tell you about my future, so I'm going to tell you about my past.' She was on to something even then. If she was around today she would agree with Toni Morrison who said that all the future is in the past. The blueswomen were themselves direct descendants of the Voodoo Queens of the 1800s. As Sally Placksin explains:

Little is known of the earliest queens – Sanité Dédé, Marie Saloppé and the most famous and long-lived queen, Marie Laveau. Free women of color, glamorous and haughty showwomen, shrewd and clever business women, the queens began to accrue power in the early nineteenth century, after the Louisiana Purchase (1803) eased the formerly strict rulings against voodoo in New Orleans . . . After 1817, New Orleans' Place Congo (or Congo Square) was the site where the most public voodoo ceremonies took place. On Sunday afternoons, slaves would gather there to chant and to dance the tamer dances that represented six African tribes . . . Sidney Bechet had grown up hearing about the music played at Congo Square. 'Improvisation', he called it . . . Not only were the voodoo queens and women an important force in the black community . . . the secret voodoo cults also provided an early ground on which women of both races mixed, for some of the women who attended the secret rites were white. Newspaper accounts considered it scandalous when white women were discovered at the ceremonies. (One man actually committed suicide the day after his wife was found at a meeting . . .)[9]

Ma Rainey was the matriarchal head of the original blues family. Every one of the women blues singers must have been influenced by her. Although she did not teach Bessie how to sing (Bessie knew how already; she had a style and a voice of her own), Ma Rainey probably influenced the way that Bessie and the other classic blueswomen dressed, and she almost certainly helped groom Bessie for a life on the road, teaching her dancing steps and the importance of performing her songs, as well as simply singing them. Ma Rainey's grandmother was a stage performer during the Reconstruction. If Ma followed in her grandmother's footsteps, all those classic blues singers who shared the surname of Smith – again as if they were part of the same family – followed in hers. There was Mamie Smith who sang the famous 'Crazy Blues', which, in 1920, was the first commercial classic blues hit. There was Clara Smith, Trixie Smith, Laura Smith and Bessie Smith. They were all unrelated, though often when they were playing together the posters would proclaim, 'The Smith Girls Are in Town'. Sometimes Bessie Smith and Clara Smith – who later recorded 'Far Away Blues' and 'I'm Going Back to My Used to Be' together – pretended to be sisters. The common surname Smith is obviously a legacy of slavery. It created the illusion that large numbers of black people were related because they so frequently shared the surname of their master.

The classic blues singers were not afraid of addressing any issue in their songs. They were far more dynamic in their lyrics than female vocalists today or indeed than any of those singers who followed them. Dinah Washington, Sarah Vaughan, Ella Fitzgerald all sing of milder stuff. The blueswomen sang songs about subjects the Spice Girls would run a mile from (with Sporty Spice in the lead). Sally Placksin again:

> Her subject matter ranged from Everywoman's lovesick, lonesome blues to the havoc wreaked against the men who did her wrong; from anguished desolation to murderous revenge. She counseled women to 'Trust No Man' – not even their own – and advised men to be careful talking in their sleep; she delivered ultimatums, and she dealt frankly and boldly with homosexuality, lesbianism, sadomasochism and sexual violence in her songs.[10]

When they did sing those songs about no-good men, women in the audience would go wild and shout out in agreement. These songs were 'telling it like it is'. They were greeted with such delirious enthusiasm because women identified with the lyrics. Here were songs about ordinary people and ordinary problems. Songs about men, every type of no-count, no-good man it was possible to imagine. A long trail of them, always cheating and letting women down. The classic blues singers sang about life, and ordinary people loved them for it. They sang about illness, death and graveyards as well. No subject was taboo. Memphis Minnie even had a song about meningitis called 'Memphis Minnie-jitis Blues':

Mmmmmmmmm,
The meningitis killing me.
Mmmmmmmmm,
The meningitis killing me.
I'm bending, I'm bending, baby,
My head is nearly down to my knee.

The classic blues singers sang songs that offered wonderful pieces of advice to other women; they were way ahead of their time, giving much more revolutionary advice than today's agony aunts! Ida Cox's 'Wild Women Don't Get the Blues' is exemplary:

You never get nothing by being an angel child,
You better change your ways and get real wild
'Cause wild women don't worry, wild women don't have
    the blues.

There was Bessie, nineteen years old, travelling with the Mother of the Blues. The audience came from miles around; they rode mules or travelled on trains, packed in like rice. The journey to the blues tent must have had nearly as much of a charge of atmosphere as the concert itself.

There they all were, jammed in with other people travelling to hear the same thing, to get the same hit from the blues, coming from all over to experience together the knockout voices of the blueswomen. They came from the river settlements, from cornrows and lumber camps. They travelled miles to get to the hotspot to hear the blues. Sterling Brown captures the mood in his wonderful poem, 'Ma Rainey':

When Ma Rainey
Comes to town
Folks from anyplace
Miles aroun',
From Cape Girardeau,
Poplar Bluff,
Flocks in to hear
Ma do her stuff,
Comes flivverin' in,
Or ridin' mules,
Or packed in trains,
Picknickin' fools . . .
That's what it's like
For miles on down
To New Orleans delta
An' Mobile town,
When Ma hits
Anywheres aroun'.

The tents filled. Even they were segregated – black people on one side and white people on the other. Ma Rainey would come on with greasepaint and powder on her face, wearing an elaborate rhinestone gown, a twenty-dollar gold-piece necklace, flashing those infamous gold and diamond teeth. A mouthful of blues. Ma Rainey's shows went on for two hours. Like the other blues performers, she did not simply sing, she danced and shimmied 'vulgarly'. She was wildly flamboyant and extrovert, a real performer. She got her audiences so worked up and wild that it wasn't so much a call and response as a shout and a shag. Her shows even included acts of freakishness. Ma Rainey loved the double entendres of blues talk; she liked to tease her audiences about bird liver and pig meat. Her repertoire of metaphors for men's sexual anatomy was probably greater than any other blues singer's. Her blues were the lewdest of the lot.

Ma Rainey was also a lesbian. Even though she married Pa Rainey on 22 April 1904, she was still a lesbian. Will Rainey was an older vaudevillean. Together they put on shows and tours for years. It was well known that Ma was a lesbian. Ma and Pa could do just about everything together, touring an act of comedy, singing and dancing, but they couldn't have sex. Amongst the many myths and legends of the Empress are different accounts of Ma Rainey kidnapping Bessie Smith and forcing her to join the Moses Stokes show. Some say she got her henchmen to go down to Charles Street and throw the girl into a big burlap bag to bring her back to her. Others say she carried out the kidnapping herself. At every crucial juncture in Bessie Smith's life there are wild stories, but this is the very best, as Albertson says:

Legend has it that Ma Rainey literally kidnapped Bessie, that she and her husband forced the girl to tour with their show, teaching her in the process how to sing the blues. It's a colorful story, but almost certainly just a story. 'I remember one time when we were in Augusta, Georgia,' says Maud Smith [wife of Bessie's brother Clarence]. 'Bessie and Ma Rainey sat down and had a good laugh about how people was making up stories of Ma taking Bessie from her home, and Ma's mother used to get the biggest laugh out of the kidnapping story whenever we visited her in Macon. Actually, Ma and Bessie got along fine, but Ma never taught Bessie how to sing. She was more like a mother to her.[11]

The person who first made up this story would never have thought of a motive. Why would Ma Rainey have wanted to do it? Maybe the story came about because of Ma Rainey's renowned lesbianism. Maybe the gossips thought it figured a girl like Ma Rainey would need a companion. Perhaps it is all down to the notion, the prejudice, that lesbians coerce, force, mislead, kidnap, rather than fall in love. But it is still a good story. Years later, whenever Ma Rainey and Bessie Smith met up, they had a good laugh about it.

*There would have been no ransom, that's for sure. No note
except a blue note. A getaway wagon, not a getaway car.
And the hideout was the Moses Stokes Travelling Show.
The Moses Stokes Travelling Show. How high the stake?
How much did Ma Rainey want Bessie Smith to sing the
blues? Would she have given her life?*

*It happens in the dead of night with Bessie asleep in
the middle of Viola and Tinnie. Ma Rainey arrives in
the wagon with Pa. Ma Rainey is so bored of Pa. Pa
never even tries to get it on with her. The names Ma and
Pa are so asexual. It is never possible to imagine a wildly
passionate relationship between a Ma and a Pa. (It is
the same with the Broons. Ma and Pa Broon have heaps
of kids but you can never imagine how they got them.)
Pa Rainey knows he is a laughing stock but something
about those gold teeth of Ma's makes him do whatever
she says. A man can be ruled by a woman with gold teeth.
He knows his wife wants this young girl Bessie Smith and
there ain't a single thing he can do about it. He aids and
abets her. He waits outside, smoking a pipe, whilst Ma is
inside kidnapping. It is not for nothing that Ma and Pa
are called the Assassinators of the Blues. They take the
role very seriously. Before coming tonight for the young
Bessie Smith, they have both intoxicated themselves with
hashish. Pa has to help Ma on with the stocking. It is one
from the current show. He pulls it over her face. Its purpose
is to disguise those gold teeth. Ma Rainey's teeth are the
most famous teeth in the whole of America. Even with*

the stocking on, Pa glimpses a frightening flash of gold.
He tells his wife to shut her mouth, to do the deed with no
words. Then he lights up his pipe again. Ma Rainey will
not bungle it. She wants the girl too bad. The things she's
heard about this girl. Skin saw her on the corner of Ninth
Street just last week singing like her soul was on fire. Skin
told Ma how he had stood transfixed on the spot for four
hours, how he hadn't even noticed the rain coming down,
how he only moved when he suddenly realised the girl had
gone. Ma had pulled his collar and said, 'Where did she
go?' But the useless pig meat knew nothing. Ma had to
assign somebody else to track down that girl's cabin. A girl
that could sing like that was worth having. Ma Rainey
opens the door quietly and picks up the girl in the middle in
her arms. Tinnie and Viola don't even turn in their sleep or
grind their teeth some more. Nothing. And this Bessie she's
heard so much about weighs some but Ma's arms are strong
as a lesbian's. Her feet don't exactly manage a straight
path to the cabin's door, but this singing girl is sleeping like
a log. Ma wants to kiss her beautiful black sleeping face,
those nice peachy lips she has, but she stops herself. There
will be time enough for all that when she gets her back to
Moses Stokes. She'll need to make sure no one else gets to
her first though. This blues girl will be worth it. Worth it
when she teaches her those shouts. She is going to teach that
girl to sing the blues in the morning – and come night-time
that girl will be singing the blues like she been singing them
all her life. Her voice will make her want her all the more.

And when they are both finished with Rabbit's Foot and Cotton Blossoms, Ma Rainey will devour her. She will get her going till she don't know where she's heading for.

Everyone knew that Ma was a lesbian. Ma even sang songs with overtly lesbian themes:

I went out last night with a crowd of my friends
It must have been womens 'cause I don't like no men.
'Prove It on Me'

Maybe Ma being so open about her sexuality encouraged Bessie to do likewise. Bessie was never hung up about her relationships with women; neither was Ma. I like to picture the pair of them on the wagon on the road, laughing and giggling at some foolish remark some foolish man just said. I like to see Ma's mouth wide and grinning, flashing bits of gold. There's the dust on the road, and the road is bumpy and travelling isn't always fun. But sometimes the journey would pass quick with Ma and Bessie gossiping and making up blues together. Bessie was a novice. Her brother Clarence had suddenly changed her life. One moment she was singing on Ninth Street, the next she was off with Ma and Pa Rainey, the Assassinators of the Blues, on the road, travelling through the South and the Midwest.

Bessie Smith is now known as a blues singer. But in the days when she performed live, she was multi-talented and versatile. Her dancing was as good as her singing. She liked acting too, dressing up in various costumes. She appeared as a singer, dancer and male impersonator in her own show, *Liberty Belles Revue*, at the 91 Theatre, Atlanta, 1918–19. (I bet she made a handsome man.) She worked with the Charley Taylor Band in local clubs in Atlantic City and she was in New Jersey in the early 1920s with the Charles Johnson Band in Paradise Gardens. She worked in Horan's Madhouse Club in Philadelphia from 1920 to 1923. She appeared with Sidney Bechet in the musical comedy show *How Come* at the Dunbar Theatre in 1923. Sidney Bechet has turned out to be one of the most perceptive commentators about Bessie Smith. He claimed to have had a love affair with her. Whether he did or not is unproven, but he certainly, early on, managed to grasp her complexity:

> She had this trouble in her, this thing that would not let her rest sometimes, a meanness that came and took her over. But what she had was alive . . . Bessie, she just wouldn't let herself be; it seemed she couldn't let herself be.[12]

Having travelled with so many troupes, she already had a large following before she made her first record. Her experiences on the road were forming the tough and uncompromising woman she was to become. She had so many encounters with racism, from black people as well as white, who thought her skin colour was just too dark, that it is easy to see how she developed a toughness to survive.

Doing the clubs, Bessie eventually became so successful and had such an enthusiastic following that she could insist on being the only one on the bill who would sing the blues. She came to regard the blues as her territory. She said to Ethel Waters (who she called 'Long Goody'):

> You ain't so bad. It's only I never dreamed that anyone would be able to do this to me in my own territory and with my own people. And you know damn well that you can't sing worth a —— [13]

The Empress was a prima donna. But then she had that voice, that talent. She could cast a spell on her audience. She could hypnotise them. She had them under her blues.

Picture her: 1921, 1922, bigger now, older now, wearing fancy clothes. She's made big money; she's drunk bootleg; she's had sex with women and men. But the little girl's face is still just behind the woman's. Pentimento. When she gets up to shout the blues, to barrelhouse, to transform her audience, part of her is still standing on Ninth Street under the moody sky. The part of her that is still standing on Ninth Street is the part of her that hypnotises the audience in the 81 Club in Atlanta, Georgia. This power she has, she was born with. It is nothing new to her. It is not really a surprise. She was just waiting for it to find a bigger place than a street corner and it did. It found a dress, a dress she thought would do. She wasn't really all that bothered about what she wore anyhow. (Except for the shoes. The leather always felt important on her feet after all those years of being barefooted.) She felt fat and clumsy in every damn dress whether it was cotton, linen or kiss-her-black-ass silk. It found a club, then it would find, in a few years, a record, then it would break the record. She wears fancy dresses now on the stage, after Ma Rainey. Ma Rainey's flamboyance set the trend for the blueswomen. I've never seen a photograph of a woman singing the blues in a wee grey slip.

When Bessie sings in these clubs with the audience literally spellbound, captivated, totally enthralled, she is not really paying them any attention. She is deep, deep inside herself in the place where the blues come from. She has shut out the clapping and the stomping and the shouting. It is quiet in the 81 Club, people's faces are like stills. They flash before her sometimes in the middle of a song. Like something she might see travelling on a train very fast; seen for an instant then gone. She catches the odd glimpses of one shining face and another troubled one and then she returns to herself. Sometimes she's sure she saw her sister Tinnie right in the middle of that sea of faces in the 81 Club. Once she even saw her dead mother standing for a single moment behind the bar. After singing the blues, she knocks down one drink, then another, full of a terrible longing for herself, for her family, for Chattanooga. The whole theatre is full of a restless longing. She gets people that way. She is right under their skin. She is closer than God.

She hasn't made a record yet. But she has tapped the source of her own power – her voice – and she knows it is like nobody else's and she doesn't want the distraction of anybody else playing with the blues. No, she is the blues. She is the blues itself. As far as she is concerned, she is the rightful Queen of the Blues and nobody else can sing worth a damn. She knows the timing. She's got the timing just right. Doesn't need to articulate it or even to think about it. It's all in the length of her pause. It's the way she hangs on to those notes when they are gone. Like she's hanging on to the little girl she was, or the very back of herself, or her grandmother's long, large hand. She is full of longing, full of trouble, restless, wandering up and down the long arms of the clock. When she sings on stage, part of her is travelling, reaching back into every hurt that's ever happened. Her voice is a poplar tree singing. Swaying in the Southern breeze.

Who knows better than we,
With the dark, dark bodies,
What it means
When April comes alaughing and aweeping
Once again
At our hearts?

Angelina Weld Grimké, 'At April'

THE TRUNK AND THE NO-GOOD MAN

Bessie Smith married twice. The marriage to her first husband, Earl Love, lasted only one year before he died. Nothing is known about Earl Love except that he came from a prominent black Mississippi family. In 1922 Bessie Smith met Jack Gee. On 7 June 1923 he obtained their marriage licence from the clerk at the Orphans' Court, Philadelphia County. They then went on to the house of a Reverend C. A. Tindley, a prolific composer of gospel songs, the most famous of which was 'We Shall Overcome', and Jack Gee and Bessie Smith were pronounced man and wife. But Bessie never became Bessie Gee. Throughout her career, she was known as Bessie Smith.

Their relationship lasted six years and obsessed Bessie for several more. The key to understanding Bessie Smith lies in understanding the pull and power that men like Jack had. Jack was a big strong man who had no special talents or skills of his own. He loved power and from the outset of his relationship with Bessie he tried to take over her life and control her. Her wild excessive drinking drove him crazy and he often beat her up for getting drunk. Being a famous blues star did not stop Bessie from being a battered wife. Jack frequently beat her up very badly, knocking her downstairs and threatening to kill her. The few people on the road who went to her defence were beaten up too.

Jack would beat up any chorine who got between him and Bessie. When she travelled with her performing girls and prop boys and her band, everyone was terrified of Jack's sudden appearances.

Jack Gee was a nightwatchman, an illiterate, who had been turned down by the Philadelphia Police Department; yet he often pretended that he was a policeman. Bessie's Smith husband, a policeman? The image of her in feathers, plumes and ermine coat, and him in his imaginary policeman's uniform, arresting people at her gigs, is hilarious. The joke was on Jack; he couldn't see it. If he had been a real policeman he probably would even have enjoyed arresting his drunken wife to thwart her devilish drinking. He would have relished throwing her in a cell and telling her to stay there till morning. Jack could only fantasise. He never made the grade. Why did Jack want to be a policeman anyway? He wanted to have the legal power to intimidate people and excuse his excesses in the name of the law. He wanted a uniform. He wanted respect from the wearing of the uniform. He wanted everyone to obey his every command. He didn't get to join the Philadelphia Police Department but he did a good job of being that impromptu self-appointed cop on the road. Bessie Smith's fame gave him a vicarious power which he used unsparingly.

Apparently, Jack Gee's favourite pastime was counting Bessie Smith's money. He was good at counting. He loved it. He would often stay up until the wee small hours separating nickels from dimes, quarters from halves. Every tiny coin counted. Although Bessie disapproved of his money-counting binges, she found them amusing. She often entertained her friends mimicking the big-eyed Jack, crouched over and stroking tiny towers of change.

Bessie met Jack in 1922, when she was appearing in Philadelphia at Horah's Cabaret before she made her first record, but well after she had made a big name for herself on the road. Bessie Smith's life is so full of drama, it seems really over the top. Her life was theatre. It is impossible to separate Bessie's blues from Bessie. She is her blues. 'She was blues from the time she got up in the morning until she went to bed at night,' Frank Walker – who later became her manager – asserted.[14]

On her very first date with Jack somebody shot him and nearly killed him. (Wasn't that enough of a sign for her?) So, on her first date, she is not stood up, or given a dizzy, as we used to say in Glasgow, but her date is shot. *Her date is shot.* You can't fuck a man with a bullet through him, but you can visit him in hospital and tend to his wounds. Bessie visited Jack daily for five weeks and became close to him. (Closer probably than if they had fucked that first night.) The strange intimacy that comes from hospital visits, the complicit vulnerability, the knowledge of the daily routine, the familiarity of every need, would have given Bessie a false sense of Jack's character. She would have been moved by his need for her. His violence would have been checked under those hospital sheets.

When he came out of hospital, Bessie already had her

first recording date for 'Downhearted Blues'. He pawned his watch so that he could buy her a red dress for the occasion. A gift that she never forgot. It was probably the single thing that tied her to him for all those years. Nobody had ever bought her anything before. No man had ever pawned his watch for her before. She took the dress as a sign of love. She loved the dress. When she sang that first song wearing the red dress that Jack bought, she probably felt lucky. Who knows, if she was superstitious (which she was apparently), she might have felt she owed her entire recording career to that one dress. Ludicrous as it may seem, there must be something that explains his enormous hold over her. There she was, an independent woman who had wealth, fame and success, tied to a man who was beating her up and trying to control her life. Did she believe he loved her because he pawned his watch? Did she love him because he pawned his watch? What was it about Jack Gee that made Bessie stick with him for so long? She was not faithful to him, but she was obsessed with him. She did everything to try and make him happy. After she started making lots of money with her recordings, he gave up his job as nightwatchman. She bought him a good gold watch. (Maybe that's why he pawned that watch; he knew she'd buy him a better one.) She bought him expensive cars if he so much as looked in their direction. She sent him off frequently to Hot Springs when he was having his 'breakdowns'. She even wrote a song about Hot Springs.

Some come here crippled, some come here lame
If they don't go away well, we are not to blame
Hot spring water sure runs good and hot.

According to Maud Smith, Bessie's sister-in-law: 'Jack never had no nervous breakdowns. . . We were the ones who should have had nervous breakdowns. He wasn't sick. Bessie would do anything for him, so she'd give him all the money he wanted, and tell him to go and take a rest.'[15] Perhaps being ill, or pretending to be ill, was also a way of controlling Bessie. If he couldn't stop her wild partying ways by beating her up, he could go to Hot Springs and make her visit him. Perhaps he wanted her to feel guilty at the state he was in. Or maybe it reminded him of their wooing when he was in hospital recovering from the gunshot.

Jack Gee applied for a marriage licence as 'Downhearted Blues' hit the market. Irony rarely comes so rich:

Gee, but it's hard to love someone, when that someone
    don't love you.
I'm so disgusted, heartbroken too, I've got those
    downhearted blues.
Once I was crazy about a man, he mistreated me all the
    time.
The next man I get has got to promise me to be mine all
    mine.
Trouble, trouble, I've had it all my days,

61

Trouble, trouble, I've had it all my days,
It seems like trouble is going to follow me to my grave.

This song was prophetic; no need for a glass ball when you've got the blues. At this point the marriage with Jack was reasonably happy, full of the promise of the future. But the real future of 'Downhearted Blues' was lying in wait and it wasn't long before Jack started mistreating Bessie. Once begun, the cycle of destructiveness seemed unbreakable. The relationship was always on the brink of crisis. They were always breaking up and making up. They were addicted to each other even though they must have known they were bad for each other and totally incompatible. Bessie was a big drinker; Jack was teetotal. Bessie was wildly sociable; Jack was antisocial. Bessie liked wild parties; Jack hated them. Bessie was generous with her money; Jack was mean. Bessie was talented; Jack was not. What did they have in common? Singing the blues on the road became a vital means of escape from the tyrannical Jack. But she couldn't stop him popping up all the time. Jack Gee cramped her style. She could not enjoy partying, drink or promiscuous sex with him around. It is strange that she married him in the first place; home life never did seem to make her very happy. She did not even spend much time or money making a nice home for them. They had an apartment at 1236 Webster Street in Philadelphia, but it lacked the atmosphere of a home. There was no fancy furniture. 'There was nothing special about it,' said Ruby Walker, Jack's niece.[16]

The majority of Bessie Smith's blues concern that 'dog-gone man of mine'. Although they were not all written by Bessie, they could easily all have been about one man: Jack Gee. The famous lines, 'I've got the world in a jug, The stopper's in my hand', at the end of 'Downhearted Blues' are still powerful. The mixture of enigma and plain talking is typical blues. What do they mean? Are they positive or negative or both? Those lines were so popular because people identified with them. People would sing those lines to themselves as if they aptly captured some truth about their own life. All the conflicts, wars and sorrows, the pains and the troubles of the world are in Bessie's jug, but the stopper's in her hand. She has some power over it then. Or is she just singing about the world of men, the world of mistreating men? And has she cast herself in the role of Everywoman? The world's experiences contain her own; she is the woman in the street who can never get away from man trouble – it will follow her to her grave. But she has that stopper in her hand. Maybe it is men she can put a stop to if she wants to, a stop to all that heartbreak. Maybe all the lying, cheating, dirty no-gooders are in that jug. A blues song, like a poem, opens itself up to multiple interpretations. There's the words and there's the music contradicting the words, adding irony and changing the meaning.

Jack tried to control Bessie's affairs but he never was her business manager. Maud Smith said in conversation with Chris Albertson, scathingly, some years later:

Jack couldn't even manage himself. He would always have signs saying, 'Jack Gee presents Bessie Smith' and he would call himself a manager, but he couldn't even sell a ticket. He could count money and he could ask for money, but that's about it.[17]

So why did Bessie Smith, the Empress, the Queen of the Blues, waste her time and much of her life with a living example of the dirty no-gooders she sang about? On the one hand, there's this strong woman who would fight anyone who said one wrong word, who stood up for herself and others, who refused to be patronised by white people or intimidated by club owners, managers or recording companies; on the other, there's this battered wife who lived in fear of her husband discovering any of her wild parties, who kept him on, despite the numerous times he beat her up, who was so frightened of him she was unable to get rid of him. In the end, Jack Gee broke Bessie down so much that her friends and family could hardly recognise her. A dancer in the Moaning Low Company said, 'She wouldn't cry, she'd just sit there, staring . . . sometimes I just couldn't believe that this was the same woman. That man really broke her down, strong as she was.'[18]

Bessie never recovered from his affair with Gertrude Saunders, Irving C. Miller's most glorious 'brownskin beauty'. (There is a bitter irony here since it was Miller who years before had turned Bessie down for being 'too dark'.) Gertrude Saunders was well known in Miller's

1926 show, *Red Hot Mama*. She was also famous for her performances in *Liza* and *Shuffle Along*. In 1929, six years after their marriage and 'Downhearted Blues', Jack used Bessie Smith's money to finance Gertrude Saunders in her own show, then on tour in Columbus, Ohio. Bessie found out because a man turned up at her show, *Steamboat Days*, and showed her a copy of an article in the *Amsterdam News* which recounted Jack's considerable success with the Gertrude Saunders show. True to form, Bessie was off like a shot, hiring a cab all the way to Columbus, accompanied by her constant companion, Ruby Walker. She tracked down his hotel in thirty minutes and went to have a piece of him. The hotel room was a shambles. There was blood, guts, pieces of furniture and feathers flying everywhere. 'Downhearted Blues' came into being for real – *Once I was crazy about a man, he mistreated me all the time.* This wasn't a blues song. This was real life. Question: what is the difference between the blues and real life? Answer: truth is stranger than the blues.

Nobody knows when Jack began his relationship with Gertrude Saunders; and Gertrude Saunders herself denied they ever had a relationship at all, claiming that Jack was 'just an ignorant darkie, but he had a good business head on him, and he was perfect for Bessie; those two belonged together.'[19] It is certain that the marriage to Jack ended in that hotel room. It was never the same afterwards. So Jack moved on to the Gertrude Saunders show, moving out of Bessie's life. Bessie Smith's adopted son, Jack Junior, claims that Gertrude did have a relationship with Jack Gee senior. According to him, Jack even brought Gertrude round to Bessie's house when she was away on the road. But Jack Gee's biggest betrayal of Bessie Smith concerned not Gertrude Saunders, but Jack Junior.

It was while she was on tour that Bessie first met her adopted son-to-be. Every time she went through Macon, Georgia, she would stop to spend time with the boy she called 'Snooks'. His mother was a niece of Margaret Warren, one of Bessie Smith's chorus girls. She had promised Bessie that if she ever ran into trouble she would let Bessie adopt her son. In the spring of 1926 Jack Gee and Bessie legally adopted Snooks, renaming him Jack Gee Junior. He was six years old when he was brought to Philadelphia on Bessie's railroad car. He joined them on the last round of that tour. Bessie Smith was overjoyed to have this son in her life. She resolved to cut out her drinking and try and create a stable family home for him. She sent for her family, who were still living in Chattanooga, so that they could help her with him when she was on tour. She bought two houses close together, 1143 and 1147 Kater Street. One was for Viola and her family and the other for Lulu and Tinnie and their families. At this period in her life, she was at her happiest. She had her family up from Chattanooga. She had enough money to buy them houses. She had a son. She showered everybody with gifts. Having a son to look after gave her a real purpose in life and she took it very seriously. She wanted him to get a good education, said she would buy him anything if he became a lawyer. Having Jack Junior gave Bessie Smith another identity. She was proud of being a mother.

Perhaps out of a desire to get revenge on Bessie, or perhaps just to make his destructive impact on her life long-lasting, Jack Gee ensured that Bessie lost not just him but their son. After leaving Bessie, he kidnapped Jack Junior and put him into care. He had no real interest in his name-sake but he knew how much the boy meant to Bessie. Jack Junior, tracked down by Chris Albertson, tells the story:

> One day when Mama wasn't home and he was living with Gertrude [Saunders] at her house, he came and got me. He told me to get in the car and said that he wanted to take me somewhere. So I got in the car and wound up at the SPCC (Society for Prevention of Cruelty to Children). They kept me down there for about two weeks – he told the people that Mama let me stay out all night, which was a lie.[20]

Jack Junior escaped from the SPCC only to find himself back with his father, whose partner made him sleep in the basement with her brother. The brother got him up the next morning and made him clean the hall. It must have been such a shock for Jack Junior to go from living with Bessie Smith for five years as her pride and joy, with a more than comfortable lifestyle, to cleaning the hall in the basement of a strange house at the age of eleven. After Bessie's death in 1937, Jack Gee Senior took all of her money, claiming that they still had a close relationship. Her son, who should have inherited the lot, was left poor. Jack Gee made sure he was never mentioned and that no writer knew of his existence. It wasn't until Chris Albertson's book came out in 1971 that the rest of the world knew of his existence.

The blues that Bessie sang often strangely anticipated her life. She became her blues. Or perhaps she always was her blues. So many of the tragedies she sang about actually happened to her after she sang the songs and not before; she didn't just sing songs that reflected her own experience, she sang songs about experiences that were lying in wait for her, further on down that old blues road. Her blues were premonitions. Some of them she wrote herself, but the majority were written by other people. It is not clear how much freedom she had in choosing her songs. But it is surely no coincidence that the next song she recorded after her devastating break-up with Jack was the quintessential 'Nobody Knows You When You're Down and Out', written by Jimmy Cox.

This song, more than any other, is the song that most people associate with Bessie Smith and it remains tremendously popular. She recorded it on 15 May 1929, accompanied by an accomplished five-piece band, and Columbia released it on 13 September. When she sang, 'Then I began to fall so low, / I didn't have a friend and no place to go, . . . It's mighty strange, without a doubt / Nobody knows you when you're down and out,' the truth of those words resonated throughout the North and the South. She hummed half the song, a haunting solo by trumpeter Ed Allen, 'Mmmmmmmmmmmmmmm when you're down and out', expressing the feeling of it even more movingly than words could do. Jack was gone in May of the same year. Her loss comes through; what other song at this point in her life could possibly have put it better? Or worse – since the pain in her voice is almost unbearable? And it prophesied Bessie's future. She was once a woman who gave everything to friends and family, buying them houses and expensive gifts. But when she fell so low, there was hardly a soul to help her, except Richard Morgan (her lover for the last six years of her life, but he comes later). Even her own sisters, when she started making less money in the late 1920s and early 1930s, resented the drop in their allowance and barely spoke to her. Everyone had grown to depend on her making big money, so much so that they almost despised her for not pulling it in any more. No wonder the last lines are so resonant:

So if I ever get my hands on a dollar again
I'm going to hold on to it 'til them eagles grin.

Bessie Smith's life, as a poor girl in a ramshackle cabin, to Queen of the Blues, to the fallen betrayed queen of a dying music, has Shakespearean dimensions, a shattering combination of tragedy and comedy. Jack Gee's betrayal of Bessie signalled the very beginning of the downward spiral; not long afterwards, the blues themselves became victims of the Wall Street Crash and were almost killed by the Depression. Bessie experienced its effects firsthand. Theatres closed down; others replaced their stage shows with talking pictures. People didn't have the same kind of money to buy blues records. Columbia terminated her contract in 1931, two years after the Crash. People still turned out to hear her, from Mobile, Alabama, to New Orleans to Dallas. But a concession Bessie had to make during these years was to share the bill with other blues singers, whose talents didn't match hers and whose fame did not last as hers has. A couple of years after the Crash, even live audiences were sorely affected. People queued for bread, not blues. Money was tight, and for those who could afford it the blues started to go out of fashion, to be replaced by jazz. The tastes of the Northern audiences changed dramatically: they wanted faster, more sophisticated voices, voices that could swing. All the downhome people who had found in the blues a cure for homesickness were now firmly rooted in the North and ready to move on in their cultural tastes. The bottom fell out of the blues recording industry.

Bessie Smith's voice could and did swing; it is one of the greatest jazz – not just blues – voices of the twentieth century. With her penchant for timing and her way of dipping down notes and hanging on to others, she almost certainly would have survived the Depression and gone on to sing standard jazz songs. Her style of dragging over a word or syllable into the next bar was copied by many others. Her genius for timing set the trend for jazz singers for generations. She knew how to hang up a note and when to let it down. Humphrey Lyttelton is convinced her recording of 'Nobody Knows You' 'stamps Bessie Smith as a supreme jazz singer':

> This is the standard which Bessie Smith set for all jazz singers to follow, using improvisation in its fullest sense on the melody of a song to express a deeper meaning than that of the words on their own. It was an example which, a decade later, enabled Billie Holiday to make remarkable music out of the popular ditties of the calibre of 'I Cried for You' and 'Back in Your Own Back Yard', not to mention 'Oooooooh, What a Little Moonlight Can Do'.[21]

But back then, she was known of and thought of as a blues singer. In the 1930s a new generation of jazz women came to the fore: Billie Holiday, Mildred Bailey, Ella Fitzgerald, Connee Boswell. So Bessie Smith lost her man and lost her power to earn really big bucks almost simultaneously. She was broken by Jack's betrayal, lost a lot of her fighting spunk and was often seen crying. (Ruby had never seen her cry before.) Bessie recorded 'Shipwreck Blues' in late 1931, although it wasn't released until July 1932. Columbia only ordered four hundred copies to be made – a far cry from the 780,000 copies of 'Downhearted Blues' sold in its first six months, making the fate of 'Shipwreck Blues' doubly ironic.

> It's rainin' and it's stormin'
> On the sea,
> It's rainin' and it's stormin'
> On the sea,
> I feel like somebody has shipwrecked poor me.

Not only did Bessie lose her man – something she had sung about so often in her recording career – but she lost him to a 'glorious brownskin' girl. 'Downhearted Blues' should have served as a warning: 'It seems like trouble is going to follow me to my grave.' Jack Gee did break her heart; she never really recovered from his betrayal. Only for the first couple of years of their marriage was she happy, although fanatically busy recording songs and going on tour. (She was so happy because she was away from him so much and the thought of him and having a husband at home was much better than the reality. The actual homecoming and the wearing of the housecoat and slippers was a disappointment.)

Jack Gee didn't just play a destructive role in Bessie Smith's life; he played a destructive role in her afterlife. His greed for her money has had a long-lasting and devastating effect on the way in which we remember Bessie Smith, from his refusal to give over the money raised at benefits to buy her a gravestone, to his refusal to co-operate with a potential biographer, as recorded by Chris Albertson:

Jazz critic Rudi Blesh has recalled how close he came to writing a Bessie Smith biography in the late nineteen-forties, when the available sources of information were more numerous. Bessie's sisters, Tinnie and Viola – then still alive, living in Philadelphia – had agreed, along with Jack Gee, to co-operate in piecing together Bessie's life. The two sisters had a trunk full of rare photographs, letters, sheet music, and other items that had belonged to Bessie . . . Just as the work was to begin, Jack Gee decided he wanted more money. His unrealistic demand made the project impossible, reignited his long-running feud with the two sisters, caused them to withdraw, and led to the complete disappearance of the trunk and its valuable contents.[22]

This is the kind of story that makes biographers weep. A whole trunk of stuff. It's shattering to imagine all the information that would have been available had the trunk been handed over. Jack almost certainly was threatened by his wife's success. Why else hijack the biography? Why did Jack prevent that book? Was it just greed, or could he not bear the story of Bessie Smith being told? Did he consciously or subconsciously stop that book from being written? The policeman in him got rid of the evidence. What was in that trunk?

*Before they died, Tinnie and Viola sent it on a ship
heading for Scotland. They had seen pictures of Scotland
and liked the look of the country, those big goddam
mountains. Could send Jack up one of those and he'd never
come down. Ben Nevis. They didn't want Jack to get his
hands on that trunk, ever. He was a dirty no-gooder and
Bessie should have known better. She ought never have
married the skunk. A note to this effect is found sixty years
later, signed by both of them. Tinnie's writing is perfect
calligraphy and Viola's is big childish print.*

*Inside: an old photograph of the shack in Chattanooga
where Bessie was born. Nobody is standing outside it. It is
just the shack. It is dark and leaning to the left. The window
is partly boarded up. The door is flung open like somebody
just left in a hurry. It's been taken on a bright day. The
bright sunlight has blanched out some bits of the roof.*

*An early daguerreotype of Bessie Smith's mother and her
father. She favours her mother, got her nose and her chin and
especially her eyes. But she has inherited the shape of her
father's face. Her mother and father are standing side by side,
and the electricity between them is still there in the image.*

*An old poster advertising the two Assassinators of the
Blues in the Moses Stokes Travelling Show, yellow and
curled at the edges.*

*Another picture, of Bessie herself standing on the corner
of Ninth Street singing.*

*A baby tooth, next to a wisdom tooth, which has its long
roots still, also attached, wrapped in newspaper.*

A bottle of bootleg liquor and a pint glass with a lipstick imprint of the lips of the Empress. A horsehair wig – shiny black hair that once long ago ran all the way down to the round shoulders of Bessie Smith. A strand of pearls and imitation rubies. A satin dress. Headgear that looks like a lampshade in someone's front room with lots of tassels hanging down. A plain dress with beaded fringes. A Spanish shawl. A skullcap with beads and pearls sewn into it. Feathers. Ostrich plumes. An ermine coat. A giant bottle of Coca-Cola.

A notebook in Bessie's handwriting of all her own blues compositions.

A letter full of curses, again in Bessie's handwriting, to the manager of the 91 Club in Atlanta. An original record of 'Downhearted Blues'. A reject selection of the songs that were never released. A giant pot of chicken stew still steaming, its lid tilted to the side. A photograph of Ethel Waters; underneath the sophisticated image Bessie has written: 'Northern bitch. Long goody. Sweet Mama Stringbean. 1922'. A photograph of Ruby. Bessie has written all over her face words that are impossible to decipher. Another photograph of Ruby, untouched, in a sexy pose and wearing a polka-dot dress with separate sleeves and a string of pearls. A jar of Harlem night air.

A blues for Ruby. A blues so raunchy it will become a lesbian classic. Every lesbian singer will make a recording of it. Sometime in the future, technology will be so sophisticated, a recording will be made of Bessie

Smith singing 'Ruby's Blues'; they will have succeeded in cutting a whole new blues by piecing together words from her previous songs and connecting them digitally. Bessie's version will outsell anyone else's, including k. d. lang.

A conical horn. A diary of the road, written by Bessie. Contains much hot gossip and many lavish curses. Every player gets a mention. Clarence Williams has more swear words attributed to him than anyone else, except Jack Gee, because he stole Bessie's money too. There are foul nicknames too rude to quote here for Carl Van Vechten, Gertrude Saunders, Jack Gee, Frank Walker, Clarence Williams and brownskin girls everywhere. All the clubs have code names. The 81 is nicknamed 'The Black Ass Club'. A long list of every woman Bessie Smith ever had sex with. A list of all the women Ma Rainey had sex with.

An old street map of Goose Hollow with faded pencil drawings of all three stores. A cross by Bessie's cabin.

A photograph of Snooks, taken in Macon, Georgia, in 1925. Another picture of him a year later, getting onto Bessie's railroad car. Bessie's written 'My son' on it.

A pillow from Bessie Smith's bed with her smell still on it. The dress she wore for the recording of 'Nobody Knows You When You're Down and Out' with her odour round the rim of the oxters. A photograph of the Smiths when Bessie was four. Andrew's top two teeth missing. Viola long and skinny. Tinnie holding Bessie's hand. Clarence looking serious. Maud holding her stomach laughing. One of them is missing.

At the bottom of the missing trunk is something that is very old and shrivelled up. It is a pig's foot. The pigfoot is swimming in water. It is the Tennessee River. A backdrop of a bright full moon floats on the river along with sheets of blues music that have been made into tiny boats. The music of Clarence Williams, Fletcher Henderson, Buster Bailey, Coleman Hawkins, Joe Smith, Louis Armstrong. The original sheets of music inside the missing trunk, folded into tiny boats and floating on the brown water of the Tennessee. The backdrop she always used, magnolia trees, in silhouette set against an orange sky, floats on the saltless water of the Tennessee. Andrew's guitar rows itself down the water. A couple of shipwrecked blues musicians sit inside it, pulling the strings apart to keep their heads above water. The voice of the young girl on the corner recorded at the very beginning of wax recordings. Somebody took her into a strange barn where she had to sing, 'Swing Low, Sweet Chariot, Papa's going to carry you home,' into an odd object that looked like an animal's horn.

Her Bible with the exact date of her birth: 15 April 1894. A marriage certificate with the date of her birth — 1894, Chattanooga. A legal certificate of adoption, naming Bessie Smith as Jack Gee Junior's mother. A divorce certificate dated 19 October 1931.

Her death certificate: 26 September 1937. Obituaries from the Boston Globe *and the* New York Times *and the* St Louis Record. *The steering wheel of the $5,000 Cadillac she bought Jack. A letter from the hospital where*

Bessie Smith died, claiming to have treated her straight away, signed by Dr W. H. Brandon, M.D.

A piece of Route 61. Her bed from her Pullman. A tiny Pullman porter in his uniform, perfectly preserved. Ma Rainey's gold fillings.

A lock of Ruby's thick black hair.

Pour O pour that parting soul in song,
O pour it in the sawdust glow of night,
Into the velvet pine-smoke air to-night,
And let the valley carry it along.
And let the valley carry it along.

Jean Toomer, 'Song of the Son'

She came out on the stage in ostrich feathers, beaded
    satin,
and shone that smile on us and didn't need the lights
    and sang.

Robert E. Hayden, 'Homage to the
Empress of the Blues'

WAX

Why the blues are part of me – almost religious, like
a chant. The blues are like spirituals, almost sacred.
When we sing the blues, we're singing out our
hearts . . . our feelings.

Alberta Hunter

The fact that we ever got to hear any of the classic blues
singers at all is a total historical accident. Mamie Smith
was the first to record a blues song, 'Crazy Blues'. The
Okeh Record Company were trying to find Sophie Tucker
to record some songs. Her voice was mellow and she was
white, but they couldn't contact her, so they decided to take
a risk with Mamie Smith, a black vaudeville singer who
didn't have half of Bessie's on-the-road followers. The
first record Mamie Smith made with the Okeh label was
not a blues. But for the second, her gutsy manager, Perry
Bradford, persuaded the Okeh company to let her sing the
blues backed by a black band. The result was 'Crazy Blues',
which was cut on 10 August 1920. It was a huge hit, selling over a hundred thousand copies during the first month
of its release. Little business sat up and wanted to be big
business.

The blueswomen had taken the blues from backrooms
in backstreets first onto the stage, and now onto the record.

The blues were changed by the blueswomen, no longer a folk music sung by folk in the fields, calling and responding to one another while they worked, or while the sun set, folk who wore poor country suits that hadn't been made for them, or vests with holes in them. We have all seen the pictures. In fact, the picture has come back recently in stereotyped romanticisation of the old bluesmen. Now they are being used to sell beer in television and cinema ads, sitting on that sepia porch with the bad-fitting brown suits to sell a bottle of beer. These old bluesmen are considered the genuine article while the women are fancy dress. The poorer the bluesman looks on that run-down porch, the more authentic his blues. The image of the blueswomen is the exact opposite of the bluesmen. There they are in all their splendour and finery, their feathers and ostrich plumes and pearls, theatrical smiles, theatrical shawls, dressed up to the nines and singing about the jail house. The blueswomen are never seen wearing white vests or poor dresses, sitting on a porch in some small Southern town. No, they are right out there on that big stage, prima donnas, barrelhousing, shouting, strutting their stuff. They are all theatre. This combination of theatre and truth is at the heart of the blueswomen. They might be dressed up as divas, queens and empresses, but they are still telling it like it is. The audience never doubts the truth of the humour or the truth of the sadness, as Sally Placksin writes:

Women shouted recognition when the singer told about the way her man mistreated her, they shouted confirmation when Bessie Smith sang her 'Young Woman's Blues' or when Ma Rainey majestically dumped her man in 'Titanic Man Blues'. Some nights the shouts turned into silence as the congregations sat spellbound in the mystical presence of Ma Rainey or 'Miss Bessie.' Like all artists, musicians, actors and show people in those years, the blues women were considered by many, even in the black community, the 'lowest of the low'.[23]

It is all there in the blues: believable and fanciful at the same time. The opposite of social realism. Realism with a string of pearls thrown in. Grimy life with fancy feathers. Poverty and pain with a horsehair wig. These blues sisters of Bessie Smith all knew how to dress, how to reach out to a full audience and let the blues rip.

The 'classic' blues singers followed in the royal footsteps of the Voodoo Queens and they signalled the emergence of a new type of record. They even gave themselves royal names: Clara Smith was 'Queen of the Moaners'; Bessie Smith was the 'Empress of the Blues'; Mamie Smith was the 'Queen of the Blues'; Ida Cox was the 'Uncrowned Queen of the Blues'; Ma Rainey was the 'Golden Necklace of the Blues' as well as being the 'Songbird of the South' and 'Mother of the Blues'. (Ma Rainey's famous neck-lace was made from gold coins of different sizes, from two–dollar–fifty pieces to heavy twenty–dollar eagles. She often kept her necklace in bed with her in case of theft.) These names were used in the billings for the concerts through the South, but they were also used by the people, the blues admirers and fans. Columbia may have promoted Bessie Smith as 'Queen of the Blues', but it was the people who called her 'Empress', the people who travelled all over to get to hear those women sing their blues. The people who crowded the streets and were moved on by the police in the heyday of the blueswomen, when hundreds were left waiting outside. Those people would have said, 'I went to see the Empress,' and everybody would have known who they were talking about. The classic blues singers were the first and only American black royal family. Ruby Walker said:

Bessie was a queen. I mean, the people looked up to her and worshipped her like she was a queen. You know, she would walk into a room or out on a stage and people couldn't help but notice her – she was that kind of woman, a strong, beautiful woman with a personality as big as a house.[24]

All things are possible: the poor girl from Chattanooga can put on a silk gown and transform herself into an empress; she can wear a lampshade–fringe crown. There is the perception, on the one hand, of the blues as lowlife (the view of middle–class jazz fans and critics) and on the other

hand, the blues as high life, royalty (for classic blues singers and their fans). This combination can't be bettered: the
result is a black working-class queen. No ordinary queen
who has inherited somebody else's lineage quite by chance,
but a diva with style, daring, panache, imagination and
talent. A queen who knows how to shimmy. A queen who
can send herself up. A queen who can holler and shout. A
queen who knows what it is all about. A Queen of Tragedy;
a Queen of Bad Men; a Queen of Poverty; a Queen of the
Jail House. A queen who understands and has been through
herself everything that other ordinary people, particularly
ordinary *women*, have been through. A Queen of the Folk.
No wonder the classic Bessie Smith and the other Blues
Queens were so loved.

The touch of class, artistry and imagination that went
into the look and style of the classic blues singers is often
devalued or misunderstood, Alan Lomax writes:

With few exceptions, only women in show business,
women of questionable reputation, women who
flaunted their loose living, publicly performed the
blues – women like Mamie Smith, Bessie Smith and
Memphis Minnie. The list isn't very long. These
female blues singers toured the black vaudeville circuits
or performed in city nightclubs . . . They did not sing
in the street or play in jukes and barrooms, where they
would inevitably be subjected to sexual advances of
every sort . . .[25]

Many people perceived the blueswomen to be vulgar, crude, lewd, common, rough, raucous, lowlife. The subject matter of the songs, the double entendres, the kitchen man, the butcher man and the jelly roll didn't help the image either. There is lots of 'rough' in the songs of the classic blues singers. Songs like Alberta Hunter's 'I Want a Two-fisted, Double-jointed, Rough-and-ready Man'; Ida Cox's 'Wild Women Don't Have the Blues'; Ma Rainey's 'Rough and Tumble Blues'; Victoria Spivey's 'I Got Men All Over This Town'; Bessie Smith's 'Dirty No-Gooder Blues'. The classic blues singers took men as their central subject and wrote songs about what swines they were, how they cheated, lied, deceived and beat you up. Men in women's blues songs do not look good; they do not look good at all. The odd good man has got to be held on to at all costs because the blueswomen recognise what a rarity he is – 'Don't Fish in My Sea'. In sending men up, mocking and deriding them, the classic blues singers were revolutionary. They took control of their own image, and their songs relentlessly told the truth about no-good men. Male blues songs about women are actually quite mild by comparison. Michele Wallace in *Invisibility Blues* believes that classic blues singing is 'critical to defining the spectrum of possibilities for black women beyond servility and self-abnegation'.[26]

Although the classic blues singers performed on the vaudeville circuit and the TOBA circuit, often appearing at the same theatre halls or in the same tent shows to notices that proclaimed they were in town, it was years before their voices were actually recorded. The wax versions were a long time coming, considering Bessie Smith had been on the road and drawing a crowd since 1912 and Ma Rainey since 1902. In fact one of the earliest references to a woman singing a blues is to be found in *Songsters and Saints*:

Among the earliest recollections with a specific date was that of John Jacob Niles who, in 1898 heard Ophelia Simpson, known as 'Black Alfalfa'. She did the current ragtime things but was most effective in the native blues. Earlier she had killed her man, one Henry 'Dead Dog' Simpson, who worked at a fertilizer plant on the Ohio River near Louisville. After a brief period in the 'Stony Lonesome' jail, Ophelia worked for Dr Parker's Medicine Show where 'she cooked, helped mix the tape-worm eradicator and shouted in the oleo'. It was there that Niles heard her sing:

'I ain't got not a friend in dis town
Cause my New Orleans partner done turned me down
Po gal wishin' for dat jail-house key,
To open up de door and let herself go free.'[27]

'Nobody Knows You When You're Down and Out', 'Jail House Blues' and many other Bessie Smith classics perhaps owe their debt to Ophelia and Dead Dog Simpson. Or maybe not. The blues that the blueswomen sang – and

Bessie was only one of them – all conveyed this epic sense of life. They feature characters who are representative rather than extraordinary. The man in the blues songs could be anyman just like the woman singing about him could be anywoman. The blueswomen could sing, act, dance, hip-shake, shout, perform, draw a crowd, and at the same time be singing about ordinary life. Black women began to sing about their own personal problems, about the everyday; anything could crop up in a blues. The subject of Bessie Smith's first recording hit was a mistreating man. On wax, he is Bessie's first in a long, long line of mistreating men, mistreating Mama all the time.

There are many stories about how Bessie Smith came to make her first record. Frank Walker – who was in charge of the Columbia record company's 'race list' of records by black artists – tells it like this:

> I don't think there could have been more than fifty
> people up North who had heard about Bessie Smith
> when I sent Clarence Williams down South to get
> her . . . I told Clarence about the Smith girl and said,
> 'This is what you've got to do. go down there and find
> her and bring her back up here.'[28]

This is apparently untrue, since only two weeks earlier she was being rejected by the Okeh company in New York. Clarence Williams only had to go as far as South Philadelphia to bring her to Frank Walker. The idea of

Clarence Williams being sent all the way to the ramshackle cabin in Chattanooga and bringing a poor black girl with an already legendary voice straight into a recording studio belongs to the American Dream.

According to Frank Walker, the girl that Clarence brought back from the Deep South was no dream singer. 'She looked anything but a singer. She looked about seventeen – tall and fat and scared to death – just awful.'[29] In fact this was 1923 and Bessie was twenty-nine. She had already been on the road for eleven years and had built up huge audiences in the clubs in the South who threw extra money to tip her. Ethel Waters said Bessie's shouting brought worship wherever she worked. She was earning $50 to $75 a week – 'big money for our kind of vaudeville'. And Ma Rainey and her like would have influenced Bessie's style of dress. But maybe she did arrive in the studio looking like a teenager, tall and fat and scared to death.

Bessie Smith's records were made very early on when recordings were done acoustically. Her voice would travel through the large conical horn; at the other end the stylus would wait to be activated by the sound. The stylus made the appropriate impressions while cutting a groove in a thick wax-like disc. (The old phonographs, with their big horn, simply reversed the procedure to provide playback.) The thick wax discs would then be processed into metal masters from which test pressings were made. The record company would then choose which version of the song they thought to be the most successful. (Editing was not technically possible in those early days.) Bessie travelled throughout the South and the Midwest with a portable mock-up of this strange-looking acoustic recording equipment. She'd explain to her audiences how records were made by singing into the big horn as if she were actually recording. She chose many of her live songs from her recorded material in order to boost record sales.

It is incredible that Bessie's first recording was not made until 1923, since she auditioned for recording companies as early as 1921. The Black Swan company, which proudly presented its discs as 'The Only Genuine Colored Record – Others Are Only Passing for Colored', turned Bessie down. This was the company that was founded that same year by composer W. C. Handy, since dubbed the 'Father of the Blues' – but he was never married to Ma Rainey! It was the first black-owned record company and was sold to the public as such. However, Bessie Smith's voice was the wrong colour even for them. They wouldn't take the risk. They thought her voice too 'rough'. (For rough, read 'black'.) The very qualities that make her voice still live on today – rough, raw, harsh – were loathed by Black Swan. They preferred the smooth tones of an Ethel Waters. From news reports at the time, it appears that the Emerson Company and the Okeh label also judged her style too 'rough'. Bessie Smith's voice and skin were not the only aspects of her that the recording companies considered too dark or rough. There was her behaviour. One story has it that she failed a test with Black Swan Records because she interrupted a song with, 'Hold on, let me spit!' and the president of the company immediately ended the audition.[30]

Bessie, even then, was no stranger to being turned down for being too rough or too dark. Irving C. Miller, who threw her out of his chorus line in 1912 for being too black-skinned, told Humphrey Lyttelton, 'She was a natural singer even then – but we stressed beauty in the chorus line and Bessie did not meet my standards as far as looks were concerned.' The advertising slogan of Miller's shows speaks volumes: 'Glorifying the Brownskin Girls'. (It was this same impresario who had no hesitation in booking Bessie's rival, Gertrude Saunders, years later.) Light-skinned black women were favoured for a very long time.

I remember being in a school show when I was eight. We were doing the dances Bessie could do so brilliantly, such as the Black Bottom and the Charleston. I remember not being able to get my steps right and my teacher saying, 'I thought you'd be good at this. I thought you people had it in your blood.' The comment made me lie awake at night, wondering what was in my blood. I don't know what the girls from Balmuildy Primary were doing learning such lewd dances in any case. I find it difficult to connect the history of the Charleston with my Scottish primary school. (An early example of multiculturalism.) As Humphrey Lyttelton explains:

In the early years of the century the population of the San Juan Hill area [of Harlem] was boosted by an influx of migrants from the South, some from Alabama, some from the part of South Carolina and Georgia that centered around Charleston. From the testimony of both The Lion and James P., it is clear that these latter people, known as Gullahs or Geechies, had a powerful effect on the style of the local piano men . . . 'They danced cakewalks and cotillions; by this time we had learned to play the natural twelve-bar blues that evolved from the spirituals . . . the Gullahs would start out early in the evening dancing two-steps, waltzes, schottisches; but as the night wore on and the liquor began to work, they would start improvising their own steps and that was when they wanted us to get-in-the-alley, real lowdown . . . it was from the improvised steps that the Charleston dance originated.'[31]

Bessie Smith was born, lived and died in a country obsessed with colour. A country that could never shake off the legacy of slavery, of segregation. The systems of slavery and racism in the United States ensured that a person with even a 'drop' of 'black blood' was considered black. Racism could not tolerate any deviation from 'pure whiteness'. Racism required white people and black people to keep absolutely separate. Any 'mixing' of the two races was regarded with suspicion. This obsession with colour and blood, and the colour of blood, manifests itself in countless ways, not least in the culture of the people. You get a sense of a country's laws by listening to the music, reading the literature and looking at the art. As late as the Second World War, the American Red Cross, though well aware that all human blood is the same, kept blood plasma segregated by race. As explained by Geneviève Fabre and Robert O'Meally in *History and Memory in African-American Culture*, here is a country that, not content with separating the living, had to separate the dead:

> There were even segregated railway compartments for the deceased. In 1835 a railroad was planned for the transport of corpses and mourners from St Claude Street to the Cemetery at the Bayou St John in New Orleans. The city council stipulated that the contractor had to provide separate cars for white, free colored and slave corpses, with fees ranging from fifty cents for slaves to three dollars for whites.[32]

The recording companies in the early 1920s could not have anticipated a blues boom. Black Swan, the company that promoted black artists, wanted its singers to possess a light tone, a light touch, a light shade. The blues were dark, rough, brutal and real, and all the major companies underestimated their marketability. But suddenly the blues were big business.

On record, Bessie's blues made an incredible and immediate impact. People who had migrated from the South to the North rushed out to buy her records, because her phrasing and words reminded them of home. Downhome talk. She filled people with awe and longing. With a sense of themselves. Black people identified with her songs. She was like a preacher in the way she affected people. Guitarist Danny Barker explains:

Bessie Smith was a fabulous deal to watch. She was a pretty large woman and she could sing the blues. She had a church deal mixed up in it. She dominated a stage. You didn't turn your head when she went on, you just watched Bessie . . . She just upset you. When you say Bessie – that was it. She was unconscious of her surroundings. She never paid anybody any mind . . . She could bring about mass hypnotism. When she was performing, you could hear a pin drop.[33]

Bessie Smith lived an epic life. Her life could be painted on a broad canvas. She was big in stature, size and influence. She was *the people*. This ability to write songs, to record songs and perform songs that touched the heart of everyone who listened to her is what has kept her alive for so many years. She was her time. She totally reflects her time. She had some pair of lungs and could fill a massive hall with her voice. She didn't need a microphone. She just belted those blues out. As Buster Bailey said, 'There was none of this whispering jive.' (A hilarious description of someone like Betty Carter.) The fascinating thing about the voice of Bessie Smith, for all its blueness, is its total lack of sentimentality. She can sing unnerving, sad songs without a note of self-pity. It is the very flatness of her voice, sing-ing about tragedies, that so moves us. It is not in any way the voice of a victim. It is unrefined. It is not sweet, sugary. It is not smooth. So why did the record companies initially reject her? Her voice was that of an ordinary working-class black woman. And they didn't think that the voice of an ordinary working-class black woman could or would ever sell records. They were, of course, wrong because they underestimated the amount of ordinary working-class black women that would buy Bessie Smith and the other blueswomen's records. They didn't realise that ordinary black people would so identify with the blues that the blues would seem to be the singing version of their lives. It was this that made the records of Mamie Smith, Bessie Smith, Ma Rainey, Alberta Hunter, Ida Cox, Victoria Spivey and others so popular and, with Bessie Smith at least, so long-lasting.

It took two days for Bessie to conquer her nerves when she first turned up on 15 February 1923 at the Columbia studio to record 'Downhearted Blues' and 'Gulf Coast Blues'. Arriving, 'tall, fat and scared to death', the Queen of the Blues did not make the wax the first day of record-ings. She tried 'T'ain't Nobody's Business' nine times and failed, and she tried 'Downhearted Blues' twice and failed. Maybe it was her nerves or maybe it was that strange conical horn which stuck out from a drapery-cov-ered wall. The effect must have been quite sinister. Bessie probably didn't trust the damn thing. Maybe she thought she'd lose herself in that big hole. Maybe she thought they ought to be able to record her differently. Maybe her head was full of bad memories of all the previous times she had been in recording studios and been rejected. She might have been at a total loss without her live audience throwing money, shouting and stomping for more. She might have found the studio cold. She might have been stone-cold sober. Imagine her in that studio in 1923, wearing the red dress that Jack pawned his pocket watch to buy, making her first ever record. Could she have realised the significance of her break into wax? Did she fear she'd lose her talent through having her live voice recorded and transmuted to wax? A waxwork? A dummy? She possibly mistrusted the whole technological thing, such as it was then. She might have felt she was being had. But she soon got the hang of it. Humphrey Lyttelton says: 'The singing that was transmitted to wax was, from the outset, mature, steeped in harsh experience and formidably commanding.'[34]

The sales of 'Downhearted Blues' – three quarters of a million copies in six months – far exceeded the sales of any other blues record. The black public were eager to purchase records through mail-order catalogues, record stores in black neighbourhoods or even through the Pullman porters. The blues sold both in the North and in the South and became part of the record companies' 'race records' series, as Paul Oliver explains:

The term 'Race records' for issues directed solely to the black purchaser was in use by Okeh as early as January 1922. Several other major record companies began to issue Race series: Columbia commenced in 1921 while Paramount . . . merged with the only label to have black ownership, Black Swan, and commenced its Race series in 1922 . . . By the end of 1922 Race records were being distributed in many Northern cities and as far south as Alabama.[35]

In the South the blues sold to black and white people; in the more 'liberal' North, they just sold to black people. It is possible to have been white in the North in the 1920s and never have known that blues records even existed. This is because in the North, advertising of so-called 'Race records' was restricted to the black press, and the distribution of the records took place only in black areas. Southerners, though, became part of the 'race market'. White and black people, though segregated, crowded into those tents to hear the blues. Bessie Smith gave many performances in 'whites only' theatres where she changed neither the style of her music nor its contents. She refused to 'water herself down'. It is interesting to me, though, that she should have played in those theatres at all. I would have thought she would have refused as a matter of principle. But maybe that is naive; maybe she had no choice.

Bessie was at the height of her popularity after her first records were released. Everywhere she went it was the same story. People had heard Bessie's records and now they wanted to hear her in person. Five records were now on the market: 'Downhearted Blues', 'Gulf Coast Blues', 'Aggravatin' Papa', 'Beale Street Mama' and 'Baby, Won't You Please Come Home'. Her reputation had grown beyond all expectations. Columbia were dazzled by her success. Bessie was now making $350 a week from personal appearances. During her tour in June 1923, streets in Atlanta were blocked and hundreds of fans were unable to gain entrance to the theatre. The crowds, wild and hysterical, pushing and crushing, fought to get to her next performance as ten policemen tried to keep things under control. At this early stage in her recording career Bessie Smith already had the kind of star status that we associate with today's pop icons.

Once Bessie had recorded her version of a song, it practically wiped out all others. Ida Cox wrote 'Graveyard Dream Blues' and recorded it with Lovie Austin's Blues Serenaders in June 1923. Three months later, Bessie Smith recorded the same song for a January release, but Columbia decided to rush it out for competitive reasons. Paramount then fought back with Ida Cox releasing yet another version of 'Graveyard Dream Blues' after Bessie's. In this way, the record companies pitted one blues singer against another in the search for total domination of the market. Bessie Smith sang a lot of songs that had previously been sung by other blueswomen because in those days there was no copyright over songs. 'Chicago Bound Blues' was another Ida Cox number that Bessie then recorded. 'Downhearted Blues', her first release, was originally an Alberta Hunter number.

In 1923 Bessie recorded song after song after song. Columbia signed her up for a lucrative eight-year contract. Between 1923 and 1931, she recorded 160 songs with Columbia, roughly twenty a year. Bessie Smith wrote and composed thirty-seven of these blues. She was right there at the beginning of the recording industry, in and out of studios like there was no tomorrow. Any time she came back from a tour, she was whisked into a studio to record another number. During that period, she worked alongside some of the best musicians of her day: Fletcher Henderson, Louis Armstrong, Fred Longshaw, Jack Teagarden, James P. Johnson, Coleman Hawkins, Joe Smith. But the most exciting combination musically was Bessie Smith and Louis Armstrong, in those sessions they recorded on 14 January 1925. 'Reckless Blues', 'Cold in Hand Blues', 'Sobbin' Hearted Blues' and 'You've Been A Good Ole Wagon' were all recorded that day. It has turned out to be one of the most memorable dates in the history of the blues. Louis Armstrong's horn understood Bessie perfectly. They are totally at ease with each other. There is all the awesome complexity of perfect harmony between the two of them. Her big blues voice and his big horn.

Neither Bessie Smith nor any of the other classic blues singers received royalties for their songs back then. They were paid per usable side; the amount varied depending on their popularity. At her peak, Bessie could earn as much as $200 per usable side. She was the best paid of all the classic blueswomen. She was on a rollercoaster and that rollercoaster was her blues. Five of her records were on the market, and her reputation had grown beyond all expectations. But success would not last. After the Wall Street Crash of 1929 and the Depression, a new combo style of blues became fashionable. 'Urban blues' or 'Chicago blues' then dominated the scene from the mid–1930s through the 1940s. The likes of Muddy Waters and Howlin' Wolf took off; the classic blues singers were replaced by men with acoustic guitars.

I dream of a place between your breasts
to build my house like a haven
where I plant crops
in your body
an endless harvest
where the commonest rock
is moonstone and ebony opal
giving milk to all of my hungers
and your night comes down upon me
like a nurturing rain.

Audre Lorde, 'Woman'

RUBY ON THE ROAD

My best friend, Gillian Innes, loved Bessie Smith. We spent many hours in Gillian's bedroom, imitating Bessie Smith and Pearl Bailey. Various objects served as microphones from hairbrushes to wooden spoons. At the age of twelve, singing Pearl Bailey's 'Tired of the life I lead, / Tired of the blues I breed, / I'm tired, mighty tired of you,' or Bessie Smith's 'You've got to give me some, please give me some,' was a way of expressing our wild emotions for each other. The one who was singing looked directly at the other, getting completely into the mood of the blues, making her body movements correspond to the words. The one who was watching killed herself laughing. Gillian's box bedroom floor was our centre stage. She had her own old gramophone in the corner. We'd rush to it as soon as the number was over to play it again and swap the singer. The singer would then jump onto the bed and become the whole audience. We fancied each other singing those blues. She could make one eyebrow raise itself way above the other. I was besotted with her and the blues. First love; I was positive Gillian would grow up to be a famous singer. My life at that age was charged with all the intense devotion and loyalty of a schoolgirl crush. I could barely breathe. The air in her box bedroom was thick with secrets. The door firmly shut. Our own private performance. Gillian lent me a book about Bessie when we were thirteen or so. I never found that book again, but I remember reading about Bessie travelling in her Pullman, which I then imagined as a sort of wagon, like one of those wagons on the Oregon trail. I imagined it to have little windows with red curtains bunched to the side. I read in this book that Bessie had sex with women on the road. I could barely contain myself. Who did she have sex with? What were the women's names? She got to sing the blues. She had a beautiful haunting voice. She got to travel in the Deep South. And she got to have sex with chorus girls on the road in her own Pullman. What more could a girl want? It must have been a bumpy ride. I remember the name Ruby. The name Ruby coming up all the time.

Ruby Walker was Jack Gee's niece. Her relationship with her aunt, Bessie Smith, was undoubtedly one of the most crucial in Bessie's life. It was possibly even more significant and complex than Bessie's relationship with Jack himself. As far as we know, Bessie Smith never had sex with Ruby, though their relationship has all the complexity of a sexual one. They were companions from before Bessie made her first record in 1923 until Jack and Bessie's marriage broke up in 1929. Bessie spent far more time with Ruby than she did with her husband. It is certainly true to say that Ruby Walker knew and understood Bessie better than anybody. She was all things to Bessie: her travelling companion, her confidante, her keeper. At various points in their nomadic showbiz life, Ruby acted as decoy, as a red herring; she would lie for her, spy for her, risk her own neck for her. Her most fascinating role in the life of Bessie Smith was as the keeper of her lesbian secrets. Bessie kept nothing from Ruby. She knew every intimate detail of Bessie's travelling promiscuity. But she kept schtum. On numerous occasions, she ran away with Bessie to avoid the wrath of Jack.

The first time Ruby heard Bessie Smith sing it changed her life. 'I didn't know it then,' she said, 'but that's what happened.' It was in New York, February 1923. Bessie and Jack were staying in Jack's mother's house on 132nd Street between Fifth and Lenox Avenues. Above 132nd Street was a Harlem full of black people, and the home of the writers, artists and musicians of the Harlem Renaissance. Harlem had a dazzling night-life which would have suited Bessie Smith down to the ground. The young Duke Ellington described Harlem as having 'the world's most glamorous atmosphere'. Bessie spent the days at 132nd Street rehearsing for her first recording. Ruby Walker first heard Bessie Smith sing in Jack's mother's living room.

Ruby was bowled over by Bessie's voice. After that initial meeting, she persuaded Bessie to let her come with her on her tours. Not only did she want the glamour and excitement of Bessie Smith's life, she wanted to *be* Bessie Smith. So Bessie took Ruby on the road, teaching her simple dancing steps until Ruby became part of her show, performing in the intervals when Bessie was making one of her elaborate costume changes. Ruby had very high hopes for herself as a singer, but unfortunately she never managed to become a star in her own right. She recognised Bessie's genius and for the most part contented herself by hanging out with it, in the hope that some of it might rub off on her.

Ruby travelled with Bessie from Cincinnati to Chattanooga, from Pittsburgh to Philadelphia, from New York to New Orleans, from Birmingham to Baltimore, from Detroit to Dallas. Ruby didn't just get to see a lot of America with her blues guide, she also discovered that Bessie knew just about every nightspot and every small bar in the country. Initially, they travelled from city to city by car, bus or train until, in 1925, Bessie's brother came up with the idea of buying her own personal railroad car. It was custom-made by the Southern Iron and Equipment Company in Atlanta and painted bright yellow with green lettering. Her fans could tell when Bessie had arrived the minute they saw the yellow train pull into their station. Bessie's Pullman became as notorious as she was. When the Empress arrived in town, she did it with the style and glamour that people associated with her. The train was as vital to her image as the rhinestones or the plumes. Ruby loved the train. It was seventy-eight feet long. It had seven staterooms – each one could sleep four – and a lower level that could accommodate up to thirty-five people. It had a kitchen and a bathroom with hot and cold running water. The long corridor carried the show tent's central pole. A room at the back of the train carried the canvas, along with the case-loads of peanuts, crackerjacks and soft drinks that Bessie sold on the road. The whole crew lived on the train.

They didn't have to split up when they hit a town; the train was big enough for all of them. Maud Smith said to Chris Albertson:

> I'll never forget it. It was Clarence's idea to buy the car, and it was delivered to us in a small town near Atlanta on a sunny Monday. Everybody was so excited, and we laughed and carried on as we walked through the car and examined every corner. And what a difference it made – some of the towns we hit didn't have hotels for us so we used to have to spread out, one staying here, another one there. Now we could just live on the train.[36]

Travelling life on the train was easier than before. For one thing, they didn't have to confront racism on the road so often, or stay at bed-and-breakfast dives, way out of town. In the 1920s, no black person, not even a superstar, could stay at even a third-rate hotel, never mind a decent one. The only hotels that would take in black people were bed-and-breakfast places and rooming houses. It was possible to be as famous as Bessie Smith and still be the victim of racism. America might be proud of its blues tradition now, but it certainly did not help blues performers when they needed it. Touring the racist South was a nightmare for the blues performer. The Ku Klux Klan were at their height in the 1920s, exactly the same time as the heyday of the classic blueswomen. (Interestingly, the Klan's decline began in the

1930s too.) But in the Pullman, there was privacy, autonomy and plenty of space. You could have your own food,
often cooked by the Empress herself. Bessie Smith made
stews for her entire troupe, pigs' feet or other Southern
specialities. She prepared the food in the galley of the railroad car. Sometimes the musicians would be drafted in to
peel potatoes. Sometimes, when the weather was fine, they
would have a picnic somewhere near the track where the
railroad car was stationed.

The sleeping quarters were tight. Everyone knew everyone else's business. Ruby knew all of Bessie's business.
Bessie frequently had sexual relationships with women on
the road. There was Marie, Lillian and countless others;
none were hidden from the other musicians, prop boys
and crew. But nobody spilled the beans. Everybody pro-
tected her when Jack came. She threatened to throw them
out of her show if anyone so much as breathed a word to
Jack. Everyone was so terrified of muscular Jack, with the
powerful and unrepentant fists, that no one would have
said anything anyway.

On the road, away from Jack, Bessie had the freedom to
be herself. She could drink good moonshine as much as she
liked. She could go to buffet party after buffet party and
watch all manner of lewd sex acts. She could be promiscuous, having sex with young men and young women on
the road. She could go out after her shows and not return
'till the break of day'. She tried to keep it all secret from
her ex-nightwatchman husband, but he often caught her

red-handed by turning up unexpectedly and surprising her. On one particular occasion, on closing night after the show, everyone changed into their nightgowns and pyjamas, as was their custom, parading from room to room in a hotel they were staying in, eating Bessie's food and drinking homemade gin. They wound up in Bessie's room on the first floor. Marie, a young ballet dancer, was wearing bright red pyjamas, a gift from Bessie. (Bessie often bought the women in her life red gifts. She bought Ruby a pair of red dancing shoes.) Marie was dancing around in her pyjamas to Bessie's shouts. Ruby passed out. She was always the first to pass out at such parties. The next thing she knew Marie was tearing down the corridor, followed by Bessie at her heels. Jack had caught Bessie in bed with Marie. The two of them huddled together in a corner of Ruby's room, hiding from Jack. He passed by the door, shouting, 'Come out of there, I'm going to kill you tonight, you bitch.' Jack went off into the street looking for Bessie. Bessie waited a while and then gathered her whole troupe together and rushed for her Pullman. Many of them left their belongings in the hotel, they were in such a hurry to escape Jack. They all got on the train. The Empress was still in her pyjamas as the train pulled out of Detroit, heading towards Columbus, Ohio. 'That's how I lost the only fur coat I ever had,' Ruby told Chris Albertson.[37] (Needless to say, the fur coat was a gift from Bessie.)

But Jack caught up with them, as he always did. His life, policing his wife, must have been pretty exciting too. He would stalk them, tracking them down until he managed to have a piece of Bessie. It probably wasn't jealousy any longer that fuelled him, but a desire for power and control. His wife's life was completely out of hand. There wasn't much he could do about it. He caught up with Bessie in Columbus and came into her dressing room before the show to tell her that he was going to beat her up afterwards. Bessie rushed everybody through the grand finale, changed into ordinary clothes and dashed for the station with Ruby, where the two of them caught the next train for Cincinnati. The rest of the musicians were dumped on the road. (Bessie frequently abandoned her musicians, leaving them penniless in Detroit or Dallas. She had a reputation for treating her musicians badly, in contrast to her generosity towards friends and family.)

When Bessie took off like this, to escape the wrath of Jack, she always took Ruby too. Not her current woman lover, but Ruby. Only Ruby; everyone else could go to hell. No wonder the other chorus girls were jealous of her. (Once a chorus girl shredded a pair of new dancing shoes that Bessie had just bought Ruby.) Why did she not dump Ruby too? Ruby was Bessie's right-hand woman. She was the one who planned, who covered for her, who lied to Jack for her. She was the one who would get on the phone to Jack the morning after a night before to tell him that Bessie was unwell. Sometimes, Bessie would actually throw herself down the stairs so that she could have genuine injuries in an attempt to avoid being beaten up by Jack.

Ruby's undying admiration for Bessie is not puzzling. But was Ruby admired by Bessie? Did she desire Ruby all those years? Did she see Ruby as unobtainable because she was Jack's niece and Jack would have gone wild at the combination of the two of them? Did Jack originally tell Ruby to tour with Bessie so that she could spy on her for him? Perhaps Ruby was feeding Jack information about his wife, until she fell for Bessie herself. The fact that their relationship was never consummated, while Bessie had frequent relationships with other women, is interesting. Many of the women Bessie travelled with had lesbian relationships. Nobody was hung up about it. Yet Bessie did not want Ruby to have sex with anybody. If Bessie couldn't have her, nobody else could. She told Ruby she would throw her off her show if she caught her having sex with anybody.

Perhaps Bessie liked Ruby's devotion without sex. Perhaps she needed her in a different way; a constant presence to give her life security, stability, a sense of belonging. Perhaps she knew that Ruby loved her more than anyone else did. More than her own sisters. Perhaps Ruby reminded her of her sisters when they were young. Bessie enjoyed fooling around with Ruby, joking and laughing. She enjoyed shopping for Ruby, stunning her with one lavish gift after another. It's quite possible that without Ruby, Bessie wouldn't have managed to get her show on the road at all.

Ruby was Bessie's accomplice. She aided and abetted Bessie's lesbianism, covering for her and inventing stories for her. Was she in love with Bessie herself? Everything she says about Bessie is drenched in admiration. 'No one messed with Bessie Smith . . . She was a strong woman with a beautiful strong constitution and she loved a good time.'[38] She even compares Bessie to a real queen.

Ruby Walker idolised Bessie Smith. She loved the way she looked; the way she sang; the way she drank; the way she partied; the way she cussed; the way she fought; the way she danced; the way she spent money. Ruby loved everything about Bessie . . . She loved her so much she wanted to be her. But Bessie could be violent if she strayed. Once Ruby tried sleeping with a young man that Bessie was also sleeping with. When Bessie found out she beat Ruby up. The boy in question was called Agie Pitts, a handsome young dancer from Detroit. Bessie caught Ruby in the dressing room with Agie Pitts and flew into a rage. She told Ruby that she was going to break her out of the habit of trying to be Bessie Smith. She told her she was going to let her know that she ain't Bessie Smith. She beat Ruby up so badly that Ruby's screams attracted the police and all three of them, Ruby, Agie and Bessie, ended up in jail. The dressing-room scene is like an early black version of *All About Eve*.

I caught that girl red-handed trying to be me. She was sitting in my dressing room looking at herself queerly in the mirror. It made my blood boil. She was trying to look sad. She was so in herself that she didn't even notice me. And she did this weird thing with her eyebrows, bringing them in on each other, like they was commiserating with each other over somebody's death. I don't think I ever look that bad. But there she was in the mirror trying to be me. Then she closed her eyes and sang 'Mama's crossed the danger line'. She must have sensed me right behind her. The real fucking thing. 'Cause when she opened her eyes, Ruby did startle. I say going up behind her all quiet like and pulling her long hair to the side, 'What on earth you up to, girl? Eh?' She say nothing. She chuckles some. Which is aggravating. I grab hold of that lump of black hair some more. She thinks she's cute 'cause her hair is practically straight as her nose. I get a hold of it and I just pull lumps of the stuff out. I will destroy you, bitch, before I let you be me, I say and walk out. But when I'm in the corridor, I hear her start up again, softly, this time singing, 'St Louis gal, look what you done done,' and I go back in and she's wearing my dress and she's dancing, swaying side to side like I do. I go up to her and I hold her hips and she takes me into her dance and I kiss her. It's the first time I have ever kissed her. I don't think I have ever had a kiss like it in my life. We lost all time in that kiss. We was dreaming, slow and soft. Her lips full and wet, moving with me, tracing my lips, finding my tongue. It was all so slow, so slow. We could have become something else in that kiss. I forgot the room, and where I was. I closed my eyes. I don't usually close my eyes, but the one time I ever kissed Ruby Walker, I closed my eyes. It was like kissing myself.

Many of the women performers that Bessie travelled with were lesbians. Some were even married lesbians. Boula Lee, a chorus girl, was the wife of the Harlem Frolics music director, Bill Woods. Boula Lee fancied Ruby Walker and once thought she saw another woman making a pass at her. She attacked Ruby, telling her that she shouldn't mess around with 'them other bitches'. Then she leaned forward and scratched Ruby's face. Right on cue, Jack turned up, demanding an explanation for his niece's scratched face. Bill Woods, totally unaware of his wife's preference for women, retorted, 'It was one of them bulldaggers who is after Ruby.' The irony was too much for Bessie Smith to take. She said, 'What do you mean, one of them? It was your wife.' When Jack caught up with Boula, he threw her down the stairs.

At that time, Bessie was having a serious relationship with Lillian, another of the women in her show. It was January 1927 and the troupe were in St Louis at the Booker Washington Theatre. Ruby and Lillian were sharing a room. Bessie entered the room and kissed Lillian in front of Ruby, from whom she had nothing to hide. Lillian, embarrassed, tried to pull herself away, admonishing Bessie for her public display. Bessie told her that she had twelve women on the show and could have 'one every night' if she wanted to; she then ignored Lillian for several days. Lillian took the cold shoulder for three nights. But on the fourth night, she did not turn up at the theatre. Maud Smith noticed a letter sticking out from under Lillian's door. It was a suicide note.

Without taking time to read the note, Bessie, with Ruby and Maud at her heels, ran next door to the hotel. When they reached Lillian's door, they smelled gas. Bessie tried to force the door, panicked, rushed downstairs and got the proprietor. When he let them in, they found Lillian lying across the bed, unconscious. The proprietor had to break the windowpanes: Lillian had nailed the window shut. She was taken to the nearest hospital . . . 'From that day on,' says Ruby, 'she didn't care where or when Bessie kissed her – she got real bold.'[39]

When Bessie went on tour in the South or the Midwest, it wasn't so much a tour of the blues as a tour of the bars. Bessie knew every backstreet and every bar there was to know in any given town. There wasn't a bar anywhere that hadn't made the Empress's acquaintance. Ruby was constantly impressed with her repertoire. Bessie also knew every buffet flat. Buffet flats, also known as goodtime flats, were small, private establishments offering illicit pastimes from gambling to sex shows, to a variety of sex acts. They were a variation on whorehouses, with a more upmarket clientele. They provided 'important persons' with their sexual equals, judging for themselves who would go with whom. You could always get good bootleg at a buffet flat. Bessie went to one in Detroit while she was performing at the Koppin Theatre. She took five of her girls along with her. Before she went, she said her mantra: 'Don't tell Jack.' Ruby describes the scene:

It was nothing but faggots and bulldykers, a real open house. Everything went on in that house – tongue baths, you name it. They called them buffet flats because buffet means everything, everything that was in the life. Bessie was well known in that place.[40]

Ruby's tone here is slightly lofty, as if she herself were above it all. Yet she went along. Everywhere Bessie Smith went, Ruby Walker went too. She admired Bessie's penchant for the night-life, Bessie's supreme talent for finding a good drink. The Empress was famous for liking the rough bootleg stuff. She was still drinking the bad stuff even when the good stuff became legal again. Like just about everything else in life, Bessie liked it rough. In Atlanta, the garbageman sold alcohol in the streets. He went up and down neighbourhoods with a huge can on wheels, shouting 'Garbageman. Garbageman.' People would run out with their glasses in hand and get them filled. Bessie didn't buy from the garbageman, however, because she drank a great deal more than just a glass. She went to a man who lived under the viaduct who sold bad liquor by the half gallon.

I'm a young woman and ain't done runnin' round,
I'm a young woman and ain't done runnin' round.
Some people call me a hobo, some call me a bum,
Nobody knows my name, nobody knows what I've
    done.
I'm as good as any woman in your town,

I ain't no high yeller, I'm a beginner brown.
I ain't gonna marry, ain't gonna settle down,
I'm gonna drink good moonshine and run these browns
   down.

'Young Woman's Blues'

Ruby enjoyed the wild life with Bessie on the road. Hiding from Jack. Hanging out in bars and buffet flats. Learning the odd dance step in small moments between one hangover and the next. She liked the way that Bessie sang. The way every blues seemed to relate directly to her life. There was a line for everything: 'There'll be a hot time in the old town tonight'. Bessie chronicled her own life. Ruby, knowing the Empress better than anybody, recognised the landmarks. The woman who ran the buffet flat in Detroit is large as life in 'Soft Pedal Blues':

There's a lady in our neighbourhood who runs a buffet flat
And when she holds a party she knows just where she's at.
Have all the fun, ladies and gentlemen,
But please don't make it too long.

There's a wonderful 'Yahoo!' in the middle of this song, screamed out after she declares how drunk she is: 'I've got them soft pedalling blues.'

On the road, more of Bessie's talents went into her secret lesbian life (secret from Jack Gee only) than into her shows. Maybe it wouldn't have been quite so exciting if the danger

of Jack discovering her had not been so present. Bessie carried out her lesbian relationships so openly that we all know
about them still. She didn't mind anyone else knowing so
long as they didn't tell Jack. She kissed Lillian in front of
people. She didn't care what anybody said about her, ever.
But she barely mentioned her own lesbianism in her songs.
(Other blues singers, Ma Rainey amongst them, sang openly about lesbianism. But Bessie mostly sang about men.)
Homosexuality is seen as something puzzling and strange
in 'Foolish Man's Blues', and apart from that song there
is hardly a reference to her own bisexuality in any of her
other blues:

> There's two things got me puzzled, there's
> Two things I don't understand.
> There's two things got me puzzled, there's
> Two things I don't understand;
> That's a mannish-acting woman, and a
> Skipping, twistin', woman-acting man.
>
> 'Foolish Man Blues'

Did she marry a man because she feared her own lesbianism? Did she marry Jack because she wanted somebody to
keep her in order, to make rules and stop her from drinking? Was she simply self-destructive, lavishing her love
and her money on a man who could never return either,
who ultimately ended up stealing both? It is interesting
to consider what Bessie Smith's life would have been like

if she had not been married to Jack, whether or not she would have had a more constant relationship with Lillian or Marie. Perhaps she thought that men were for marrying and women were for having on the road. Otherwise, why was she married in the first place? She was frequently unfaithful to Jack with both sexes. Chorus girls. Younger men. When she went home to Jack after an exhaustive tour – booze, sex and brawls – she'd try and cut out the drink, play the role of demure wife, quiet and calm, padding about in her slippers, talking to neighbours over the fence, cooking Jack big stews for his dinners. But it never lasted long. Wild Bessie always wanted out: 'I'm a young woman and ain't done runnin' round.' She was a wild woman and no Jack Gee was ever going to successfully tame her. Even after Jack left her, she didn't stop having relationships with men. She would have been free by then to concentrate solely on a relationship with a woman, but she didn't. Until the end of her life, she continued to have involvements with both sexes, to 'swing' both ways. It wasn't just jazz that could swing.

Ruby Walker's intimate relationship with Bessie Smith lasted the same length of time as Bessie's marriage to Jack. But when Jack left Bessie in 1929 for Gertrude Saunders, Ruby left too. She claimed that Jack forced her to join him and it is certainly true that she, like everyone else, feared him. Even so, Ruby's departure was a big betrayal. Bessie did not recover from the double blow. Ruby was present when Jack beat Bessie up in Columbus and she witnessed

Bessie's huge distress, but that did not stop her leaving too. After Jack left, he returned for a brief while to help Bessie work on her show at the Wallace Theatre in Indianapolis. Bessie was drunk all the time, totally devastated. She went off, leaving the whole troupe and not even taking Ruby with her this last time. So Jack tried to save the engagement by substituting Ruby for Bessie. They even padded her up to look fatter. Bessie heard about Ruby playing the part of the Empress and came back and ran Ruby off the stage. Perhaps Bessie stopped trusting Ruby right there and then. Ruby had always been ambitious for herself. Life with Bessie Smith did not give her many opportunities to be famous, as she had hoped. Bessie was content with Ruby doing dance routines, but she never wanted Ruby to sing. Ruby and Jack left to join Gertrude Saunders' show again, leaving Bessie to sing to herself: *Ain't nobody's blues but mine*.

I like to imagine that Bessie's great love of her life was Ruby Walker. I like to think that Ruby on the road was in love with Bessie all along. When she betrayed Bessie to join the Gertrude Saunders show, she never forgave herself. Her life had no meaning without the Empress riding her through the backstreets of Harlem in the dead of night. There was nothing to live for without Bessie's blues.

Bessie,

I know you are reading this saying 'I ain't never heard of such shit!' But I'm going to do as I want to anyway, as you would say, and write this letter. I am sorry I am here with Jack. He made me come. You know we was always frightened of him. He threatened me if I stayed in your show. Anyhows I have just made one big mistake. The biggest mistake of my life. This is Ruby talking. You know I don't bullshit. I know you better than anybody. You are still reading and laughing some despite yourself. I can picture you holding on to this letter with one finger, holding it some distance from yourself, your face grinning that wide grin of yours when you whole mouth opens up. I like to kiss you when your mouth opens up like that. I like to see the way you eyes glint when I dance. You were surprised how good I was at dancing. I don't know why I left you. I guess I thought you never loved me as much as I loved you. I remember the first time I heard you sing. I just stood there and watched you and my whole life changed. I didn't know it then, but that's what happened. My whole life changed because of you, Bessie. I won't never love anyone like I loved you and I want you to know it. So if you have kept reading this letter of mine, you know it now. I don't expect a reply. Though I'd like one. I'd like one more than I can say. But I'll leave that down to you. I am yours, even though I'm with Jack, I'm yours. The man has a power over people. You know that better than anybody, don't you. Why don't you come and get me and take me home. Why don't you come. I wait and I watch out of this window here but you ain't never appeared outside. I have tried to make you, using the power of my mind. I imagine you coming for me in a smart car, dressed up in your blue and white satin *dress with the hoop skirt, your favourite. I see you with a deep red shawl round your shoulders and pearls round your neck. You've got a gift for me. You always was crazy about buying folks gifts. I open it up and inside the little jewellery box is a ring with a ruby stone. Come and get me. Don't you dare ever forget me.*

Ruby xxx

It is dangerous for a woman to defy the gods;
To taunt them with the tongue's thin tip,
Or strut in the weakness of mere humanity,
Or draw a line daring them to cross;
The gods own the searing lightning,
The drowning waters, tormenting fears
And anger of red sins.

Anne Spencer, 'Letter to My Sister'

TALES OF THE EMPRESS

But the stories of her portentous rages, tempestuous
love-affairs, wild bouts of generosity, epic binges
and not infrequent recourse to devastating fisticuffs
– all of which stretch back as far as her public career
itself – leave us with the impression of a totally
untamed creature who remained, to the end of her
days, immune from the restraints of manners or
convention.

Humphrey Lyttelton, *The Best of Jazz*

Bessie Smith's life lent itself to legends. Her life so aptly
corresponded with her times that each seems emblematic of
the other. She was successful and high in the roaring, manic
1920s; she was depressed and low in the down-to-earth-with-
a-crash 1930s. At various points in her life, she exhibited
perfect timing. She was there when blues made their shift
from the backrooms and onto the stage; she was there at
the beginning of the record industry; she was there at the
beginning of the 'race records' promotions; she was there at
the beginning of the jazz era. She was in Harlem during the
Harlem Renaissance. She was with Louis Armstrong when
he was just a wee boy, practically. She became poorer when
the blues started to die. She sang 'Nobody Knows You When
You're Down and Out' at a time that ensured it would be

a classic forever, right after the Wall Street Crash. She was killed in a car crash in the Jim Crow South. Even her death was memorable and, sadly, beautifully timed. It is arguable that she might not have been so incredibly influential had she not died at precisely that moment in time.

A long time ago, far away from the Tennessee River and the small ramshackle cabin, lived a woman who married a greedy man. The woman was actually Queen of the Blues. The man's name was Jack Gee. Now Jack Gee knew his Queen could make him a lot of money and money was the thing that made that fellow happy. He could sit stroking it and counting it all through the night and into the morning. One day the fellow Jack Gee discovered that another man, by the name of Clarence Williams, was stealing his wife's money. Well, at daybreak he and his Queen went off to the city to see the man who had stolen his Queen's money.

The thief was the man who played the piano while his Queen sang the blues. That man had signed a piece of paper that gave him half of the Queen's earnings. He made an extra three hundred and seventy-five dollars. Jack Gee had smoke coming from his nostrils he was so angry. But the Queen now, she was not a woman to mess with. She didn't need a husband to stand up for her. The minute the thief Clarence Williams sees them storm into his midtown Manhattan office, he loses his voice, loses his power of speech. The sight of Jack and his muscles and gigantic fists sent the thief crawling underneath his desk. Jack yelled,

*'Come out of there, you dirty no-good cheating bastard' —
words which could have accurately described Jack himself,
a few years later. Williams, shaking and his bottom lip
trembling uncontrollably, crawled out; he looked like a
large soft insect. Before the two men could say boo to a
goose, the Queen jumped on Williams and pounded him
to the floor with her clenched fists. She kept on battering
him until Jack and she got what they wanted – a release
from the bit of paper that bound her to Clarence Williams.
But the strange thing was late that same night the Queen
discovered that it was not just the thief who was stealing
from her, but her own husband, the one that was lying there
sleeping in her big bed. That man could not feel a pea under
his sheets; but he could always tell right away if she placed
a dime under the mattress.*

Having been initially wary of the blues, Columbia now
knew they were on to a good thing. Bessie was selling like
pigs' feet. After they had 'sorted things out' with Clarence
Williams, Jack Gee and Bessie Smith went to see Frank
Walker at Columbia Records, who became Bessie's manager. He wrote Bessie a cheque for $500 and struck a new
deal. It guaranteed her $125 for four songs. But with-
out saying anything, Frank Walker struck out the royalty
clause. Bessie and Jack knew nothing about royalties, so
they didn't realise that Bessie was being had yet again. For
Bessie, completely unaware of the deception, $500 was
more money than she had ever dreamed of.

*Let's talk about one time. There was this woman, Bessie
Smith, who was Empress of the Blues. One night the people
were out lining the street waiting to hear a big concert with
lots of blues singers at the New Star Casino. There was
Mamie Smith, Sara Martin, Edna Hicks, Eva Taylor
and Clara Smith. All of a sudden, there was a terrific
commotion on the stage. Bessie threw a chair at somebody.
Then she was dragged out of there by three men. She was
cussing and carrying on. The three strong men had a hard
job of it, dragging her across the floor. She was big and
powerful. She could cuss worse than a sailor. She never did
appear that night with the other blueswomen.*

*Well, first up was Mamie Smith. She tried to sing
'Crazy Blues', but the minute she opened her mouth she
froze on the spot and her voice went dead. The same thing
happened to Sara Martin. When Clara Smith got on,
she was going to sing 'Far Away Blues' when all of a
sudden she just started flying up and up and up. She went
right through the roof of the New Star Casino and ended
up in the sky. Some say that's her, next to the Plough,
still burning bright and still claiming 'She Going Back
to Where She Used Ta Be'. Now Bessie, well, she was a
different matter. She made it out through the front door.
Not a soul in sight. Jumped in a plush car that was waiting
for somebody else. Got the driver to take her to a buffet
flat. At the buffet flat she saw a woman perform tricks she
wouldn't have thought humanly possible. There was one fat
woman performing tricks with a lit cigarette. Now that's*

what I call an educated pussy, Bessie thought to herself,
before downing a pint of good moonshine.

The Empress often refused to share a bill with any other
woman who was singing the blues. She gave this order to
'the men who ran No. 91', as Ethel Waters recalled:

I agreed to this. I could depend a lot on my shaking,
though I never shimmied vulgarly and only to express
myself. And when I went on I sang 'I Want to Be
Somebody's Baby Doll So I Can Get My Lovin' All the
Time'. But before I could finish this number the people
out front started howling, 'Blues! Blues! Come on,
Stringbean, we want your blues!'
  Before the second show, the manager went to Bessie's
dressing room and told her he was going to revoke
the order forbidding me to sing any blues . . . There
was quite a stormy discussion about this, and you
could hear Bessie yelling things about 'these northern
bitches'. Now nobody could have taken the place of
Bessie Smith.[41]

What did the acknowledged 'Empress of the Blues' have
to fear from any other woman singing the blues? There was
no other voice like hers. Other blueswomen have often said
so. 'Nobody could sing the blues like Bessie,' admitted May
Wright Johnson, while being sure to tell us: 'She wasn't a
friend, mind you. She would come round to the house. But

she was very rough.'[42] So why did she refuse? Perhaps it was nothing to do with insecurity, or feeling threatened, or competitiveness; perhaps the sound of others singing the blues before she was ready to go on simply put her off her own work. Perhaps she knew she was unique and wanted her talent to be taken seriously, or perhaps she just had a huge ego and didn't like the idea of anybody else being celebrated for the kind of songs that she regarded as her own. Perhaps she thought, 'I am the greatest.' It is interesting there has been so much disapproval of Bessie Smith for her reaction to other blues singers; in fact, she had good friendships with most of them, had a good laugh with Ma Rainey whenever they met up, affectionately called Ethel Waters 'Long Goody', and recorded songs with Clara Smith. She was happy to be friendly with her so-called 'rivals'. I don't believe she saw them as rivals at all. I think she knew her own talent and did not want it compromised, or the impact of it weakened by having other blues on the same bill.

There are many tales about Bessie's generosity. If she were around today, she would probably be labelled a shopaholic, a binge-buyer, addicted to giving people gifts. She bought cars for Jack on impulse. She would just need to see him looking adoringly at a car, and she'd buy it, even if she was told it was a showpiece Cadillac and not intended for sale. She paid $5,000 for that Cadillac in cash. A lot of money in the 1920s. She was almost as generous to her sisters, buying them a house and giving them an allowance. But what was she really trying to buy and why did she have to buy expensive gifts for people so often? It is easy to see why Bessie Smith spent her money like she did, with no mentality of 'saving up for a rainy day'. The way she spent her money mirrored her own personality. She liked to put her hand down. Money just became another expression of her impulsive, party-loving, binge-drinking generosity. Her attitude to money wasn't any different to her attitude to sex. If she liked something, she bought it. If she wanted somebody, man or woman, she had them. She was totally free in this sense, completely uninhibited, and never once stopped indulging her own desires. I suppose you could call her hedonistic. But when you consider her background and the poverty that she came from, you can understand why she so enjoyed showering people with gifts, why it was so important to her to spend so much on her family. I can picture her, lying on her bed, awake at night, dreaming about what she was going to buy her family. It must have been like a childhood wish come true. Lying awake on a bed with lovely soft, clean sheets, in a house with running water, a warm house in Philadelphia, a long long way from Chattanooga, lying there dreaming about what to buy her sisters. From poor to rich, like a fairy tale. Who would have believed that one day she'd be able to buy anything she wanted. Except love, of course. She couldn't buy love. Perhaps she thought she could. Perhaps she imagined that if she bought Jack a new Cadillac he would love her like no other. She loved him for buying her that dress with the money from his pawned watch. She never forgot that moment when he arrived home with that red dress in that big paper bag. He'd got her size right and everything. She liked that dress so much she wanted to sleep in it. It was the first time in her life that anybody had ever bought her anything. Presents mean more to people who have been poor. Maybe she thought everyone would be like she was with that dress Jack bought her. But Jack and her sisters started to take her wild generosity for granted; they started to expect the gift, to wait for it and to turn sour if it didn't just magically appear as she walked through the door.

She also used her money to help friends out of trouble. When she heard an old friend, Buster Porter, a man she had worked with, was in a Cleveland hospital and unable to pay his medical bills, she grabbed a cab during one of the breaks in her show and went to him. She gave the sick entertainer $50. She bailed friends out of jail with her money. Once she had to rescue Ma Rainey, who had found herself in a compromising position with the Chicago police. She'd been having a wild party with a group of young women; the party got so noisy that a neighbour called the police. The cops arrived just as Ma and the girls were in the middle of an orgy. There was pandemonium as everyone scrambled madly for their clothes and ran out of the back door, wearing each other's knickers. Ma Rainey, clutching someone else's dress, was the last to leave by the back exit. But she fell down the stairs. The police caught her, accused her of running an indecent party and threw her in jail. Bessie, who must have been more than sympathetic to

Ma's predicament, having run many such parties herself, bailed Ma out the following morning.

The party life was never better than in the Harlem of the 1920s. It could be said that home entertainment never had it so good. These parties were known as 'rent parties' or parlor socials. You could get into one for anything from ten cents to a dollar. The other guests were ordinary, working-class people: tradesmen, housemaids, laundry workers, seamstresses, porters, elevator 'boys'. But writers and artists and singers loved to go along too. On a Saturday night in Harlem, the music pounded out of the open windows. There was always an upright piano, a guitar, a trumpet and sometimes a snare drum. Rent parties originated in the South, where rents were so high that people had to organise such socials to pay their landlords. You needed no social standing to throw a rent party. All you needed was a piano player and a few dancing girls. Jervis Anderson describes a typical rent party:

Drinks were bathtub gin and whiskey. Food was fried fish, chicken, corn bread, rice and beans, chitterlings, potato salad, pigs' feet. 'Whist' games were actually poker and dice. Music, in the parlance of the day, was 'gut bucket', played by some of the masters and students of Harlem stride piano. Dancing – the Charleston, the black bottom, the monkey hunch, the mess around, the shimmy, the bo-hog, the camel, the skate and the buzzard – went on till the break of day . . .

You were not regarded as much of a jazz pianist unless, wherever else you appeared, you played the rent-party circuit. You earned your spurs not only by sending the dancers into flights of ecstasy but also by 'cutting,' or outperforming, rival piano players.[43]

Duke Ellington, Bill Basie – not yet Count – a young Fats Waller and Bessie Smith enjoyed these rent parties. Fats was already fat and playing the piano like nobody else. Count Basie remembers Fats at those parties. He was a daily customer, 'hanging on to his every note, sitting behind him all the time, enthralled by the ease with which his hands pounded the keys and his feet manipulated the pedals'. Did Bessie get into them chitterling struts or was she too interested in the pigs' feet and the seamstresses? Some of those piano battles lasted five hours; by then the Empress would be out of her head on bathtub gin, not giving a damn for the boys' 'carving battles'.

One of Bessie's best-known songs, 'Gimme a Pigfoot', written by Leola 'Coot' Grant and Wesley Wilson and performed with Jack Teagarden and Benny Goodman in 1933, is about rent parties:

Up in Harlem ev'ry Saturday night
When the high brows git together it just too tight.
They all congregates at an all-night strut
And what they do is tut-tut-tut.
Old Hannah Brown from 'cross town

Gets full of corn and starts breakin' 'em down.
Just at the break of day
You can hear old Hannah say –
Gimme a pigfoot and a bottle of beer,
Send me again, I don't care.
I feel just like I want to clown,
Give the piano player a drink because he's bringing me
    down.
He's got rhythm, yeah, when he stomps his feet
He sends me – right off to sleep.

The last lines of this Harlem party song sum Bessie up
– 'Slay me 'cause I'm in my sin, / Slay me 'cause I'm full
of gin.' The gutsy way she sings that 'yeah' is like nobody
else. She drags that *yeah* out of herself. She knew how to
let herself go; didn't give a damn what anyone thought of
her. She knew how to give herself up to the mood: 'Send
me 'cause I don't care, / Slay me 'cause I don't care . . .'
Drunk and full of the high of that strutting piano, the
place Bessie was sent to was more wonderful than any
other. It was free of Jack, free of the pressure to record
or perform herself, she could completely lose herself at
those parties. And Bessie did want to lose herself. Jack was
very judgemental of her going to or throwing any type of
party. If he caught her partying, he'd beat her. The parties
were an escape from Jack; they were also forbidden fruit,
or forbidden pigs' feet.

Say that the Empress went out one night to the house of a man called Carl Van Vechten. Say it was a nice house, a smart house. Say he had a pretty wife. Now Carl Van Vechten was one of Bessie's many white admirers. He was rich and he was a journalist whose interest was the black writers and musicians of the Harlem Renaissance. There was no time like the twenties in Harlem if you liked to have fun. Van Vechten held parties that were a bit different from the rent parties; at his parties, you could meet the New York cultural elite. 'Dicty blacks' that Bessie loathed. They didn't like her either. That woman knew crude language inside out. She could cuss up a storm. She could make clouds in the sky appear dark and heavy with her bad language. The rain could pour down and it still couldn't put out her tongue. She wasn't scared of nothing, least of all cussing. She was never quiet awhile. She just came right out and said what she thought. No, those well-to-do blacks believed the Empress swearing reflected badly on them. Alone, they could hold up a mirror to themselves and see the wild face of that roaring violent woman staring right back. It frightened them to think they and she could be connected in any way. She was the same colour. 'That woman is bad for the race,' they said and walked away.

Heard tell about 1928. The Empress arrives with Porter Grainger, the composer of her current show, Mississippi Days, *dressed up to the nines in ermine and dripping with jewels. Right aways she realises she is on alien territory. There's a whole sea of white faces staring at her and the*

polite white handshake of Van Vechten is no comfort. She is
out of her depth, and she sure as hell is not going to drown.
Whenever the Empress is out of her own territory, she
defends herself with her own aggression. Van Vechten offers
her a martini; she demands a whiskey. She tells Ruby, at
her side at all times, to pull her damn mink coat off. Van
Vechten, of course, wants her to sing in his own home. He's
got Grainger and he's got a piano and he's got the New
York elite. What more could a man like Van Vechten, who
published a novel in 1926 entitled Nigger Heaven, *want?*
Now that was a dream come true. As soon as Bessie sings
her blues, after downing practically a pint of whiskey, she
is back to herself. Only music and drink can bring herself
back. The white faces in Carl Van Vechten's chic apartment
light up. Polite society can still enjoy 'Jail House Blues',
'Sorrowful Blues' and 'Bleeding Hearted Blues'. The
sadder the song, the better the hit. They could shake their
sad white heads when Bessie sang 'St Louis Blues'. Here
was somebody from the heart of the poor black community
dressed up in furs, drinking whiskey in between blues,
and singing her whole black heart off. They were moved.
This was a real experience. Polite society drank a bit
more than usual that night. Not pint measures, mind you.
Not whiskey. Martinis all the way through 'I Ain't Got
Nobody'.

I ain't got nobody, nobody,
Nobody cares for me.

That's why I'm awfully sad and lonely,
Won't some good man take a chance with me?
I'll sing good songs, all the time,
Want some gal to be a pal of mine.
I ain't got nobody, nobody
Ain't nobody cares for me.

As Bessie goes to leave the tiny wife of Carl Van Vechten says, 'You're not leaving without kissing me goodbye,' and leans forward to kiss her. Bessie doesn't even think. She's got fists for words like these. She knocks the small woman down, saying, 'Get the fuck away from me! I ain't never heard of such shit.' Polite society shook its small white head, horrified. And who had ever seen a hostess knocked down like that before at such a do? Van Vechten helped his wife up and escorted Bessie to the elevator. The patron of black arts went back into his room and wiped his brow with a big white handkerchief. Then he said to his company, 'Well, that's the blues for you.' His tiny wife jumped up and down on his feet. 'I'll give you the blues,' she said. At which point, polite society left. There was a queue of them waiting at that same elevator. Well, the elevator boy had a laugh that night. He laughed and laughed and laughed.

The story of Van Vechten's party is told more than any other story about Bessie. Outside of the time and place, it is perhaps difficult to imagine the enormous impact a black woman knocking down a white woman (and rich white, too) would have had. But Bessie had no real sense of her own danger or the consequences of her violent, thoughtless actions. She just loathed being patronised by white people. She hated it when any member of her travelling troupe, from her boys in the band to the prop boys and chorines, sucked up to white people. She never did. Perhaps she just couldn't stomach the contradictions of finding herself at such an elitist soirée, and the only way to resolve the conflict was to behave completely differently from anyone else there so that she need never fear being like 'one of them'. Or perhaps she didn't think about it at all.

Bessie Smith did make a habit of reacting swiftly to anything that annoyed her with her fists. Violence and she were no strangers. Jack beat her up and she beat up other people. Anybody, from a chorine in her show, to a piano player like Clarence Williams, to Carl Van Vechten's wife, if they upset Bessie, could receive blows just about anywhere on their body. Why did she respond with violence so often? She never used words to fight. She had a fearsome temper and when she got riled there was no cooling her down. When she was wild she hated words. Only the action of fists could do. She was completely out of control. Once, having heard that Jack was unfaithful to her, she tried to shoot him from a moving train. There are countless tales about the violence of the Empress, but none actually enlighten us as to what was going on inside her mind. The point about violent people when they lose their temper is that they don't have any mind. Whether she thought about her own violence afterwards is not known either. But she always used her fists. The blues she sang also contained lots of violence: 'St Louis Gal, I'm going to hammer you, I mean manhammer you [*sic*].' In that song, she wants to beat up the woman that has stolen her man, which again anticipated real life when she beat up Gertrude Saunders for stealing Jack Gee.

Sometimes her resort to fisticuffs was admirable. Like the time when she fought off the Ku Klux Klan single-handedly. This was the twentieth-century Klan, organised in 1915 by 'Colonel' William J. Simmons, a preacher and promoter of fraternal orders. The Klan was fuelled by patriotism and a romantic nostalgia for the old South. In the 1920s its membership exceeded four million nationally. Profits rolled in from the sale of its regalia. A burning cross became its symbol. White-robed Klansmen participated in marches, parades and night-time cross burnings all over the country. Bombings, shootings, whippings and lynchings were carried out in secret. The Klan exercised a reign of terror over Southern black people. But Bessie Smith was not cowed even by the Ku Klux Klan.

Now here it is. It was a hot day that's just cooled down. The moon is way up in the sky. The tents are up for the blues. People have arrived from all over on mules, trains, in cars, wagons. They have come to hear the woman they have been playing on their phonographs. The one who can tell it like it is. The one who speaks their own tongue. On the way to the tent, past the train station, they have

*caught a glimpse of her famed train, bright as the yellow
sun was in the sky, with its great green letters. That train
is something. They are not arriving in quite that style. But
they have picnics with them. Enough food and drink to feed
the blues. Chicken legs. Pigs' feet. Cornbread. Tomatoes.
It is July 1927 in Concord, North Carolina. The whole
tent is buzzing and it's hard to find a place to sit down. It's
difficult to see where one person ends and another begins.
The excitement is making the canvas flap back and forth
even though there is no wind. The big pole in the middle
of the tent looks unsteady. Some folks have bought drinks
from the Empress's vendor. Crackerjacks. The moment
finally comes. The one they have journeyed all the miles
for. The single opening moment when that voice breaks out
like an animal in heat. The Empress just howls them blues
out. She starts off with 'Florida Bound Blues'.*

I got a letter from my daddy,
He bought me a sweet piece of land;
I got a letter from my daddy,
He bought me a small piece of ground;
You can't blame me for leavin', Lawd,
I'm Florida bound.

*Then she goes on to strut her stuff, when one of her
musicians, overhot, goes out and finds a bunch of hooded
men about to drop her tent. The sight of them in their white
robes makes the musician dizzy. They've already pulled*

up several stakes. He rushes back and warns the Empress. The Queen commands her prop boys to follow her round the tent. She doesn't tell them why. So when they see the Klansmen they make a speedy retreat. The Empress, on her own, confronts the Klansmen: 'I'll get the whole damn tent out of here if I have to. You just pick up those sheets and run.' The Klansmen, shocked, stand and gawp whilst the Empress shouts obscenities at them until finally they disappear into the darkness. 'I ain't never heard of such shit,' says the Empress, walking over to the prop boys. 'And as for you, you ain't nothing but a bunch of sissies.' Then she goes right back into that same tent for her encore.

Sorted.

The Empress was often knocking people down or beating them up with her own fists. But in the wild stories recounted about her, there are also some wonderful one-liners. In Atlanta at the 81 Club in 1925, two years after the release of her first records and at the height of her massive popularity, the manager, Bailey, requested that all stars go in through the back door, which was rat-infested. But the Empress grabbed Ruby's arm and led her past the box office into the theatre through the front door. Bailey caught up with them just as they entered the auditorium. 'If they see you before the show, they won't find you as interesting,' he said. 'I don't give a fuck,' replied the Empress. 'If you don't like it you can kiss my black ass and give me my drops.'

The story of Bessie Smith is not just the story of the blues but the story of a woman who was not in control of her life. There was the constant drinking. She started drinking heavily when she was a child. Sometimes she was so drunk concerts had to be cancelled. Other times she was propped up. The speedy resort to violence probably had a lot to do with bathtub gin and white lightning. It is ironic that a lot of these violent outbursts took place in the time of Prohibition; like the time when a chorine in her show cut up the new pair of dancing shoes Bessie bought for Ruby. Bessie threatened her and she fought back, throwing a bottle across the room. Bessie asked her to go and she refused. There was one hell of a fight. The chorine ended up totally dishevelled and covered in blood. She called the police and pressed charges against Bessie. Ruby had to get the money for her bail. Bessie was often arrested by the police for her violence.

It happened again when she discovered Jack had given Gertrude Saunders some of her jewellery. She saw Gertrude in the street and knocked her unconscious. Another time, according to Albertson, Bessie spotted Gertrude out with Jack and got a hold of her. 'You motherless bitch,' she called Saunders – an interesting insult since Bessie was motherless herself – and dragged her through the mud. Gertrude screamed that she was going to get a gun and kill Bessie. 'I'll make you eat it, bitch,' was Bessie's answer, 'and every time I see you, you yellow bitch, I'm going to beat you. One of these days when you are up on stage, I'm gonna be in the audience and I'm comin' up and grabbin' you off.'[44]

Some people said Gertrude Saunders did indeed purchase a gun for defence against Bessie. Another time when Bessie attacked Gertrude she was actually given a summons.

The timing of those attacks signalled the beginning of her downward spiral. In 1929 Jack left. And by this time Bessie's recordings were only selling about eight thousand each – a far cry from the sales in 1923 of 'Gulf Coast Blues' and 'Downhearted Blues'. The Empress became a victim of the Depression; Bessie's luck was running out. In the same year she made the film *St Louis Blues*, based on W. C. Handy's song. A two-reel short. Seventeen unforgettable minutes. The only film footage in existence of the Empress. It was all about a woman suffering from an unfaithful lover. I saw it once, when I was in my teens, on the television. I remember the shock of the grainy monochrome image of my heroine appearing in this sad tale of woe. There she was, a tall, beautiful woman, driven to drink by her feckless lover. She catches her lover at it, marching through a crap game, and throws her rival out of the room. Then she herself is thrown to the floor by the faithless lover. On the floor she starts singing 'St Louis Blues':

> I hate to see the evening sun go down,
> I hate to see the evening sun go down,
> It makes me think I'm on my last go-round.
> Feeling tomorrow like I feel today,
> Feeling tomorrow like I feel today,
> I pack my grip and make my getaway.

St Louis woman with her diamond rings,
Pulls my man around by her apron strings.

The lover reappears; they embrace; he steals a wad of notes from her stocking, pushes her into the bar and struts out the place. He could have been Jack Gee.

Ironically enough, 'St Louis Blues' was turned down by every major Tin Pan Alley publisher. W. C. Handy had to set up his own publishing company to issue the song. 'It did not become popular for years, until vaudeville and revue singers began using it in their acts.'[45]

It is impossible to know how many of the stories of Bessie Smith's violence are true, and how many are made up, or embellished and exaggerated out of all proportion. Gertrude Saunders, years later, maintained that she and Bessie had never come to blows, and that she felt sorry for Bessie because her love of Jack was so intense and yet Jack treated her so badly. Today her so-called binge-drinking or heavy drinking would simply be called alcoholism, and all her wild, tempestuous behaviour would be explained by that addiction. Her terrible mood swings, her violence, her insecurity, her bad language, the fact that she never went anywhere without corn liquor in her purse, all point to her being a serious alcoholic. Why did she need to drink so much? The stories are good crack, but she was out of control. She was running. She frequently drank herself unconscious and woke up to beat people up. She can't have felt all that good about herself. After her violent outbursts, she probably felt remorse. Who knows? Maybe she felt justified. Maybe she felt whoever it was that was receiving her blows at any given time deserved it. Maybe she didn't think about violence at all. She just let rip and then it was over until the next time.

That there are almost as many stories about Bessie Smith's violent, drunken bawdiness as there are blues is interesting in itself. It is not just her blues that fascinate, but her wild, drunk, promiscuous, generous, cussing personality. The legends and myths of the Empress will live on as long as the blues survive. Whether they are true or not. 'All stories are true.'

Preach 'em blues, sing 'em blues,
Moan 'em blues, holler 'em blues,
Let me convert your soul.
Sing 'em, sing 'em, sing them blues,
Let me convert your soul.

'Preachin' the Blues'

THE BLUES

It was the real thing: A wonderful heart cut open
with a knife until it was exposed for all to see.

Carl Van Vechten

Whatever Bessie Smith sang automatically became
a blues.

Paul Garon

Whatever pathos there is in the world, whatever
sadness she had, was brought out in singing – and
the audience knew it and responded to it.

Frank Schiffman

Bessie Smith's blues tell the story of her life better than
biography or autobiography ever could. Bessie Smith's
life and her blues are as opposite sides of the same coin.
Many weirdly anticipated events in her life. She sang
'Nobody Knows You When You're Down and Out' before
the Depression hit her and her friends started to leave her.
'Jail House Blues' she sang in 1923, before she'd ever been
arrested for any violence. The lyrics probably came right
back to her though, when she was in jail for real. 'Good
morning blues, blues, how do you do, / Say I've just come
here to have a few words with you.' She was not singing

about some distant imaginary character. Even when she didn't know it yet, it turned out she was singing about herself. Her blues are autobiography. They favoured almost exclusively the first-person narrative. She is so intimate with her lyrics because she knows what they are saying. She knows her life. The key to understanding the complex personality of Bessie Smith is all in her blues.

'Baby Doll', 'Backwater Blues', 'Blue Blues', 'Death Valley Moan', 'Dirty No-Gooder Blues', 'Dixie Flyer Blues', 'Don't Fish in My Sea', 'Foolish Man Blues', 'Golden Rule Blues', 'Hard Time Blues', 'He's Gone Blues', 'Hot Springs Blues', 'In the House Blues', 'It Makes My Love Come Down', 'Jail House Blues', 'Jot 'Em Down Blues', 'Lonesome Desert Blues', 'Long Old Road', 'Lost Your Head Blues', 'My Man Blues', 'Pickpocket Blues', 'Pinchbacks, Take 'Em Away', 'Please Help Me Get Him Off My Mind', 'Poor Man's Blues', 'Reckless Blues', 'Rocking Chair Blues', 'Safety Mama', 'Shipwreck Blues', 'Soft Pedal Blues', 'Sorrowful Blues', 'Spider Man Blues', 'Standin' in the Rain Blues', 'Sweet Potato Blues', 'Telephone Blues', 'Thinking Blues', 'Wasted Life Blues', 'Young Woman's Blues'.

These are just the ones she wrote herself. Some of them uniquely capture specific moments in her life. Jack's nervous breakdowns in 'Hot Springs Blues'. Her own reaction to the postwar boom that favoured white people is in 'Poor Man's Blues':

> Now the war is over, all man must live the same as you;
> Now the war is over, all man must live the same as you;
> If it wasn't for the poor man, Mister Rich Man, what
> would you do?

'Backwater Blues' is an example of her writing a blues herself to reflect the experience of ordinary working-class people. She went to a small town near Cincinnati which was flooded. She had to step off the train into a little row-boat. Her audience shouted for her to sing 'Backwater Blues', but this was a song she did not know. Perhaps they were suggesting she should write it. 'Backwater Blues' is one of her most powerful compositions, probably because it came out of this strange experience, and because there was a demand for a song that did not yet exist. She recorded it on 17 February 1927 with James P. Johnson.

> I woke up this mornin', can't even get out of my do';
> I woke up this mornin', can't even get out of my do';
> There's enough trouble to make a poor girl wonder
> where she wanna go . . .
> Backwater blues done caused me to pack my things and
> go;
> Backwater blues done caused me to pack my things and
> go;

'Cause my house fell down, and I can't live there no
  mo'.

The blues she sang and the blues she wrote often contained elements of burlesque, music hall and vaudeville
which reflected her background as a young girl who had
first joined a travelling troupe in 1912. A lot of her blues
were raunchy, bawdy, double-entendre-filled, sexy songs,
as well as tragic, painful and depressing. As W. C. Handy
put it:

> The blues came from the man farthest down. The blues
> came from nothingness, from want, from desire. And
> when a man sang or played the blues, a small part of the
> want was satisfied from the music. The blues go back to
> slavery, to longing. My father, who was a preacher, used
> to cry every time he heard someone sing, 'I'll See You
> on Judgement Day.' When I asked him why, he said,
> 'That's the song they sang when your uncle was sold
> into slavery in Arkansas. He wouldn't let his masters
> beat him, so they got rid of him the way they would a
> mule.'[46]

Bessie's blues moved people. Her voice just got to them.
Perhaps she reminded them of the past, of losses, of longing. Something in her voice went way back into a deeper
past. Her voice seemed to contain history, tragedy, slavery, without self-pity. It had the ability to stretch beyond

even the lyrics of her blues into something more complex. Her blues were universal, but also deeply personal. They allowed her to express the whole range of her complex personality: the wild promiscuous drunken side ('T'ain't Nobody's Business', 'Gimme a Pigfoot', 'It Makes My Love Come Down', 'He's Got Me Goin', 'Young Woman's Blues') and the depressed, insecure, lonely side ('I've Been Mistreated and I Don't Like It', 'Nobody Knows You', 'Wasted Life Blues', 'Baby Won't You Please Come Home', 'Empty Bed Blues').

Relationships, principally relationships with men, are the key troubles in her blues. Not poverty, not poor health, not alcoholism, not floods, not death, not gambling – although she did write and sing blues about all these subjects. As Albert Murray points out:

In fact in the 160 available recordings of Bessie Smith . . . the preoccupation is clearly not all with hard workmasters, cruel sheriffs, biased prosecutors, juries and judges, but with the careless love of aggravating papas, sweet mistreaters, dirty no–gooders, and spider men.[47]

Spider men. They sound awful. She wrote 'Spider Man Blues' herself too. In some of her blues, she begs for the man to change himself, plaintively; and in others she states she's had enough. She's not putting up with it any more. 'I Ain't Gonna Play No Second Fiddle', 'Safety Mama'. Her songs do not reflect the fact that she had just as much sex with women. There is no reason to presume, however, that every character she sings about is actually a man. When she sings songs like 'Empty Bed Blues', she could be missing a woman just as easily as a man.

The men in her life, principally Jack Gee and Earl Love, her two unsuccessful husbands, as well as the men she had relationships with on the road, consistently let her down; she did not believe that a good man could exist. Her blues reflect this. 'Dirty No-Gooder Blues', 'Foolish Man Blues', 'My Man Blues', 'Spider Man Blues', 'Please Help Me Get Him off My Mind'. Not just one man, but all men are bastards as far as these songs go. They are not just bastards, they are bad for you. Men, in Bessie's blues, destroy and ruin women, they waste their lives; in order for a woman to survive she needs to get away from men. She needs to stand up to them, to not take any more of it, and to get on a train to Chicago. 'I'm a good ole gal but I've just been treated wrong' ('Lost Your Head Blues').

There are times when she puts up with it:

I swear I won't call no copper
If I'm beat up by my papa.
T'ain't nobody's business if I do.
                    'T'ain't Nobody's Business'

But these lines are in direct contrast to:

I've been mistreated and I don't like it,
There's no use to say I do.

'I've Been Mistreated and I Don't Like It'

Or these:

Daddy, Mama's got the blues
The kind of blues that's hard to lose
'Cause you mistreated me
And drove me from your door . . .
Mistreating Daddy, mistreating Mama all the time . . .
Mistreating Daddy, Mama's crossed the danger line.

'Mistreating Daddy'

The songs she wrote herself about men are much more revealing. The lines are more direct and the pain cuts deeper:

Lord, I wish I could die because my man treats me like a
    slave,
Lord, I wish I could die my man treats me like a slave.
As to why he drives me, I'm sinking low, low in my
    grave.
He's a hard drivin' papa, drive me all the time,
Drives me so hard I'm afraid that I'll lose my mind. . .
And when the sun starts sinking, I start sinking into
    crime.

He takes all my money and starts to cry for more.
I'm goin' to the river feelin' so sad and blue,

I'm goin' to the river feelin' so sad and blue,
Because I love him 'cause there's no one can beat me
like he do.

'Hard Drivin' Papa'

Bessie's 'Hard Drivin' Papa', was first recorded on 4 May 1926. Six months later, she's still writing about a man who wants her money in 'Lost Your Head Blues': 'I was with you baby when you didn't have a dime.' Jack Gee obviously features highly in her own songs. She was sorely let down by him. 'Since you got money, it's done changed your mind.' She wrote this song three years before she actually left him: 'I'm gonna leave baby, ain't gonna say goodbye.' Once again, the blues saw into her future. Jack Gee couldn't handle having money. It totally corrupted him; he was no longer the nice man who had been shot on a date, but a money-grabbing chancer. The fact that his behaviour made the Empress suicidal is obvious in her blues. 'I'm going to the river, feeling so sad and blue.'

Bessie Smith's blues show how perceptive she was about people and herself, about her own complicity. They also reveal a fear of losing herself, that a man can drive her so hard, 'I'm afraid that I'll lose my mind.' But even worse than losing her mind is the death wish at the beginning of the song. 'Lord, I wish I could die because my man treats me like a slave.' Addressed to the Lord, this blues is like a plea to get out of life, to be free. 'Treats me like a slave' would have been a more shocking line then than it is now for African-American people hadn't been out of slavery for all that long. The man in this song gives her no freedom; there is no escape; no respect; no autonomy. Just like the girl in 'Backwater Blues', 'There ain't no place for a poor old girl to go.' There is a pervasive sense of Bessie being totally trapped in her destructive relationships with men. There she was, with more money than any black woman in the country, making her own records, writing her own songs, yet still being battered by her talentless husband. It really says something about the power of sexism. The Empress is 'treated like a slave'.

It is this complexity that beats at the heart of Bessie Smith's personality. The woman who didn't take any rubbish from anybody, who cussed and beat up many people, and the woman who was beaten herself, who had her money stolen and was humiliated by her husband. No matter how rich or famous she became, she never resolved that contradiction. She was never totally in control of it. There is the sweet man and there is the sour man. 'I fell in love with a sweet man once, he said he loved me too.' Then there are the men who are both, the sweet mistreaters. Her blues never offer resolutions. They explore conflicts but they don't resolve anything. Like good stories, they let you enter with your imagination and participate in the conflict. Nothing is pat or finished, except jelly rolls, as Will Friedwald states:

Smith also encompassed a tantalizing mixture of what was past and what was to come . . . With Smith, the two seemingly incongruous attitudes are compatible, a sort of tender invective. Smith sings about love without a trace of sentiment, and of sex without guilt. She has an amazingly realistic attitude toward life and love . . . devoid of self-pity. As far as jazz, blues and popular music are concerned, Smith was the first fully three-dimensional recording artist, the first to use the new medium to express a complete personality.[48]

There is so much sauce in her blues as well, so much feisty cheek and humour. One of my early favourites, 'Kitchen Man', is a wonderful example. It was written by Andreamenentania Paul Razafinkeriefo (Andy Razaf), who was the nephew of Queen Ranavalona III of Madagascar. He also wrote, with Fats Waller, 'Honeysuckle Rose' and 'Ain't Misbehavin'. The double entendres in 'Kitchen Man' are practically pornographic, though I didn't really understand them back then. Often Bessie Smith deviated from the words written down, probably because some of them were unfamiliar. In 'Kitchen Man' she sings 'foor men' for 'footmen', which suggests she hadn't heard of a footman.

I love his cabbage gravy, his hash . . .
I can't do without my kitchen man . . .

Like the way he warms my chops,
I can't do without my kitchen man . . .

Oh his jelly roll is so nice and hot,

Never fails to touch the spot,
I can't do without my kitchen man.

His frankfurters are oh so sweet,
How I like his sausage meat,
I can't do without my kitchen man.

Oh how that boy can open clams,
No one else can touch my hams,
I can't do without my kitchen man.

When I eat his doughnuts,
All I leave is the hole
Anytime he wants to,
Why, he can use my sugar bowl . . .

Only when food is mentioned in Bessie Smith's blues are sexual relationships with men any fun – raunchy, low-down, wild-about-that-thing fun. The rest is a bum ride. Doughnuts and jelly rolls, clams and hams aside, what she sings most about is a lack of love. The majority of Bessie's blues are about looking for love. Love is something she is constantly searching for, but never quite finds: 'I wanna be somebody's baby doll / So I can get my lovin' all the time.' She is desperate for love. Possibly, her mother dying when she was so young intensified this quest for love. When she wrote 'Reckless Blues' in 1925 her mother had already been dead for twenty-three years. Despite her powerful looks, there is a perception of herself in her own songs as unattractive. No wonder, with impresarios turning her down for being 'too dark'. In her blues, she believes other women are beautiful, but she still deserves love.

> When I wasn't nothing but a child,
> When I wasn't nothing but a child,
> All you men tried to drive me wild.
>
> Now, I am growing old,
> Now, I am growing old,
> And I got what it takes to get all of you men told.
>
> My mama says I'm reckless, my daddy says I'm wild.
> My mama says I'm reckless, my daddy says I'm wild.
> I ain't good-lookin' but I'm somebody's angel child.
>
> Daddy, Mama wants some loving,
> Daddy, Mama wants some loving,
> Hurry pretty Papa, Mama wants some lovin' right now.
>
> 'Reckless Blues'

The line 'I ain't good-lookin' but I'm somebody's angel child' is wishful thinking because both her parents were long dead. If her mother were still alive, how much she would have been loved. Maybe the line 'Mama wants some loving' is ambiguous and she is actually talking about wanting love from her mother. She wrote this song when she was thirty-one. Already she had a sense of getting older. The song looks back at herself when she was younger. The way she hangs on to the 'now' for so long in 'Now [beat, beat, beat], I am growing old' is extraordinary.

Although the main issue in her blues is love and lack of love, a significant number of her songs were about social issues, crime and punishment, poverty and ill health, work and death. Some of the blues she sang that are less well known, and not included in any of the more recent compilations are: 'Send Me to the 'Lectric Chair', 'Homeless Blues', 'Jail House Blues', 'Sing–Sing Prison Blues', 'Work House Blues' and 'House Rent Blues'. The woman in 'Send Me to the 'Lectric Chair' asks the judge to hear her plea, but doesn't want any sympathy because she has slit her good man's throat. Then she actually asks to be sent to the electric chair, wanting to take a journey to the devil down below: 'I just killed my man, I want to reap what I sow, / Judge, Judge, hear me Judge, send me to the 'lectric chair.' She tells the judge how she sat laughing round her man whilst he was dying. The combination of the extraordinary plea with the graphically violent descriptions of the murder makes the song wildly funny. I can imagine women hearing it in 1927 and splitting their sides laughing. The music is humorous in the background: Joe Smith on witty cornet, Charlie Green on trombone. The trombone after each 'Judge, Judge', is critically timed.

I imagine her sitting down and composing her blues, sitting down because in every photograph but one that I have ever seen of Bessie Smith, she is standing up, performing. I imagine her at her kitchen table singing the words to herself till they sound right, then memorising them. She was not very literate; she'd get out a pencil and write down what she could but her memory was a necessity. She learned all 160 songs that she sang off by heart, so that when she went into the recording studio she would not slip up. She'd sing her life to herself while cooking and stirring, humming the melody till it sounded as good as the stew tasted. I imagine her trying out her new blues on Ruby. 'Ruby, what do you think of this?' and Ruby maybe making one or two adjustments. I like to see the two of them laughing at those lines, 'There's nineteen men living in my neighbourhood, / Eighteen of them are fools and the one ain't no doggone good.' Ruby saying, 'That's definitely a hit!' and pouring Bessie some homebrewed lethal concoction. Did she enjoy creating blues lyrics and compositions as much as singing them, or more? They were surely her way of working herself out, her way of exploring her own contradictions and making sense of herself. Her blues are littered with clues about how she saw herself and what she thought of herself and also how she saw the world. Luck changes, turns your life upside down. This sense of bad luck, of sudden things happening and changing your fortune, is everywhere in her blues, like a sixth sense.

'Thinking Blues', which she wrote in 1928, begins with a woman thinking about someone who has been 'nice and kind' when a letter comes from an old lover giving her the blues so bad. 'Don't you hear me baby, knocking on your door, / Have you got the nerve to drive me from your door?'

The man begs to be taken back and tries one more time. She replies, 'The good book says you've got to reap what you sow.' Again the song doesn't offer a resolution, it just explores the problem of whether or not a woman should take a man back after he has betrayed her. 'Hard Time Blues', which she wrote in 1926, also explores the dilemma of a woman who has been abandoned by her man. In this song the woman is packing her bags and leaving town.

My man says he didn't want me,
I'm getting tired of his dirty ways.
I'm going to see another brown,
I'm packing my clothes, I'm leaving town . . .

But when the good woman is gone, hard times will strike at the man and cut him down to size. Many of her blues reveal this glorious desire for revenge, even invoking the weather on her side. That man is going to get his comeuppance: 'The rising sun ain't gonna set in the east no more.' The revenge continues till it ends up with the man 'Down on your knees you'll ask for me.' Her blues are not just about being mistreated; they also detail an attempt to take control over her life. It is the man who gives the woman the blues in the first place – 'I've got the blues and it's all about my honey man' – but after that it's up to the woman to get out of it, even if that means escaping by dying – 'I'd rather be in the ocean . . . than to stay with him and be mistreated like a dog.' ('Honey Man Blues', 1926). Or, in 'Rocking

Chair Blues' (1924), she sings: 'I won't be back until you change your ways / I'm going to the river, carrying a brand new rocking chair.'

She recorded 'Black Mountain Blues' (not to be confused with 'Red Mountain Blues', altogether a happier affair) in 1930, with cornettist Ed Allen, who also played for her in 'Nobody Knows You'. 'Black Mountain' becomes synonymous with the Depression. It is a rough and ready place where there are no laws and everybody is out for themselves: 'Back in Black Mountain, a child will smack your face, / Babies crying for liquor and all the birds sing bass.' Bessie was on the long road down; she was making $500 per week now as opposed to her former $2,000. Ma Rainey wasn't doing too well either in her show *Bandanna Babies*. The audiences of the South still loved the Empress and gave her a warm welcome whenever she returned. But the South was not enough. She didn't stop being generous though and still bought presents for everybody.

On 20 November 1931, Columbia dropped the Empress. Bessie had recorded with Columbia for over nine years and had been almost single-handedly responsible for Columbia's rise to fame. The last two songs Bessie recorded for them on that date were 'Safety Mama' and 'Need a Little Sugar in My Bowl' – the latter would go on to be sung by many jazz singers, most memorably by Nina Simone years later. Columbia did not cut anything like the number of records they had cut in her heyday; now it was down to only four hundred. It was the decline and fall of the Empress. She became a has-been. She lived for a while off the earnings of her boyfriend, Richard Morgan, an easy-going Chicago bootlegger who was the uncle of musician Lionel Hampton. (Bootleggers prospered during the Depression. People always needed a good drink, especially if times were hard. A good moonshine could always cheer you up.) All over America talking pictures started to take over from theatre shows, and the vaudeville theatres closed. Ethel Waters made it in France; but Bessie Smith never left America, never reached out to a wider European audience. Had she survived, Bessie might have succeeded in the way that Ethel Waters or Alberta Hunter did.

In 1933 she recorded some songs for Okeh Records, who had turned her down ten years before. Here she was in good company, with Jack Teagarden and Buck Washington, who recorded some of her most famous songs: 'Do Your Duty', 'Gimme a Pigfoot', 'Down in the Dumps' and 'Take Me for a Buggy Ride'. 'Down in the Dumps' turned out to be the Empress's swansong. It is the only time you can hear her with swing accompaniment. Later both Billie Holiday in the 1940s and Nina Simone in the 1960s went on to sing 'Gimme a Pigfoot', although both omitted the references to reefers and marijuana that are in Bessie's version. Billie Holiday shared that same troubling quality that Bessie Smith had. According to Johnny Mercer, 'There was something about her – not just the torchy quality of her voice – that made you want to try to help her.'[49]

In 1935, Bessie appeared at the Apollo Theatre in

Harlem, owned by Frank Schiffman, who tried to persuade her to change her chorus girls because their skin was too dark and they looked gray. Bessie, however, demanded amber lights. 'If you don't want my girls, you don't want me . . . I don't give a damn, because I'm tired of wearing myself out. I can go home, get drunk, and be a lady – it's up to you.'[50] Bessie won the battle of the dark-skin girls; a little *coup d'état* of her career, possibly remembering that time in 1912 when she was thrown out of Irving C. Miller's chorus line for not being a 'glorious brownskin girl'. Ruby said, 'You never saw people give applause like they did for us girls – we broke it up!'

Two weeks earlier, Billie Holiday had made her Apollo debut. And people were already talking about the young Ella Fitzgerald. Bessie had stars on her back and she had to fight to maintain her own reputation. Luckily for Bessie, in 1936 Billie came down with a case of ptomaine poisoning and had to leave her show. Bessie Smith substituted for her. That night in 1936 at Connie's Inn helped Bessie get back on her feet again. She reached out to a wider audience. Around this time, she chummed up with Mildred Bailey. They both shared the camaraderie of fat, with routine jokes to each other about size. When they met, they would both say, 'Look I got this brand-new dress, but it's too big for me, so why don't you take it?'[51]

During this time Bessie continued to write and compose blues, although they were not recorded. Eubie Blake, a pianist and composer, remembers her coming into W. C.

Handy's office at the Apollo in April 1936 with new songs. He says they were good ideas but he was concerned that they would not make money, so they were never record- ed. We will never know the lyrics of the last blues of the Empress.

Please don't let me lose my rightful mind
'Cause them is graveyard words.

                        'Them's Graveyard Words'

MISSISSIPPI, 1937

Note: Other reports on the circumstances of this
singer's death have proven incorrect.
Sheldon Harris, *Blues Who's Who*

Someways, you could have said beforehand that
there was some kind of an accident, some bad hurt
coming to her. It was like she had that hurt inside
her all the time, and she was just bound to find it.
Sidney Bechet, *Treat It Gentle*

On 26 September 1937 at 11.30 a.m. in Ward One of
the Afro-American Hospital, 615 Sunflower Drive,
Clarksdale, Mississippi, Bessie Smith died. She had been
on her way down Route 61 to a show date, travelling with
Richard Morgan, with whom, for the past six years, she
had been very happy. Bessie's old Packard, which Richard
was driving, hit a National Biscuit Company truck parked
without lights in the town of Coahoma. Her arm was virtually severed. She was still alive when she was taken to
the Afro-American Hospital, where her arm was ampu-
tated, but she died hours later of shock, blood loss and
internal injuries.

Her funeral was held in Philadelphia on Monday 4
October. Her body had been sent, two days earlier, by

train to Philadelphia's Thirtieth Street Station, where her brother Clarence was waiting. Richard Morgan, who had survived the terrible accident, accompanied her body on the train. She was laid out at Upshur's funeral home on Twenty-first and Christian Street. When word of her death reached the black community, however, the body had to be moved to the O. V. Catto Elks Lodge on Sixteenth and Fitzwater Streets, where ten thousand mourners filed past her bier on Sunday 3 October.

Her insurance policy ensured that she had a lavish send-off the following day. Her coffin, which cost $500, was trimmed with gold and lined with pink two-tone velvet. There were forty floral arrangements. The auditorium at Catto Elks Lodge was full. The Reverend Andrew J. Sullivan presided over the service. A Mrs Emily Moten read a poem called 'Oh Life'. Somebody fainted. The crowd outside was seven thousand strong and policemen were having a hard time holding it back, just like they had in the heyday of the Empress when thousands could not get into the theatres to hear her sing the blues.

The pallbearers were hired men. They did not know Bessie. It was unusual for a dead person at a Philadelphia funeral to be carried by strangers. They walked down Sixteenth Street to the waiting hearse. A choir sang 'Rest in Peace'. The hearse did not head straight for the cemetery; it gave the Empress one last twirl round her old neighbourhood. The streets were lined with thousands of ordinary blues fans and admirers who were all devastated at the tragic death of the Empress. She was a huge star who, at the time of her death, was just beginning to make a comeback. Yet no other major stars showed up at her funeral. Ethel Waters ('Long Goody') was not there. Duke Ellington was not there. The cast of the Cotton Club Revue was not there. Louis Armstrong was not there.

It seems fitting for a woman who preferred pigs' feet parties to polite society, who never ever forgot her working-class roots, who despite her fame and fortune never changed her speech, her behaviour or her habits to conform to some notion of a star, that it was not stars and personalities who attended her funeral but ordinary people. The people who always mattered to her all along. Not the Harlem elite. All of the faces in the photographs taken of her grand exit are black. There is not a white face to be seen.

Her family attended the funeral and Richard Morgan was there. It was Jack Gee, though, who talked to reporters, claiming he had never lost touch with his wife. He later became rich on royalties paid to him from her songs. Over the years he invested in small businesses and real estate. Richard Morgan, who had been her partner for as long as Jack Gee, six years apiece, received nothing. Nor did Bessie Smith's adopted son, Jack Gee Junior.

She was buried at Mount Lawn Cemetery, Sharon Hills, Philadelphia, wearing a long silk dress that Jack Gee claimed he bought her – the second significant dress, the first being the one she wore to her first recording studio date. The newspapers said it was a farewell gift. Her family denied that he ever bought that dress and said the money from her life insurance went to pay for it.

Jack Gee, although not poor, did not buy a headstone for Bessie Smith. Nor did any of her family, despite her generosity towards them for years. It is scandalous but true that the Empress of the Blues lay in an unmarked grave until 1970. Her relatives squabbled over money. Everyone thought somebody else should be the one to pay. Bessie would probably have written a blues about that if she had been looking down from above. Her own 'Funeral Blues' or 'Nobody Bought Me a Headstone Blues'. A Bessie Smith Memorial Concert was held in Town Hall, New York City in 1948 to raise money to buy a headstone. But the stone was never bought; Maud Smith, Bessie's sister-in-law, said that Jack Gee pocketed the money. Other surviving relatives of Bessie Smith have also said that money raised from benefits was turned over to Jack Gee on at least two occasions. Chris Albertson says:

> Ruby [Walker] recalls performing at such an event
> in New York's Town Hall during the forties: 'Jack
> showed up and demanded the money. He said he'd
> have his lawyers stop the concert if he didn't get it. He
> got it, but knowing Jack he spent it on something for
> himself.'[52]

Maud Smith tells of another benefit held at Philadelphia's Blue Note Club in the early 1950s. She handed extra miles each day to get to the next city in which I was to sing so that we could get a place to eat and sleep . . . by the time I was supposed to sing I was almost dizzy.[57]

And because Bessie Smith came from the South and was killed in Mississippi, the rumours began immediately and continued for over thirty years. It is not surprising, given the racism of the segregated South and the drama of Bessie Smith's own life, that such a story should have survived many different tellings over all those years. It was initially sparked off by an article of John Hammond's in *Downbeat* magazine entitled 'Did Bessie Smith Bleed to Death While Waiting for Medical Aid?' Here was the terrible sentence that burned for many years: 'When finally she did arrive at the hospital she was refused treatment because of her color and bled to death while waiting for attention.' Once the article appeared people refused to believe any other version of events. She *had* died that way. Racism had murdered her. Racism had killed so many black people in the American South through Jim Crow laws and lynchings; Bessie Smith was seen as yet one more victim. It is easy to see why, to imagine that a whites-only hospital might have been closer by on that road where she had the accident; that they might have wasted time driving to the Afro-American Hospital; that the woman travelling in another car which became involved in the accident might have been treated more quickly because she was white. The truth is that black lives in the United States of America at that point in history were not valued, so precious time could easily have been wasted. There's no rush to save a black life. Sidney Bechet said, 'I was told the doctors, they hadn't too much concern for getting to her quick.'[58] People from Mezz Mezzrow to Edward Albee believed the story about her death. Albee was so outraged and moved that he wrote a play about it called *The Death of Bessie Smith*. It was written in 1959 and opened in West Berlin in 1960:

> JACK: Ma'am, I got Bessie Smith out in that car there.
> SECOND NURSE: I don't care who you got out there, nigger. You cool your heels![59]

The image of the racist white nurse is not a joke. It is true. Bessie Smith could have died of racism. The truth about racism has kept the story going. Sadly, for racism to have an impact on the general public it needs to happen to somebody well known enough to matter. There probably were many people who died in road accidents in the South because white hospitals refused to admit them, but not being famous, their names don't stick. Bessie Smith's death became a symbol of racism. It was a warning. Just like her blues were warnings. This could happen to you. The fact that it maybe didn't happen to her is not the point. People believed it had. And in a strange way this is fitting. There was her life, full of fabrications and embellishments and stories galore. And there's her death: more stories.

that arm back on herself if she was really alive you think.
And that is nearly your last thought. Then you think hell,
I ain't never thought death would be like this. You ain't
never thought death would get you to thinking about all of
this stuff. When you are dying you think fast on your back
and you can get in more thoughts in those dying minutes
than you've thought your whole life and you suddenly
get a chance to see yourself different. Shit, you even get a
chance to see yourself dying. Not just see yourself dying,
'cause that isn't important any more. Visual things aren't
important when the crunch comes. You feel yourself going
down, like you going down underneath the world, swirling,
doing the Black Bottom, all the way down. Your feet don't
stop moving. You start to stomp. You lying there on the road
covered in your own blood and at the same time you are
dancing down to death. And then everything goes dark but
the noise continues. You don't see your own death coming;
you hear it. It is not just one loud noise; it's a whole gang
of noise, shouting, barrelhousing, cussing, crashing, cutting
piano, stomping, snare-drumming, car-screeching, glass-
shattering, tyres-screaming, car-weeping noise. Death is
loud. Don't let anybody tell you different. You ain't never
heard such shit.

That's the way it is. It is rough and dark. Don't let
anybody tell you any different. And where does that
lonesome road take you? It takes you all the way down to
hell. You knew you was never going no place else.

Feeling tomorrow like I feel today,
Feeling tomorrow like I feel today,
I pack my grip and make my getaway . . .

'St Louis Blues'

The great thing about hell is the parties. Hell's parties are rough and dark, way, way better than Harlem's. They've got pigs' feet here you ain't seen in your wildest dreams and beer to wash them down with that you ain't tasted in your sweet life, dancing so fast you can't see your feet touch the rims of hell. They still do the camel, the shimmy, the bo hog, the skate and the buzzard here, only they don't call them that. They don't have any language in hell. No fucking words at all. It's all sensations. They feed you well and they dance you well and when you get to sing you raise hell. A lot of the old blues people are whooping it up. Well, well beyond the break of day.

THUMBNAIL SKETCH

1894 Born 15 April, Chattanooga, Hamilton County, Tennessee. Father: William Smith, a Baptist minister. Mother: Laura Smith. Born into poverty, one of eight children. Both parents died before she was nine.

1903 Sang on the corner of Ninth Street for pennies.

1912 Joined the Moses Stokes Travelling Show as a dancer. Ma Rainey and Pa Rainey were part of the troupe.

1912–21 Toured the South gathering a huge live audience. Worked with the Rabbit Foot Minstrels, the Florida Cotton Blossoms, Silas Green's Minstrel Show.

1918 Appeared in her own show, *Liberty Belles Revue*, at 91 Theatre, Atlanta as singer, dancer, male impersonator.

1920 Married her first husband, Earl Love, who died a year later.

1923 Appeared with Sidney Bechet in musical comedy *How Come* at Dunbar Theatre, Philadelphia. Turned down by Okeh Records for being 'too rough'. Made her first record, 'Downhearted Blues', with Columbia Records. It sold a record-breaking 780,000 copies in six months. Married Jack Gee, her second husband. Met Ruby Walker.

1923–31 Recorded extensively with Columbia Records on an exclusive contract, 160 songs in all. At her peak, in 1925, Columbia paid her $200 per usable side.

1925 Bought her own Pullman railroad car.
– 14 January: recorded 'St Louis Blues' with Louis Armstrong.

1926 Legally adopted Snooks, renaming him Jack Gee Junior.

1929 Separated from Jack Gee.
– 15 May: recorded 'Nobody Knows You When You're Down and Out'.
– June: appeared in film *St Louis Blues*.

1931 Dropped by Columbia. Started relationship with Chicago bootlegger Richard Morgan.

1933 Recorded 'Gimme a Pigfoot', 'Do Your Duty', 'Take Me for a Buggy Ride' and her swansong, 'Down in the Dumps', with Okeh Records, receiving just $37 per usable side.

1930–37 Earned a living touring the South again with her own show, *Bessie Smith Revue*.

1937 26 September: Bessie Smith died in a car crash on Route 61, Clarksdale, Mississippi.

NOTES

1 Chris Albertson, *Bessie*, Stein & Day, New York, 1971

2 Donald Clarke, *Wishing on the Moon: The Life and Times of Billie Holiday*, Viking, London, 1974

3 Will Friedwald, *Jazz Singing*, Da Capo, New York, 1996

4 W. C. Handy, *Father of the Blues: An Autobiography*, Macmillan, New York, 1941; Da Capo, New York, 1985

5 Richard Newman, *Everybody Say Freedom: Everything You Need to Know About African-American History*, Plume, New York, 1996

6 Albertson, *Bessie*

7 Paul Oliver, *Songsters and Saints: Vocal Traditions on Race Records*, Cambridge University Press, Cambridge, 1984

8 Sally Placksin, *Jazzwomen: 1900 to the Present*, Pluto Press, London, 1985

9 ibid.

10 ibid.

11 Albertson, *Bessie*

12 Sidney Bechet, *Treat It Gentle*, Hill and Wang, New York, 1960; Da Capo, New York, 2002

13 Nat Shapiro and Nat Hentoff, *Hear Me Talkin' to Ya: The Story of Jazz as Told by the Men Who Made It*, Dover, New York, 1966

14 ibid.

15 Albertson, *Bessie*

16 ibid.

17 ibid.

18 ibid.

19 ibid.
20 ibid.
21 Humphrey Lyttelton, *The Best of Jazz*, Robson Books, London, 1978
22 Albertson, *Bessie*
23 Placksin, *Jazzwomen*
24 Chris Albertson, Notes to *Bessie Smith: The Complete Recordings*, Columbia, 1991
25 Alan Lomax, *The Land Where the Blues Began*, Methuen, London, 1993
26 Michele Wallace, *Invisibility Blues*, Verso, New York, 1990
27 Oliver, *Songsters and Saints*
28 Shapiro and Hentoff, *Hear Me Talkin' to Ya*
29 ibid.
30 Hettie Jones, *Big Star Fallin' Mama*, Viking, New York, 1974
31 Lyttelton, *The Best of Jazz*
32 Geneviève Fabre and Robert O'Meally (eds), *History and Memory in African-American Culture*, Oxford University Press, New York, 1995
33 Shapiro and Hentoff, *Hear Me Talkin' to Ya*
34 Lyttelton, *The Best of Jazz*
35 Oliver, *Songsters and Saints*
36 Albertson, *Bessie*
37 ibid.
38 ibid.
39 ibid.
40 ibid.
41 Shapiro and Hentoff, *Hear Me Talkin' to Ya*
42 ibid.
43 Jervis Anderson, *Harlem: The Great Black Way*, Orbis Publishing, London, 1972

44 Albertson, *Bessie*

45 Philip Furia, *Poets of Tin Pan Alley: A History of America's Great Lyricists*, Oxford University Press, New York, 1990

46 Handy, *Father of the Blues*

47 Albert Murray, *Stomping the Blues*, Da Capo, New York, 1989

48 Friedwald, *Jazz Singing*

49 Clarke, *Wishing on the Moon*

50 Albertson, *Bessie*

51 ibid.

52 ibid.

53 Newman, *Everybody Say Freedom*

54 Albertson, *Bessie*

55 *Cashbox* Magazine, 1966

56 Lomax, *The Land Where the Blues Began*

57 Mahalia Jackson, *Movin' On Up*, Hawthorn Books, 1966

58 Bechet, *Treat It Gentle*

59 Edward Albee, *The Death of Bessie Smith*, First Plume, New York, 1988

60 Mezz Mezzrow in Shapiro and Hentoff, *Hear Me Talkin' to Ya*

61 Oliver, *Songsters and Saints*

62 Newman, *Everybody Say Freedom*

63 Shapiro and Hentoff, *Hear Me Talkin' to Ya*

SELECTED READING

Albee, Edward, *The Death of Bessie Smith*, First Plume, New York, 1988

Albertson, Chris, *Bessie*, Stein & Day, New York, 1971

Anderson, Jervis, *Harlem: The Great Black Way*, Orbis Publishing, London, 1972

Armstrong, Louis, *Satchmo: My Life in New Orleans*, Da Capo, New York, 1986

Bechet, Sidney, *Treat It Gentle*, Hill and Wang, New York, 1960

Clarke, Donald, *Wishing on the Moon: The Life and Times of Billie Holiday*, Viking, London, 1974

Dahl, Linda, *Stormy Weather: The Music and Lives of a Century of Jazzwomen*, Quartet, London, 1984

Fabre, Geneviève, and O'Meally, Robert (eds), *History and Memory in African-American Culture*, Oxford University Press, New York, 1995

Friedwald, Will, *Jazz Singing*, Da Capo, New York, 1996

Furia, Philip, *Poets of Tin Pan Alley: A History of America's Great Lyricists*, Oxford University Press, New York, 1990

Garon, Paul, *Blues and the Poetic Spirit*, City Lights, San Francisco, 1975

Garon, Paul, and Garon, Beth, *Woman with Guitar: Memphis Minnie's Blues*, Da Capo, New York, 1992

Handy, W. C., *Father of the Blues: An Autobiography*, Macmillan, New York, 1941; Da Capo, New York, 1985

Harris, Sheldon, *Blues Who's Who*, Da Capo, New York, 1991

Lomax, Alan, *The Land Where the Blues Began*, Methuen, London, 1993

Lyttelton, Humphrey, *The Best of Jazz*, Robson Books,
    London, 1978
Murray, Albert, *Stomping the Blues*, Da Capo, New York, 1976
Newman, Richard, *Everybody Say Freedom: Everything You
    Need to Know About African-American History*, Plume, New
    York, 1996
Oliver, Paul, *Songsters and Saints: Vocal Traditions on Race
    Records*, Cambridge University Press, Cambridge, 1984
Placksin, Sally, *Jazzwomen: 1900 to the Present*, Pluto Press,
    London, 1985
Shapiro, Nat, and Hentoff, Nat, *Hear Me Talkin' to Ya: The
    Story of Jazz as Told by the Men Who Made It*, Dover, New
    York, 1966
White, Newman, *American Negro Folk Songs*, Folklore,
    Pennsylvania, 1965

CREDITS

POETRY

In some instances we have been unable to trace the owners of copyright material, and we would appreciate any information that would enable us to do so.

14.95

956.
91
BAL

Please renew/return this item by the last date shown.

So that your telephone call is charged at local rate,
please call the numbers as set out below:

	From Area codes 01923 or 020:	From the rest of Herts:
Renewals:	01923 471373	01438 737373
Enquiries:	01923 471333	01438 737333
Textphone:	01923 471599	01438 737599

L32   www.hertsdirect.org/librarycatalogue

Scorpion Publishing Ltd
MCS

H50 117 841 2

First published in 1994 by
Scorpion Publishing Ltd and MCS
Victoria House
Victoria Road
Buckhurst Hill
Essex
England

ISBN 0 905906 96 9

Typeset by MasterType, Newport, Essex
Printed and bound in Singapore by Craftprint